Dietary Nutrients Effects on Metabolic and Physiological Adaptations: Mechanisms of Action in Health and Disease

Dietary Nutrients Effects on Metabolic and Physiological Adaptations: Mechanisms of Action in Health and Disease

Editor

Anna M. Giudetti

MDPI • Basel • Beijing • Wuhan • Barcelona • Belgrade • Manchester • Tokyo • Cluj • Tianjin

Editor
Anna M. Giudetti
University of Salento
Italy

Editorial Office
MDPI
St. Alban-Anlage 66
4052 Basel, Switzerland

This is a reprint of articles from the Special Issue published online in the open access journal *Nutrients* (ISSN 2072-6643) (available at: https://www.mdpi.com/journal/nutrients/special_issues/ Metabolic_Physiological_Adaptations).

For citation purposes, cite each article independently as indicated on the article page online and as indicated below:

LastName, A.A.; LastName, B.B.; LastName, C.C. Article Title. *Journal Name* **Year**, *Volume Number*, Page Range.

ISBN 978-3-0365-5117-3 (Hbk)
ISBN 978-3-0365-5118-0 (PDF)

Contents

Preface to "Dietary Nutrients Effects on Metabolic and Physiological Adaptations: Mechanisms of Action in Health and Disease"

Changes in lifestyle and the shift from a traditional to an industrialized/Westernized diet have drastically increased the rates of diet-related chronic diseases, such as overweight and obesity, cardiovascular diseases, diabetes, hypertension, stroke, and cancers. The physiological mechanisms of action of nutritional compounds are closely associated with their ability to recognize and modulate specific cellular enzymes as well as metabolic pathways. The identification of these targets is essential to recognize the true mechanisms of action of nutritional compounds and their direct or indirect effects on cellular processes and functions in physiological and pathological conditions.

This book is dedicated to increasing and updating our knowledge on the association between dietary patterns and diet-related diseases. It is focused on the molecular, biochemical, and physiological processes underlying the mechanisms of action of nutritional compounds and other dietary components. The book collects topics regarding the effects of nutrients on physiological functions and systems, cellular and molecular processes, energy homeostasis, and human diseases.

The book explores the need for food in addition to the interactions between health and diet, and provides a foundation of scientific knowledge for the interpretation and evaluation of future advances in nutrition and health sciences.

This book is relevant to any student or practitioner interested in how diet influences our health, including in the fields of nutrition, dietetics, medicine, and public health.

Anna M. Giudetti
Editor

Article

Paeonol, an Ingredient of Kamishoyosan, Reduces Intracellular Lipid Accumulation by Inhibiting Glucocorticoid Receptor Activity in 3T3-L1 Cells

Masayuki Izumi, Takashi Yoshida, Takashi Nakamura and Minoru Wakamori *

Division of Molecular Pharmacology & Cell Biophysics, Department of Oral Biology, Tohoku University Graduate School of Dentistry, Sendai 980-8575, Japan; izumi-masa@umin.ac.jp (M.I.); takashi.yoshida.d5@tohoku.ac.jp (T.Y.); takashi.nakamura.d2@tohoku.ac.jp (T.N.)
* Correspondence: minoru.wakamori.b1@tohoku.ac.jp; Tel.: +81-22-717-8310

Received: 26 December 2019; Accepted: 21 January 2020; Published: 24 January 2020

Abstract: Excessive triglyceride accumulation in lipid-metabolizing tissues is associated with an increased risk of a variety of metabolic diseases. Kamishoyosan (KSS) is a Kampo composed of 10 constituent herbs, and contains moutan cortex (MC) and paeonol (PN) as the major ingredient of MC. Here, we demonstrate the molecular mechanism underlying the effect of KSS on the differentiation of mouse preadipocytes (3T3-L1 cells). KSS inhibited the accumulation of triglycerides in a dose-dependent manner in 3T3-L1 cells that were induced to differentiate into adipocytes. We also found that MC and PN were responsible for the anti-adipogenetic effect of KSS and significantly suppressed the expression of CCAAT/enhancer-binding proteins-δ (C/EBP-δ) mRNA 3 days after the induction of differentiation. Thus, PN may contribute to the anti-adipogenetic property of MC in 3T3-L1 cells. In addition, PN inhibited dexamethasone (Dex)-induced glucocorticoid receptor (GR) promoter activity. Taken together, these results suggest that PN suppresses C/EBP-δ expression by inhibiting Dex-induced GR promoter activity at the early stage of differentiation and, consequently, delays differentiation into mature adipocytes. Our results suggest that the habitual intake of Kampo-containing PN contributes to the prevention of the onset of metabolic diseases by decreasing the excessive accumulation of triglycerides in lipid-metabolizing tissues.

Keywords: adipocyte; Kamishoyosan; Moutan cortex; paeonol; C/EBP-δ; glucocorticoid receptor; Kampo medicine

---

1. Introduction

Metabolic syndrome is a cluster of metabolic disorders that are associated with numerous lifestyle-related risk factors [1]. Intracellular lipid accumulation is a common feature in the pathogenesis of a variety of lifestyle-related diseases such as cardiovascular disease, fatty liver, type 2 diabetes, and dyslipidemia [2–6]. Prevention and control of lifestyle-related diseases is a major public health concern, especially in developed countries.

The 3T3-L1 cell line is one of the adipocyte cell models used to study lipid metabolism *in vitro* [7]. A mixture of dexamethasone (Dex), 3-isobutyl-methylxanthine (IBMX), and insulin (hereafter DMI) can efficiently differentiate 3T3-L1 cells into mature adipocytes [8]. Differentiation of adipocytes can be largely divided into the early stage and the late stage. Adiponectin is a bioactive factor secreted from differentiated adipocytes [9]. We focused on CCAAT/enhancer-binding proteins (C/EBPs), forkhead transcription factor 1 (FoxO1), and peroxisome proliferator-activated receptor-gamma (PPAR-γ) transcription factors as the markers of adipocyte differentiation as well as adiponectin. C/EBPs are a family of six gene members, among which C/EBP-δ and C/EBP-β are expressed at the early stage, and C/EBP-α or PPAR-γ are expressed at the late stage of adipocyte differentiation [10,11]. Knockdown of either C/EBP-δ or C/EBP-β (in combination or independently) suppresses the differentiation of primary embryonic fibroblasts into mature adipocytes and lipid accumulation [12]. The expression of PPAR-γ is induced in response to insulin, and it increases glucose uptake in adipocytes [13,14]. FoxO1 is a transcription factor containing characteristic a winged helix structure termed the Forkhead box. FoxO1 is involved in the commitment of the early stage of adipogenesis by insulin *via* Akt and FoxO1 phosphorylation by the insulin signaling [15].

Traditional Japanese herbal medicine (Kampo medicine) was originally based on traditional Chinese medicine but was adapted to Japanese culture [16,17]. Kamishoyosan (KSS, Chinese name: Jiaweixiaoyaosan, Korean name: Gamisoyosan) is a complex drug composed of 10 herbs. It is prescribed for climacteric disorder, dysmenorrhea, neurosis, and in cancer supportive therapy [18,19]. Hormones and cytokines released from adipocytes are involved in the aggravation of diseases for which KSS is prescribed [20–24]. KSS reduces lipid accumulation in human hepatoma HepG2 cells in the presence of oleic acid [25]. In Kampo medicine, Orengedokuto, inhibits differentiation of 3T3-L1 cells and Yokukansan reduces fat synthesis by reducing the expression of the transcription factor SREBP-1c and glycerol-3-phosphate dehydrogenase and increasing the expression of antioxidant enzymes *via* the transcription factor FoxO1 in differentiating 3T3-L1 cells [26,27]. One of the compounds included in KSS, geniposide, reduces lipid accumulation in 3T3-L1 cells during differentiation [28]. Another compound, paeonol, reduces lipid accumulation in HepG2 cells [29].

The molecular mechanism underlying the pharmacological action of KSS in 3T3-L1 cells during their differentiation into mature adipocytes is unclear. Here, we demonstrate that paeonol is an inhibitory compound of KSS during adipogenesis.

2. Materials and Methods

2.1. Cell Culture

The mouse 3T3-L1 preadipocyte cell line was obtained from the Japanese Collection of Research Bioresources Cell Bank (JCRB Cell Bank, Osaka, Japan). 3T3-L1 cells were cultured in high-glucose Dulbecco's modified Eagle's medium (DMEM; Sigma–Aldrich, St. Louis, MO, USA) containing heat-inactivated 10% fetal bovine serum (FBS; Thermo Fisher Scientific, Waltham, MA, USA), 30 U/mL penicillin (Meiji Seika Pharma, Tokyo, Japan), and 30 µg/mL streptomycin (Meiji Seika Pharma) at 37 °C in a humidified atmosphere of 5% CO_2. For adipocyte differentiation, cells were seeded on a 24-well plate at a density of 2×10^4 cells per well using the time course shown in Figure 1A. After reaching the confluence (0 days), adipocyte differentiation was initiated using the same medium, but supplemented with 1 µM Dex (Wako, Osaka, Japan), 0.5 mM IBMX (Wako), and 10 µg/mL insulin (Sigma–Aldrich) (hereafter DMI) for 3 days (day 0–3). The medium was then replaced with medium containing 5 mg/mL insulin for 2 more days (day 3–5) and then changed to fresh medium every 2 days (day 5–8). This differentiation induction method was termed the DMI method. 3T3-L1 cells were used up to the 4th passage to avoid cell phenotypic changes.

2.2. Preparation of Kampo Medicine and Composition of the Ten Herbs and Eight Major Components of KSS

Kamishoyosan (KSS), Hochuekkito (HET), Shoseiryuto (SST), and Goreisan (GRS) were kindly provided by Tsumura & Co. (Tokyo, Japan). Bupleuri Radix (BR), Paeoniae Radix (PR), Atractylodes Lancea Rhizoma (ALR), Angelicae Radix (AR), Poria (PO), Gardeniae Fructus (GF), Moutan Cortex (MC), Glycyrrhizae Radix (GLR), Ingiberis Rhizoma (IR), and Menthae Herb (MH) were purchased from Tsumura & Co. The extraction method of each Kampo medicine was as follows. One hundred milligrams of the powder was dissolved in 1 mL medium. The solution was shaken at 200 rpm and 38 °C for 30 min in a model BR-22FH constant temperature shaker (Taitec, Saitama, Japan). The solution was then centrifuged at 14,000× *g* for 25 min using a model 5922 device (Kubota, Tokyo, Japan). The supernatant was sterilized by using a 0.22-µm pore size membrane filter (Thermo Fisher Scientific). The eight major components of KSS were dissolved in dimethyl sulfoxide (DMSO, Wako). These compounds were: saikosaponin B1 (SSb1, Wako), saikosaponin B2 (SSb2, LKT Laboratories, St. Paul, MN, USA), paeoniflorin (PNF, LKT), β-eudesmol standard (EU, Wako), geniposide (GEN, Wako), paeonol (PN, LKT), glycyrrhizic acid (GA, Combi-Blocks, San Diego, CA, USA), and 6-shogaol (SG, Cayman Chemicals, Ann Arbor, MI, USA). L-glutamine added to the phenol red-free DMEM was purchased from Nissui pharmaceutical (Tokyo, Japan), and sodium pyruvate solution was purchased from Wako. Glucocorticoid receptor inhibitor of RU-486 (mifepristone) was purchased from Tokyo Chemical Industry (Tokyo, Japan).

2.3. Evaluation of Intracellular Lipid Accumulation

Cells were stained with Oil-Red-O (Sigma–Aldrich) as described previously [30]. 3T3-L1 cells were plated on a 24-well plate and induced to differentiate using the DMI method described earlier. The cells were rinsed with phosphate-buffered saline (PBS), fixed with 4% paraformaldehyde (PFA; Wako) for 14 min, and then stained with 3 mg/mL Oil-Red-O (in 60% isopropanol) for 10 min at room temperature. After staining, cells were washed once with 60% aqueous isopropanol and twice with PBS. After washing, the cells were observed using an ECLIPSE Ti-U inverted microscope (Nikon, Tokyo, Japan). Cell images were captured with a CCD camera (digital sight DS-L3, Nikon). Additionally, after the dye was extracted for 10 min with isopropanol, the absorbance was measured at 490 nm using a DTX 880 Multimode Detector (Beckman Coulter, Brea, CA, USA). All experiments were performed at least three times.

2.4. Assessment of Cell Viability

To investigate the effect of Kampo medicine, herbs, and major components on the cell viability, a Cell Counting Kit-8 (CCK-8; Dojindo Molecular Technologies, Kumamoto, Japan) was used according to the manufacturer's instructions. Absorbance was measured at 450 nm using the DTX 880 Multimode Detector. All experiments were performed at least three times.

2.5. Quantitative PCR (qPCR)

3T3-L1 cells were seeded in a four-well plate (Thermo Fisher Scientific). Total RNA was prepared using ISOGEN II (Nippon Gene, Tokyo, Japan). The total RNA concentration was determined by measuring the absorbance at 260 nm using a NanoDrop 2000c spectrophotometer (Thermo Fisher Scientific) and NanoDrop 2000/2000c Operating software, version 1.4.2. Using 1 μg of total RNA as a template, cDNA was synthesized by reverse transcription reaction using the SuperScript® Vilo™ cDNA synthesis kit (Thermo Fisher Scientific). Reverse transcription was conducted with a G-STORM GS482 thermal cycler (Life Science Research, Somerton, UK). The expression level of various genes regulating adipocyte differentiation was measured in the CFX 96™ Real-Time System (BIO-RAD, Hercules, CA, USA) with the KOD SYBR® qPCR Mix (TOYOBO, Osaka, Japan). qPCR was performed at 98 °C for 2 m for the initial denaturing, followed by 45 cycles of 98 °C for 10 s, 61 °C or 67 °C for 10 s, and 68 °C for 30 s, using specific primers (Table 1). The results were determined using the $\Delta\Delta C_T$ method and are shown as the fold-change relative to the control after normalizing to the expression of the 14-3-3 protein zeta/delta (Ywhaz) gene. The primer sets of Ywhaz housekeeping gene was purchased from TaKaRa Bio (Shiga, Japan). All experiments were performed at least three times.

Table 1. Primer sequences for qPCR.

Gene	Sense	Antisense	Ref.
C/EBP-α	5′-CAAGAACAGCAACGAGTACC-3′	5′-GTCACTGGTCAACTCCAGCAC-3′	[31]
C/EBP-β	5′-ACGACTTCCTCTCCGACCTC-3′	5′-CGAGGCTCACGTAACCGTAG-3′	
C/EBP-δ	5′-CTGCCATGTACGACGACGAGAG-3′	5′-GCTTTGTGGTTGCTGTTGAAGA-3′	
PPAR-γ	5′-CTGATGCACTGCCTATGAGC-3′	5′-TCACGGAGAGGTCCACAGAG-3′	
Adiponectin	5′-GCACTGGCAAGTTCTACTGCAA-3′	5′-GTAGGTGAAGAGAACGGCCTTGT-3′	[32]
FoxO1	5′-ACGAGTGGATGGTGAAGAGC-3′	5′-TGCTGTGAAGGGACAGATTG-3′	[33]

2.6. Luciferase Reporter Assay

Dual-luciferase reporter assays were conducted as described previously [34]. One day before transfection, 3T3-L1 cells were cultured in antibiotic-free and phenol red-free medium (Sigma–Aldrich) containing L-glutamine, sodium pyruvate, and 10% charcoal-dextran-stripped FBS (GE Healthcare, Salem, CT, USA). For electroporation, a mixture of 300 ng pGL 4.36 vector [luc2P, MMTV, Hygro] (Promega, Madison, WI, USA) and 3 ng internal standard pNL 1.1 PGK vector [Nluc PGK] (Promega) were used for 1.0–1.6×10^5 cells with a MicroPorator MP-100 device (NanoEnTek, Seoul, Korea). The electroporation condition was two times voltage change to 1300 V for 20 ms. After transfection, cells were seeded on a 24-well plate and cultured in the antibiotic-free and phenol red-free DMEM containing 10% charcoal-dextran-stripped FBS. One day after electroporation, the medium was changed to the medium with Kampo medicines, herbs, or their components together with 1 μM Dex for 6 h. At 30 h after transfection, cells were collected using 100 μL of Reporter Lysis Buffer (Promega). The collected cell lysates were stored at −80 °C. GR-dependent luciferase activity was analyzed with the Dual-Luciferase® Reporter Assay System kit (Promega). Emission intensity was measured using the GloMax® Discover Microplate Reader GM3510 (Promega). The results were normalized to the internal pNL 1.1 PGK control and expressed as the fold-change of mean relative intensity. All experiments were performed at least three times.

2.7. Statistical Analysis

Results are expressed as mean ± standard deviation. Comparisons between groups were made using Student's *t*-test or one-way analysis of variance (one-way ANOVA). $p < 0.05$ was considered statistically significant. Statistical analysis was performed using JMP® Pro 14.2.0 (SAS Institute, Cary, NC, USA).

3. Results

3.1. KSS Reduces Intracellular Lipid Accumulation

Mouse 3T3-L1 cells, which are preadipocytes, were differentiated into adipocytes by DMI treatment. We examined the effects of KSS and the three Kampo medicines. Kampo medicines are composed of two to ten herbs. The number and volume of herbs are changed in the prescribed formulation depending on the intensity of the chief and indefinite complaints. The 129 kinds of Kampo medicines can be divided into two groups: those with and without Bupleuri Radix (BR). KSS contains BR and is prescribed for climacteric disorder and orofacial pain [35,36]. In this study, we examined the effects of Hochuekkito (HET), Shoseiryuto (SST), and Goreisan (GRS), as the control of KSS. HET contains BR, but its prescription spectrum is different from KSS. SST does not contain BR, and its spectrum is different from KSS. GRS does not contain BR but has a similar spectrum to KSS [18,37]. As shown in Figure 1A, KSS, HET, SST, or GRS (1 mg/mL) was added to DMEM together with DMI, and the cells were cultured for 8 days. Cells were stained with Oil-Red-O to assess lipid accumulation and the degree of differentiation [12]. Only KSS reduced the Oil-Red-O staining (Figure 1B). To compare the quantity of the accumulated lipid, we dissolved the accumulated Oil-Red-O in isopropanol and measured the absorbance. As shown in Figure 1Ca, KSS significantly reduced lipid accumulation. In addition, the CCK-8 assay showed that KSS, HET, SST, and GRS had no effect on cell viability at a concentration of 1 mg/mL (Figure 1Cb).

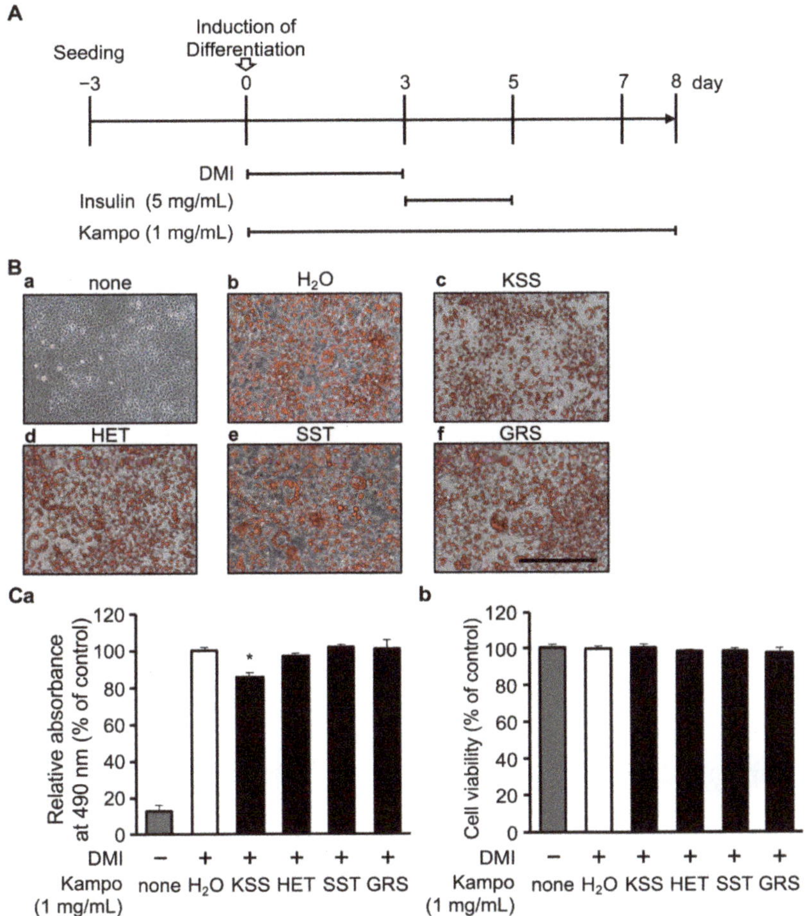

Figure 1. Kamishoyosan (KSS) reduces lipid accumulation. (**A**) Induction time course of 3T3-L1 cell differentiation. 3T3-L1 cells were seeded on a 24-well plate at 2×10^4 cells per well and cultured for 3 days. After confluent (0 day), 1 μM dexamethasone (Dex; D), 0.5 mM 3-isobutyl-1-methylxanthine (IBMX; M), and 10 μg/mL insulin (I) were added to induce differentiation for 3 days followed by additional 2 days culture with 5 μg/mL insulin (day 3–5). Thereafter, the medium was changed to normal Dulbecco's modified Eagle's medium (DMEM) every 2 days (day 5–8). Kampo medicines were administered at a concentration of 1 mg/mL after the onset of differentiation (day 0–8). (**B**) Images of Oil-Red-O staining of 3T3-L1 cells on day 8 are shown. (**Ba**) None (no DMI differentiation induction). (**Bb–f**) control and Kampo medicine administration groups ((**Bb**); H_2O (**Bc**); KSS, (**Bd**); Hochuekkito (HET), (**Be**); Shoseiryuto (SST), (**Bf**); Goreisan (GRS) with DMI differentiation induction). The scale bar indicates 200 μm. (**C**) KSS suppresses lipid accumulation. (**Ca**) Oil-Red-O dye was extracted with isopropanol to measure intracellular lipid accumulation, and the absorbance was measured at 490 nm as compared with H_2O (with DMI differentiation induction). (**Cb**) Cell viability after 8 days of culture was evaluated with the CCK-8 assay. Absorbance measured at 450 nm is shown as relative absorbance to H_2O (with DMI differentiation induction). Data are shown as the mean ± standard deviation ($n = 3$). * $p < 0.05$ vs. H_2O (with DMI differentiation induction).

Next, we examined the dose-dependent effect of KSS. 3T3-L1 cells were differentiated for 8 days in the presence of 0.1, 1, 2, 5, or 10 mg/mL KSS. The cells were stained with Oil-Red-O. KSS reduced lipid accumulation (Figure 2A,Ba) without an appreciable change in cell viability at a concentration up to 5 mg/mL (Figure 2Bb). In the subsequent experiments, a concentration of KSS of 5 mg/mL or less was used.

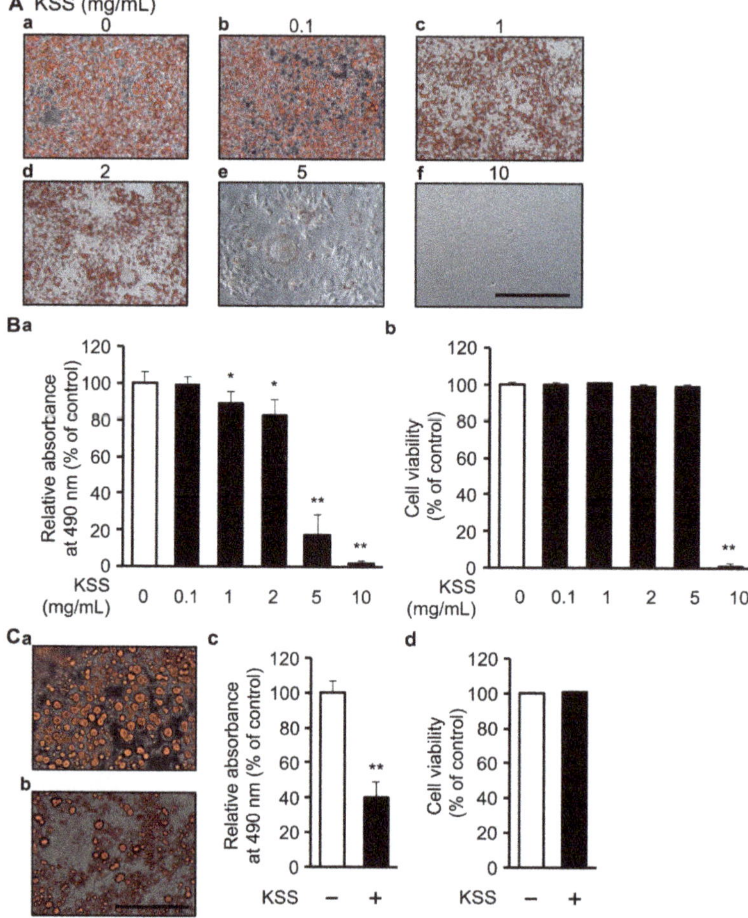

Figure 2. KSS reduces lipid accumulation in a dose-dependent manner. (**A**) KSS inhibited lipid accumulation in 3T3-L1 cells induced by the DMI method in a dose-dependent manner. Oil-Red-O staining images of cells cultured for 8 days is shown. (**Aa**) KSS 0 mg/mL, (**Ab**) 0.1 mg/mL, (**Ac**) 1 mg/mL, (**Ad**) 2 mg/mL, (**Ae**) 5 mg/mL, and (**Af**) 10 mg/mL. The scale bar indicates 200 μm. (**B**) KSS inhibited the lipid accumulation in 3T3-L1 cells cultured for 8 days in a dose-dependent manner. As in Figure 1, (**Ba**) the amount in lipid accumulation and (**Bb**) the cell viability was measured. (**C**) KSS suppressed lipid accumulation after 21 days of administration. (**Ca,b**) Oil-Red-O staining images of cultured cells are shown. (**Ca**) KSS (0 mg/mL) and (**Cb**) KSS (2 mg/mL). The scale bar indicates 200 μm. As in (**B**), (**Cc**) the amount of lipid accumulation was measured and is expressed as relative absorbance. (**Cd**) Cell viability. Data are shown as the mean ± standard deviation ($n = 3$). * $p < 0.05$ and ** $p < 0.01$ vs. H_2O.

Next, we examined the effects of KSS on lipid accumulation for longer periods. 3T3-L1 cells were differentiated for 21 days in the presence of 2 mg/mL KSS. KSS reduced the Oil-Red-O staining without affecting the cell viability (Figure 2Ca,b). KSS reduced the rate of lipid accumulation at 21 days more strongly than at 8 days (Figure 2Cc,d).

3.2. MC, Paeoniflorin, and Paeonol Inhibit Differentiation of 3T3-L1 Cells into Mature Adipocytes

We sought to identify which of the 10 herbs of KSS suppressed the accumulation of lipids (Table S1). The 10 herbs of KSS were individually applied to cells at a concentration of 2 mg/mL together with DMI, and the cells were cultured for 8 days (Figure 1A). The cells were then stained with Oil-Red-O. BR, PR, MC, GLR, and MH significantly inhibited lipid accumulation (Figure 3 and Table S1). Next, we examined the dependence on their concentration. 3T3-L1 cells were differentiated for 8 days in the presence of 0.03, 0.1, 0.3, 1, or 2 mg/mL of PR, MC, GLR, and MH. All four herbs inhibited lipid accumulation in a dose-dependent manner (Figure S3Aa,Ba,Ca,Da), and the cell viability did not change greatly, except for in the group administered 2 mg/mL MH (Figure S3Ab,Bb,Cb,Db). Each herb contains several components. Furthermore, we aimed to identify the major components of KSS that suppress lipid accumulation. The eight major components of KSS were selected based on both the 3D-HPLC (Figure S1 and Table S2) and the KSS product management information reported by Tsumura Co. Among the eight major components, herb-derived components that inhibited differentiation were BR-derived saikosaponin B1 (SSb1) and saikosaponin B2 (SSb2); MC-derived paeoniflorin (PNF) and paeonol (PN); and GLR-derived glycyrrhizic acid (GA). Herb-derived components that had no effect on differentiation were ALR-derived β-eudesmol standard (EU), GF-derived geniposide (GEN), and IR-derived 6-shogaol (SG). We excluded MH-derived L-menthol from the examination because there was no information on the 3D-HPLC (Figure S1). One of the eight major components of KSS was applied to the cells together with DMI, and the cells were cultured for 8 days.

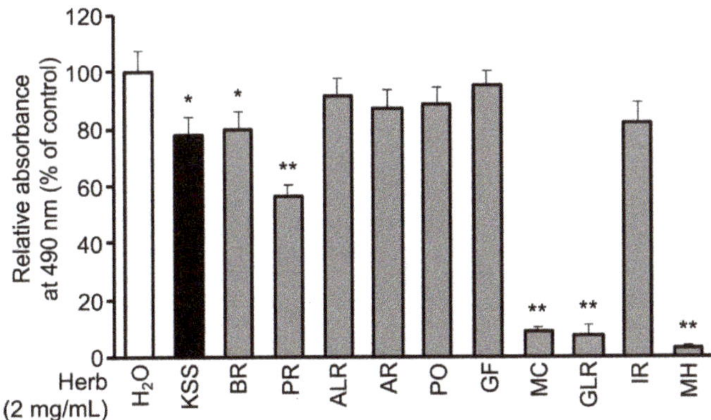

Figure 3. Bupleuri Radix (BR), Paeoniae Radix (PR), Moutan Cortex (MC), Glycyrrhizae Radix (GLR), and Menthae Herb (MH) in KSS reduce lipid accumulation. During differentiation induction by the DMI method, KSS or the 10 constituent herbs (2 mg/mL) were administered and cultured for 8 days. As in Figure 1, the amount of lipid accumulation was measured and is shown as relative absorbance. Data are shown as the mean ± standard deviation ($n = 3$). * $p < 0.05$ and ** $p < 0.01$ vs. H_2O.

The reduced number of Oil-red-O staining cell nodules observed in the treatment with 1000 µM PNF, 1000 µM PN, 300 µM SSb1, or 100 µM SSb2 compared with DMSO. Furthermore, Oil-Red-O staining was reduced in a dose-dependent manner (Figure 4A,B), but cells detached from the bottom by SSb1 and SSb2 within 2 days after the application (Figure 4A). SG at 30 µM reduced Oil-Red-O staining, 1000 µM GEN non-significantly reduced the red signal of Oil-Red-O staining, and 300 µM GA did not change the signal intensity (Figure S4A,B). EU at 1000 µM also reduced the red signal intensity, but cells detached from the bottom within 2 days after the application (Figure S4A). These results suggested that MC, PNF, and PN reduced lipid accumulation, and SSb1, SSb2, and EU were cytotoxic at the concentrations used. SSb1, SSb2, and EU were excluded from further studies.

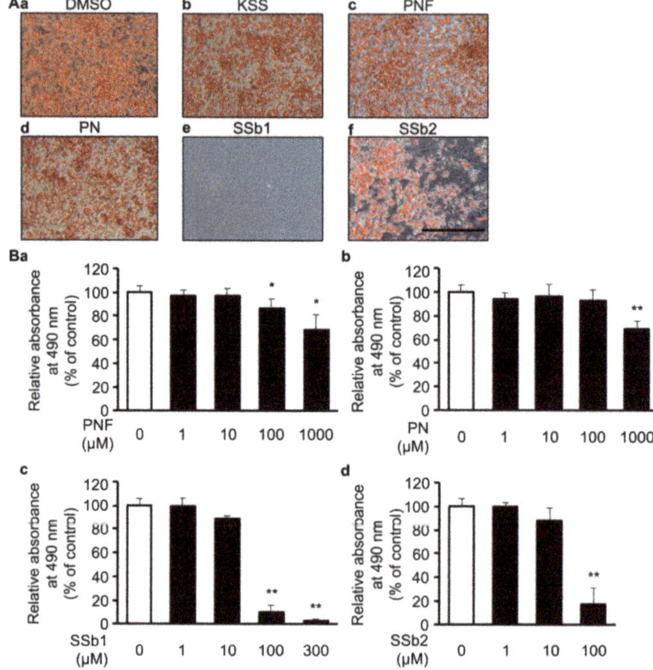

Figure 4. Paeoniflorin (PNF), Paeonol (PN), Saikosaponin B1 (SSb1), and Saikosaponin B2 (SSb2) reduce lipid accumulation in a dose-dependent manner. (**A**) (**Aa**) DMSO (DMI method only), (**Ab**) KSS 2 mg/mL, (**Ac**) PNF 1000 µM, (**Ad**) PN 1000 µM, (**Ae**) SSb1 300 µM, or (**Af**) SSb2 100 µM was added in culture media and cultured for 8 days. Images of Oil-Red-O staining are shown. The scale bar indicates 200 µm. (**B**) PNF, PN, SSb1, and SSb2 were administered during differentiation induction by the DMI method and cells were cultured for 8 days. As in Figure 1, the amount of lipid accumulation was measured and is shown as relative absorbance. (**Ba**) PNF, (**Bb**) PN, (**Bc**) SSb1, and (**Bd**) SSb2. Data are shown as mean ± standard deviation ($n = 3$). * $p < 0.05$ and ** $p < 0.01$ vs. DMSO.

3.3. KSS Inhibits Early Differentiation into Mature Adipocytes by Inhibiting the Action of Dex

In the process of differentiation of 3T3-L1 cells into mature adipocytes, various genes are expressed under strict regulation to form differentiation stages (early, middle, and late) (Jun do et al., 2011). To understand the differentiation stage at which KSS exhibited the inhibitory effect, 3T3-L1 cells were differentiated in the presence of 2 or 5 mg/mL KSS during 0–3 days (early stage), 3–5 days (middle stage), 5–8 days (late stage), and 0–8 days (Figure 5A). As shown in Figure 5B, the application of 2 mg/mL KSS during the early stage significantly reduced lipid accumulation (69.7 ± 11.3%, $n = 3$), similar to that during 0–8 days (66.7 ± 10.6%, $n = 3$). The inhibitory effect during the middle and the late stages was very low (77.5 ± 11.5%, $n = 3$ and 76.4 ± 8.0%, $n = 3$ for both). Next, we identified the target(s) of the inhibitory effect of KSS in the early stage of differentiation induced by DMI (Dex, IBMX, insulin). As shown in Figure 5C, Dex-deficient medium, IBMX-deficient medium, and insulin-deficient medium reduced lipid accumulation to 10.2 ± 6.2% ($n = 3$), 24.5 ± 13.1% ($n = 3$), and 70.3 ± 12.7% ($n = 3$), respectively, indicating that Dex and IBMX promote differentiation of 3T3-L1 cells. Moreover, 2 mg/mL KSS inhibited the lipid accumulation to 6.6 ± 2.8% ($n = 3$) in Dex-deficient medium, 13.5 ± 1.5% ($n = 3$) in IBMX-deficient medium, and 37.2 ± 4.3% ($n = 3$) in insulin-deficient medium. The reduction of the lipid accumulation by KSS was less in the Dex-deficient medium than in the IBMX-deficient medium, suggesting that KSS inhibits differentiation through Dex (Figure 5C,D).

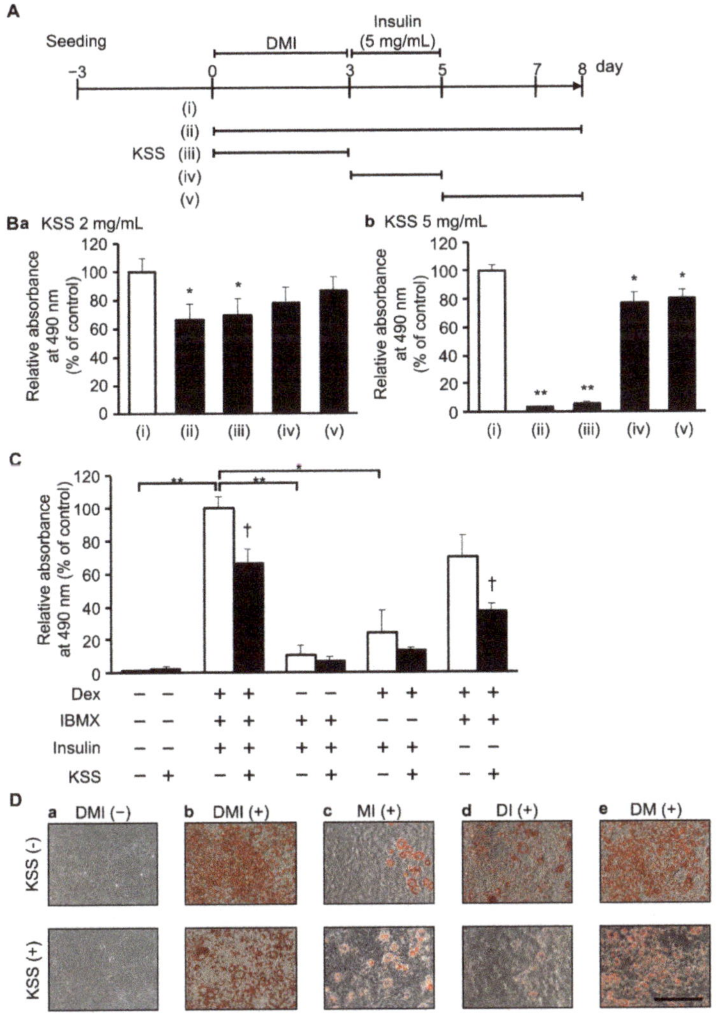

Figure 5. KSS inhibits early differentiation into adipocytes through the inhibition of Dex action. (**A**) 3T3-L1 cell differentiation stage and KSS administration time are shown. (i) No KSS administration, (ii) 0–8 days, (iii) 0–3 days (early-stage), (iv) 3–5 days (middle-stage), and (v) 5–8 days (late-stage). (**B**) KSS was administered at each differentiation stage. As in Figure 1, the amount of lipid accumulated at 8 days is measured and shown as relative absorbance. (**Ba**) KSS (2 mg/mL), (**Bb**) KSS (5 mg/mL). (**C**) 3T3-L1 cells were cultured for 3 days in the presence of two of the three types of differentiation inducers (1 μM Dex, 0.5 mM IBMX, and 10 μg/mL insulin). KSS (2 mg/mL) was administered for 8 days from the start of differentiation. The amount of lipid accumulated was measured as in Figure 1, and the relative absorbance is shown for the DMI method and KSS (-). (**D**) Oil-Red-O stained images of the cells cultured for 8 days. Upper: KSS (-), Lower: KSS (+), (**Da**) DMI (-), (**Db**) DMI (+), (**Dc**) MI (+), (**Dd**) DI (+), and (**De**) DM (+). The scale bar indicates 200 μm. Data are shown as mean ± standard deviation ($n = 3$). * $p < 0.05$ and ** $p < 0.01$ vs. (i) KSS (-) administration (**B**) or DMI method and KSS (-) (**C**). † $p < 0.05$ vs. KSS (-) (C).

3.4. KSS, MC, and Paeonol Alter the Expression of Adipocyte Differentiation Marker Genes

Since the expression levels of the genes of transcription factors C/EBP-δ, C/EBP-β, C/EBP-α, and PPAR-γ change during differentiation, they are considered to be differentiation marker in adipogenesis. We examined the expression of these four transcription factors by qPCR. KSS significantly suppressed gene expression of C/EBP-δ at differentiation day 3 (Figure 6Aa). Moreover, KSS suppressed gene expression of C/EBP-α, and PPAR-γ at differentiation day 8, no significant differences in the expression of C/EBP-β was observed (Figure 6Ab–d). Next, we examined the effect of MC and its constituents PNF and PN on the C/EBP-δ and C/EBP-β gene expression at differentiation day 3 (Figure 6B,C). MC at 1 mg/mL and 1000 µM PN suppressed gene expression of C/EBP-δ (Figure 6Ba,c). In addition, PN suppressed gene expression of C/EBP-β (Figure 6Cc). These results suggest that KSS, MC, and PN suppressed gene expression of C/EBP-δ at differentiation day 3. Supportively, the expression of adiponectin, a mature adipocyte marker and a secretion factor, was also diminished by KSS (Figure 6Da). On the other hand, KSS promoted the expression of FoxO1, that is an early marker of adipocyte differentiation (Figure 6Db).

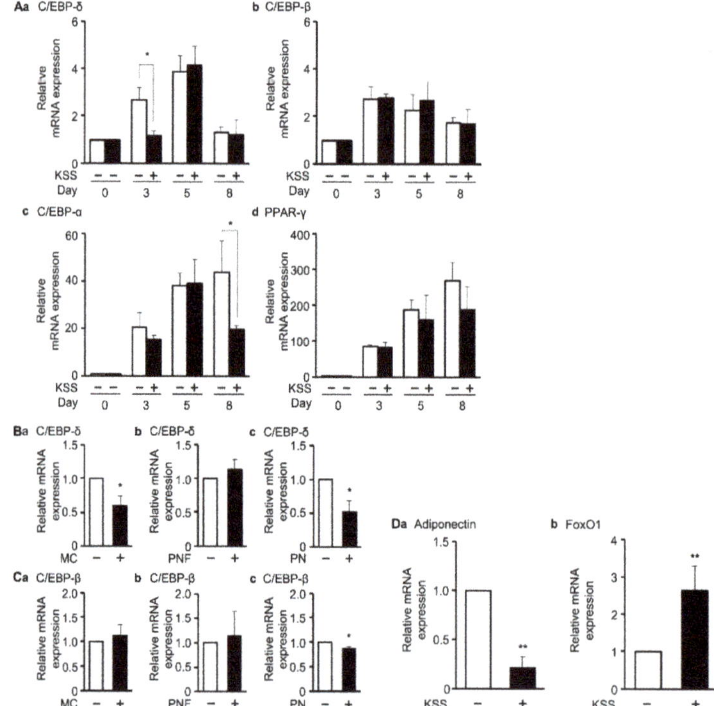

Figure 6. KSS, MC, and PN suppress C/EBP-δ gene expression. (**A**) KSS suppresses differentiation marker gene expression in a time-dependent manner. The gene expression level was quantified by the qPCR method. (**Aa**) C/EBP-δ, (**Ab**) C/EBP-β, (**Ac**) C/EBP-α, (**Ad**) PPAR-γ. (**B,C**) During differentiation induction, MC (1 mg/mL), PNF (1000 µM), and PN (1000 µM) were administered, and the expression level of the C/EBP-δ and C/EBP-β genes was quantified as in (**A**). (**a**) MC, (**b**) PNF, (**c**) PN. (**D**) Alteration of the expression level of (**Da**) Adiponectin and (**Db**) FoxO1 during differentiation induction with or without KSS. Data are shown as mean ± standard deviation ($n = 3$). * $p < 0.05$ and ** $p < 0.01$ vs. H_2O or DMSO.

3.5. KSS, MC, Paeoniflorin, and Paeonol Inhibit Promoter Activity of GR

Dex is involved in C/EBP-δ gene expression through the activation of the glucocorticoid receptor (GR) [38]. We investigated the effects of KSS, MC, PNF, and PN on the GR activity by luciferase reporter assay. The GR inhibitor RU486 was used as a GR antagonist [39]. KSS and MC inhibited Dex-induce GR promoter activity (Figure 7). Similarly, PNF and PN inhibited GR promoter activity, but the activity was weaker than that observed with RU486.

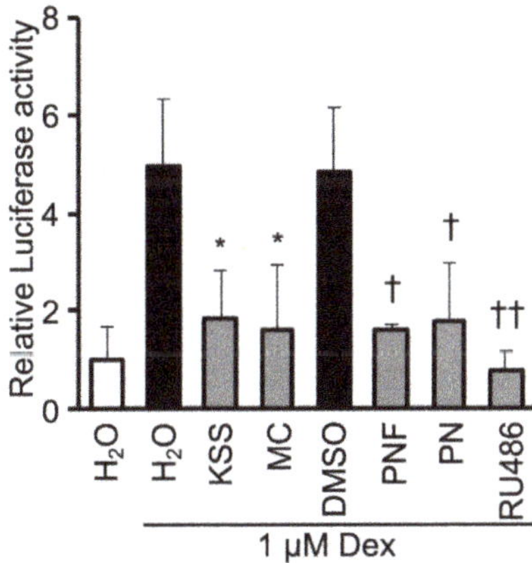

Figure 7. KSS, MC, PNF, and PN inhibit GR promoter activity. In the presence of 1 µM Dex, 2 mg/mL KSS, 1 mg/mL MC, 1000 µM PNF, 1000 µM PN, or 10 µM RU486 was administered, and a luciferase reporter assay was performed. Dex was used as a positive control for the GR promoter activity, and the relative activity was analyzed. Data are shown as the mean ± standard deviation ($n = 3$). * $p < 0.05$ vs. H₂O (Dex added). † $p < 0.05$ and †† $p < 0.01$ vs. DMSO (Dex added).

4. Discussion

In this study, we demonstrated that KSS suppressed C/EBP-δ expression by inhibiting Dex-induced GR promoter activity at the early stage of differentiation and, consequently, delayed differentiation into mature adipocytes as summarized in Figure 8. Kampo medicines are complex drugs composed of many herbs. The herbs have many components. The effects of Kampo medicines have not previously been analyzed hierarchically. We systematically analyzed the reduction of lipid accumulation caused by Kampo medicines, their herbs, and their major components. KSS inhibited the preadipocyte differentiation into mature adipocytes, but HET, SST, and GRS did not (Figures 1 and 2). MC was identified as an effective herb among the 10 constituent herbs of KSS (Figure 3 and Figures S2 and S3). Finally, PN was shown to be an effective major component among eight MC-derived components (Figure 4 and Figure S4). In addition to KSS, MC, and PN inhibited Dex-induced GR activity (Figures 6 and 7).

PN, an MC-derived ingredient, reduces lipid accumulation in the livers of high-fat diet-induced diabetic mice and improves glucose and lipid metabolism by increasing the phosphorylation level of Akt and expression of glucokinase and low-density lipoprotein receptor in human liver cancer-derived HepG2 cells [29]. In this study, PN reduced lipid accumulation by suppressing early differentiation into mature adipocytes by inhibiting Dex-induced GR promoter activity (Figures 4, 6 and 7) and the expression of the PPAR-γ gene, which is one of the late-stage markers of adipocytes. We hypothesize that PN-mediated anti-adipogenetic activity diminishes the phosphorylation of Akt by blocking GR promoter activity. Clarifying this detailed mechanism would be interesting for a better understanding of the function of PN in developing mature adipocytes.

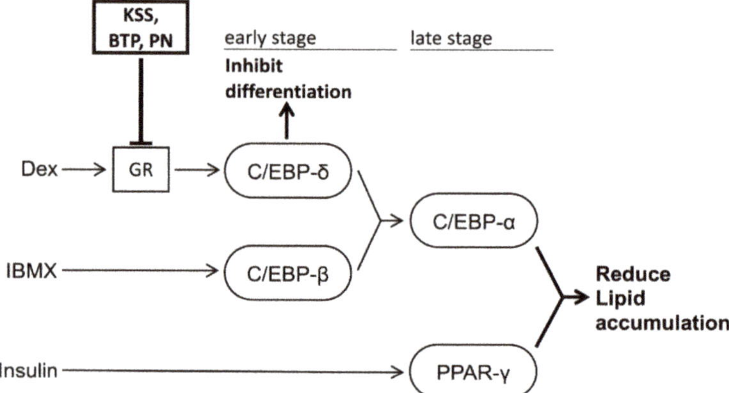

Figure 8. KSS, MC, and PN suppress lipid accumulation by suppressing Dex-induced GR activation and controlling early differentiation in 3T3-L1 cells. 3T3-L1 cells are differentiated into mature adipocytes by the addition of Dex, IBMX, and insulin (DMI method). As a result of differentiation, inducers bind to the receptor and the signals are transferred to the nucleus. KSS, MC, and PN inhibit the promoter activity of GR induced by Dex. This suppression of the promoter activity reduces the expression of the transcription factor C/EBP-δ gene and consequently inhibits the initial differentiation of 3T3-L1 cells. By suppressing early differentiation, the expression levels of C/EBP-α and PPAR-γ genes are reduced, and lipid accumulation is suppressed.

KSS as well as PN inhibited the activation of GR by Dex, which was followed by the suppression of C/EBP-δ gene expression. GR knockdown reduced lipid accumulation in adipocytes at day 7, even in the presence of insulin [40]. However, lipid accumulation and expression of the adipogenetic marker genes were intact in GR knockdown adipocytes at day 21 with the continuous supplementation of insulin [40]. KSS as well as PN reduced lipid accumulation at the late stage of differentiation (5–8 days) (Figure 5), similar to the results in the GR knockdown adipocyte model. However, KSS reduced lipid accumulation at day 21 more than that on day 8 (Figure 2). Since KSS was added at the late-stage of differentiation, it is possible that mechanisms other than suppression of Dex and GR might be involved in the long-term administration of PN.

One limitation of our study was that we could not quantify the content of the major components in the KSS extract supernatant that was used. Therefore, each drug was administered at the maximum dose at which the drug did not affect cell viability was administered. Kampo medicines and herbal medicines exert drug actions after ingestion. Information on the absorption-distribution-metabolism-excretion of Yokukansan is available [41], but there is no information on the absorption-distribution-metabolism-excretion of KSS. The blood concentration of PN, one of major components of KSS, after oral intake, is very small [42]. We will analyze the bioavailability of PN using an in vivo system and optimize the proper concentration of KSS needed to exert anti-adipogenetic activity in a future study.

5. Conclusions

In conclusion, our results demonstrate that KSS can inhibit the early stages of differentiation in mouse adipocytes and suppress lipid accumulation in 3T3-L1 cells. Paeonol is a critical ingredient that contributes to the anti-adipogenetic effect of KSS by blocking glucocorticoid receptor activity in lipid-metabolizing cells. The effects of Kampo medicine and herbal medicines have not been hitherto analyzed hierarchically. In this study, we identified the effects of the ingredients.

Supplementary Materials: The followings are available online at http://www.mdpi.com/2072-6643/12/2/309/s1. Figure S1: 3D-HPLC diagram of KSS, Figure S2: PR, MC, GLR, MH change in lipid droplets, Figure S3: PR, MC, GR, MH reduce lipid accumulation in a dose-dependent manner, Figure S4: EU, SG major components of KSS reduce fat accumulation in a dose-dependent manner, Table S1: Composition of ten herbs of KSS, Table S2: Eight major compounds of KSS.

Author Contributions: M.I. designed the study and wrote the manuscript. T.Y., T.N., and M.W. assisted in the preparation of the manuscript. M.I., T.Y., T.N., and M.W. contributed to data collection and interpretation and critically reviewed the manuscript. All authors approved the final version of the manuscript and agree to be accountable for all aspects of the work in ensuring that questions related to the accuracy or integrity of any part of the work are appropriately investigated and resolved.

Funding: This research received no external funding.

Acknowledgments: We are grateful to Akira Sugawara, and Shin Takayama, and Koh Iwasaki for advice and helpful discussion. Masayuki Izumi received a scholarship from The College Women's Association of Japan.

Conflicts of Interest: The authors declare no conflict of interest.

References

1. Haslam, D.; Sattar, N.; Lean, M. ABC of obesity: Obesity—Time to wake up. *BMJ* **2006**, *333*, 640–642. [CrossRef] [PubMed]

2. Kivipelto, M.; Ngandu, T.; Fratiglioni, L.; Viitanen, M.; Kareholt, I.; Winblad, B.; Helkala, E.L.; Tuomilehto, J.; Soininen, H.; Nissinen, A. Obesity and vascular risk factors at midlife and the risk of dementia and Alzheimer disease. *Arch. Neurol.* **2005**, *62*, 1556–1560. [CrossRef] [PubMed]

3. Vegiopoulos, A.; Herzig, S. Glucocorticoids, metabolism and metabolic diseases. *Mol. Cell. Endocrinol.* **2007**, *275*, 43–61. [CrossRef] [PubMed]

4. Kim, S.H.; Despres, J.P.; Koh, K.K. Obesity and cardiovascular disease: Friend or foe? *Eur. Heart J.* **2016**, *37*, 3560–3568. [CrossRef] [PubMed]

5. Oray, M.; Abu Samra, K.; Ebrahimiadib, N.; Meese, H.; Foster, C.S. Long-term side effects of glucocorticoids. *Expert Opin. Drug Saf.* **2016**, *15*, 457–465. [CrossRef] [PubMed]

6. Sunaga, H.; Matsui, H.; Ueno, M.; Maeno, T.; Iso, T.; Syamsunarno, M.R.; Anjo, S.; Matsuzaka, T.; Shimano, H.; Yokoyama, T.; et al. Deranged fatty acid composition causes pulmonary fibrosis in Elovl6-deficient mice. *Nat. Commun.* **2013**, *4*, 2563. [CrossRef] [PubMed]

7. Kasturi, R.; Wakil, S.J. Increased synthesis and accumulation of phospholipids during differentiation of 3T3-L1 cells into adipocytes. *J. Biol. Chem.* **1983**, *258*, 3559–3564.

8. Rubin, C.S.; Hirsch, A.; Fung, C.; Rosen, O.M. Development of hormone receptors and hormonal responsiveness in vitro. Insulin receptors and insulin sensitivity in the preadipocyte and adipocyte forms of 3T3-L1 cells. *J. Biol. Chem.* **1978**, *253*, 7570–7578. [PubMed]

9. Hu, E.; Liang, P.; Spiegelman, B.M. AdipoQ is a novel adipose-specific gene dysregulated in obesity. *J. Biol. Chem.* **1996**, *271*, 10697–10703. [CrossRef]

10. Madsen, M.S.; Siersbaek, R.; Boergesen, M.; Nielsen, R.; Mandrup, S. Peroxisome proliferator-activated receptor gamma and C/EBPalpha synergistically activate key metabolic adipocyte genes by assisted loading. *Mol. Cell. Biol.* **2014**, *34*, 939–954. [CrossRef]

11. Inokawa, A.; Inuzuka, T.; Takahara, T.; Shibata, H.; Maki, M. Tubby-like protein superfamily member PLSCR3 functions as a negative regulator of adipogenesis in mouse 3T3-L1 preadipocytes by suppressing induction of late differentiation stage transcription factors. *Biosci. Rep.* **2015**, *36*, e00287. [CrossRef] [PubMed]

12. Tanaka, T.; Yoshida, N.; Kishimoto, T.; Akira, S. Defective adipocyte differentiation in mice lacking the C/EBPbeta and/or C/EBPdelta gene. *EMBO J.* **1997**, *16*, 7432–7443. [CrossRef] [PubMed]

13. Janani, C.; Ranjitha Kumari, B.D. PPAR gamma gene—A review. *Diabetes Metab. Syndr.* **2015**, *9*, 46–50. [CrossRef] [PubMed]

14. Mi, L.; Chen, Y.; Zheng, X.; Li, Y.; Zhang, Q.; Mo, D.; Yang, G. MicroRNA-139-5p Suppresses 3T3-L1 Preadipocyte Differentiation Through Notch and IRS1/PI3K/Akt Insulin Signaling Pathways. *J. Cell. Biochem.* **2015**, *116*, 1195–1204. [CrossRef]

15. Nakae, J.; Kitamura, T.; Kitamura, Y.; Biggs, W.H.; Arden, K.C.; Accili, M. The forkhead transcription factor Foxo1 regulates adipocyte differentiation. *Dev. Cell* **2003**, *4*, 119–129. [CrossRef]

16. Iwasaki, K.; Satoh-Nakagawa, T.; Maruyama, M.; Monma, Y.; Nemoto, M.; Tomita, N.; Tanji, H.; Fujiwara, H.; Seki, T.; Fujii, M.; et al. A randomized, observer-blind, controlled trial of the traditional Chinese medicine Yi-Gan San for improvement of behavioral and psychological symptoms and activities of daily living in dementia patients. *J. Clin. Psychiatry* **2005**, *66*, 248–252. [CrossRef]

17. Takayama, S.; Arita, R.; Kikuchi, A.; Ohsawa, M.; Kaneko, S.; Ishii, T. Clinical Practice Guidelines and Evidence for the Efficacy of Traditional Japanese Herbal Medicine (Kampo) in Treating Geriatric Patients. *Front. Nutr.* **2018**, *5*, 66. [CrossRef]

18. Yamaguchi, K. Traditional Japanese herbal medicines for treatment of odontopathy. *Front. Pharmacol.* **2015**, *6*, 176. [CrossRef]

19. Huang, K.C.; Yen, H.R.; Chiang, J.H.; Su, Y.C.; Sun, M.F.; Chang, H.H.; Huang, S.T. Chinese Herbal Medicine as an Adjunctive Therapy Ameliorated the Incidence of Chronic Hepatitis in Patients with Breast Cancer: A Nationwide Population-Based Cohort Study. *Evid. Based Complement. Altern. Med.* **2017**, *2017*, 1052976. [CrossRef]

20. Herva, A.; Laitinen, J.; Miettunen, J.; Veijola, J.; Karvonen, J.T.; Laksy, K.; Joukamaa, M. Obesity and depression: Results from the longitudinal Northern Finland 1966 Birth Cohort Study. *Int. J. Obes. (Lond.)* **2006**, *30*, 520–527. [CrossRef] [PubMed]

21. Alexander, C.; Cochran, C.J.; Gallicchio, L.; Miller, S.R.; Flaws, J.A.; Zacur, H. Serum leptin levels, hormone levels, and hot flashes in midlife women. *Fertil. Steril.* **2010**, *94*, 1037–1043. [CrossRef] [PubMed]

22. Park, J.; Morley, T.S.; Kim, M.; Clegg, D.J.; Scherer, P.E. Obesity and cancer—mechanisms underlying tumour progression and recurrence. *Nat. Rev. Endocrinol.* **2014**, *10*, 455–465. [CrossRef] [PubMed]

23. Ju, H.; Jones, M.; Mishra, G.D. A U-Shaped Relationship between Body Mass Index and Dysmenorrhea: A Longitudinal Study. *PLoS ONE* **2015**, *10*, e0134187. [CrossRef] [PubMed]

24. Thanakun, S.; Pornprasertsuk-Damrongsri, S.; Izumi, Y. Increased oral inflammation, leukocytes, and leptin, and lower adiponectin in overweight or obesity. *Oral Dis.* **2017**, *23*, 956–965. [CrossRef]

25. Go, H.; Ryuk, J.A.; Hwang, J.T.; Ko, B.S. Effects of three different formulae of Gamisoyosan on lipid accumulation induced by oleic acid in HepG2 cells. *Integr. Med. Res.* **2017**, *6*, 395–403. [CrossRef] [PubMed]

26. Ikarashi, N.; Tajima, M.; Suzuki, K.; Toda, T.; Ito, K.; Ochiai, W.; Sugiyama, K. Inhibition of preadipocyte differentiation and lipid accumulation by Orengedokuto treatment of 3T3-L1 cultures. *Phytother. Res.* **2012**, *26*, 91–100. [CrossRef]

27. Izumi, M.; Seki, T.; Iwasaki, K.; Sakamoto, K. Chinese herbal medicine Yi-Gan-San decreases the lipid accumulation in mouse 3T3-L1 adipocytes by modulating the activities of transcription factors SREBP-1c and FoxO1. *Tohoku J. Exp. Med.* **2009**, *219*, 53–62. [CrossRef]

28. Kwak, D.H.; Lee, J.H.; Kim, D.G.; Kim, T.; Lee, K.J.; Ma, J.Y. Inhibitory Effects of Hwangryunhaedok-Tang in 3T3-L1 Adipogenesis by Regulation of Raf/MEK1/ERK1/2 Pathway and PDK1/Akt Phosphorylation. *Evid. Based Complement. Altern. Med.* **2013**, *2013*, 413906. [CrossRef]

29. Xu, F.; Xiao, H.; Liu, R.; Yang, Y.; Zhang, M.; Chen, L.; Chen, Z.; Liu, P.; Huang, H. Paeonol Ameliorates Glucose and Lipid Metabolism in Experimental Diabetes by Activating Akt. *Front. Pharmacol.* **2019**, *10*, 261. [CrossRef]

30. Kim, H.; Hiraishi, A.; Tsuchiya, K.; Sakamoto, K. (-) Epigallocatechin gallate suppresses the differentiation of 3T3-L1 preadipocytes through transcription factors FoxO1 and SREBP1c. *Cytotechnology* **2010**, *62*, 245–255. [CrossRef]

31. Kopinke, D.; Roberson, E.C.; Reiter, J.F. Ciliary Hedgehog Signaling Restricts Injury-Induced Adipogenesis. *Cell* **2017**, *170*, 340–351. [CrossRef] [PubMed]

32. Alonso-Vale, M.I.; Peres, S.B.; Vernochet, C.; Farmer, S.R.; Lima, F.B. Adipocyte differentiation is inhibited by melatonin through the regulation of C/EBPbeta transcriptional activity. *J. Pineal Res.* **2009**, *47*, 221–227. [CrossRef]

33. Renault, V.M.; Thekkat, P.U.; Hoang, K.L.; White, J.L.; Brady, C.A.; Kenzelmann Broz, D.; Venturelli, O.S.; Johnson, T.M.; Oskoui, P.R.; Xuan, Z.; et al. The pro-longevity gene FoxO3 is a direct target of the p53 tumor suppressor. *Oncogene* **2011**, *30*, 3207–3221. [CrossRef] [PubMed]

34. Ando, T.; Nishiyama, T.; Takizawa, I.; Miyashiro, Y.; Hara, N.; Tomita, Y. A carbon 21 steroidal metabolite from progestin, 20beta-hydroxy-5alpha-dihydroprogesterone, stimulates the androgen receptor in prostate cancer cells. *Prostate* **2018**, *78*, 222–232. [CrossRef] [PubMed]

35. Yasui, T.; Yamada, M.; Uemura, H.; Ueno, S.; Numata, S.; Ohmori, T.; Tsuchiya, N.; Noguchi, M.; Yuzurihara, M.; Kase, Y.; et al. Changes in circulating cytokine levels in midlife women with psychological symptoms with selective serotonin reuptake inhibitor and Japanese traditional medicine. *Maturitas* **2009**, *62*, 146–152. [CrossRef] [PubMed]

36. Arai, Y.C.; Makino, I.; Aono, S.; Yasui, H.; Isai, H.; Nishihara, M.; Hatakeyama, N.; Kawai, T.; Ikemoto, T.; Inoue, S.; et al. Effects of Kamishoyosan, a Traditional Japanese Kampo Medicine, on Pain Conditions in Patients with Intractable Persistent Dentoalveolar Pain Disorder. *Evid. Based Complement. Altern. Med.* **2015**, *2015*, 750345. [CrossRef]

37. Nagai, T.; Arai, Y.; Emori, M.; Nunome, S.Y.; Yabe, T.; Takeda, T.; Yamada, H. Anti-allergic activity of a Kampo (Japanese herbal) medicine "Sho-seiryu-to (Xiao-Qing-Long-Tang)" on airway inflammation in a mouse model. *Int. Immunopharmacol.* **2004**, *4*, 1353–1365. [CrossRef]

38. MacDougald, O.A.; Cornelius, P.; Lin, F.T.; Chen, S.S.; Lane, M.D. Glucocorticoids reciprocally regulate expression of the CCAAT/enhancer-binding protein alpha and delta genes in 3T3-L1 adipocytes and white adipose tissue. *J. Biol. Chem.* **1994**, *269*, 19041–19047.

39. Chivers, J.E.; Cambridge, L.M.; Catley, M.C.; Mak, J.C.; Donnelly, L.E.; Barnes, P.J.; Newton, R. Differential effects of RU486 reveal distinct mechanisms for glucocorticoid repression of prostaglandin E release. *Eur. J. Biochem.* **2004**, *271*, 4042–4052 [CrossRef]

40. Park, Y.K.; Ge, K. Glucocorticoid Receptor Accelerates, but Is Dispensable for, Adipogenesis. *Mol. Cell. Biol.* **2017**, *37*, e00260-16. [CrossRef]

41. Kitagawa, H.; Munekage, M.; Ichikawa, K.; Fukudome, I.; Munekage, E.; Takezaki, Y.; Matsumoto, T.; Igarashi, Y.; Hanyu, H.; Hanazaki, K. Pharmacokinetics of Active Components of Yokukansan, a Traditional Japanese Herbal Medicine after a Single Oral Administration to Healthy Japanese Volunteers: A Cross-Over, Randomized Study. *PLoS ONE* **2015**, *10*, e0131165. [CrossRef] [PubMed]

42. Wu, X.; Chen, H.; Chen, X.; Hu, Z. Determination of paeonol in rat plasma by high-performance liquid chromatography and its application to pharmacokinetic studies following oral administration of Moutan cortex decoction. *Biomed. Chromatogr.* **2003**, *17*, 504–508. [CrossRef] [PubMed]

Article

Whether AICAR in Pregnancy or Lactation Prevents Hypertension Programmed by High Saturated Fat Diet: A Pilot Study

Wan-Long Tsai [1], Chien-Ning Hsu [2] and You-Lin Tain [3],*

[1]   Department of Pediatrics, Chiayi Chang Gung Memorial Hospital, Chiayi County, Puzi City 613, Taiwan
[2]   Department of Pharmacy, Kaohsiung Chang Gung Memorial Hospital, Kaohsiung 833, Taiwan
[3]   Department of Pediatrics, Kaohsiung Chang Gung Memorial Hospital and Chang Gung University College of Medicine, Kaohsiung 833, Taiwan
*   Correspondence: tainyl@hotmail.com

Received: 14 January 2020; Accepted: 8 February 2020; Published: 11 February 2020

Abstract: High consumption of saturated fats links to the development of hypertension. AMP-activated protein kinase (AMPK), a nutrient-sensing signal, is involved in the pathogenesis of hypertension. We examined whether early intervention with a direct AMPK activator 5-aminoimidazole-4-carboxamide riboside (AICAR) during pregnancy or lactation can protect adult male offspring against hypertension programmed by high saturated fat consumption via regulation of nutrient sensing signals, nitric oxide (NO) pathway, and oxidative stress. Pregnant Sprague–Dawley rats received regular chow or high saturated fat diet (HFD) throughout pregnancy and lactation. AICAR treatment was introduced by intraperitoneal injection at 50 mg/kg twice a day for 3 weeks throughout the pregnancy period (AICAR/P) or lactation period (AICAR/L). Male offspring ($n = 7$–8/group) were assigned to five groups: control, HFD, AICAR/P, HFD + AICAR/L, and HFD + AICAR/P. Male offspring were killed at 16 weeks of age. HFD caused hypertension and obesity in male adult offspring, which could be prevented by AICAR therapy used either during pregnancy or lactation. As a result, we demonstrated that HFD downregulated AMPK/SIRT1/PGC-1α pathway in offspring kidneys. In contrast, AICAR therapy in pregnancy and, to a greater extent, in lactation activated AMPK signaling pathway. The beneficial effects of AICAR therapy in pregnancy is related to restoration of NO pathway. While AICAR uses in pregnancy and lactation both diminished oxidative stress induced by HFD. Our results highlighted that pharmacological AMPK activation might be a promising strategy to prevent hypertension programmed by excessive consumption of high-fat food.

Keywords: AMP-activated protein kinase; asymmetric dimethylarginine; developmental origins of health and disease (DOHaD); high-fat; hypertension; nitric oxide; nutrient-sensing signals; oxidative stress

---

1. Introduction

Hypertension is a highly prevalent disease that can have a substantial impact on the global burden of cardiovascular disease in all world regions. The developmental origins of disease hypothesis (DOHaD) suggests that the early life environment plays a key role in the early origins of hypertension [1,2]. Blood pressure (BP) is regulated by a complex process that contains key contributions from the kidney. Developmental programming of renal structure and function, namely renal programming [3], increases the risk for developing hypertension in adult life. Conversely, reprogramming intervention aimed at reversing the programming processes, preceding the onset of hypertension, to prevent or delay the development of hypertension [4].

A high-fat diet is commonly used in animal models to induce obesity-related diseases, such as hypertension [5]. Maternal high-fat intake has been reported to induce an increase [6,7] or no change [7,8] on offspring's BPs, mainly depending on the age, sex, and diverse fatty acids compositions [9]. Accumulative evidence indicates that diets containing high saturated fatty acids cause obesity/metabolic risk phenotypes, while high-fat diets based on poly-unsaturated fatty acids relates to beneficial effects [10]. Our previous study showed maternal and post-weaning high saturated fat (coconut oil-based) diets induced elevation of BP and kidney damage in male offspring at 24 weeks of age [7].

Dysregulated nutrient-sensing signals and impaired asymmetric dimethylarginine (ADMA, an endogenous inhibitor of nitric oxide synthase)-nitric oxide (NO) pathway are the proposed mechanisms underlying renal programming and programmed hypertension [11,12]. Fetal metabolism and development in response to maternal nutritional insults is mainly modulated by nutrient-sensing signaling pathways. In the kidney, several nutrient-sensing signals exist, including silent information regulator transcript (SIRT), cyclic adenosine monophosphate (AMP)-activated protein kinase (AMPK), peroxisome proliferator-activated receptors (PPARs), and PPARγ coactivator-1α (PGC-1α) [13]. Among them, AMPK is a serine/threonine protein kinase that serve as a central hub. AMPK is a heterotrimer composed of a catalytic α subunit in complex with two regulatory subunits, β and γ. All subunits exist in form of different isoforms (α1, α2, β1, β2, γ1, γ2, and γ3) [14]. AMPK is involved in BP control and renal physiology [15]. Additionally, AMPK has been shown to elicit antioxidant effects and regulate NO production [16].

HF diet was reported to reduce AMPK activity [17]. We previously observed that maternal metformin, an indirect AMPK activator, protects adult offspring against hypertension programmed by HFD [18]. Although a direct AMPK activator 5-aminoimidazole-4-carboxamide riboside (AICAR) has been reported to lower BP in adult spontaneously hypertensive rats (SHRs) [19], its reprogramming effect on programmed hypertension has not been examined yet. Thus, we hypothesized that maternal and post-weaning high saturated fat diets induced programmed hypertension via reducing AMPK and its related signals, whereas AICAR can prevent adult offspring against hypertension programmed by HFD.

2. Materials and Methods

2.1. Animal Models

The investigation conformed to the Institutional Animal Care and Use Committee of Kaohsiung Chang Gung Memorial Hospital (Permit number: 2018061302) that complies with the Guide for the Care and Use of Laboratory Animals of the National Institutes of Health. Virgin Sprague-Dawley (SD) rats (12–16 weeks old) were obtained from BioLASCO Taiwan Co., Ltd. (Taipei, Taiwan). Animals were maintained in an AAALAC-accredited animal facility, housed in a 12 h light/12 h dark cycle condition with a relative humidity of 55%. Rats were permitted ad libitum access to tap water and standard rat chow. Male SD rats were housed with individual females. Mating was confirmed by the examination of a vaginal plug.

Maternal rats received either a control diet with regular rat chow (Fwusow Taiwan Co., Ltd., Taichung, Taiwan; 52% carbohydrates, 23.5% protein, 4.5% fat, 10% ash, and 8% fiber) or high-fat diet (HFD; D12331, Research Diets, Inc., New Brunswick, NJ, USA; 58% fat (hydrogenated coconut oil) plus high sucrose (25% carbohydrate)) during pregnancy and lactation periods. In order to equal the received quantity of milk and maternal pup care, litters were standardized to eight pups per litter at birth. Because males are likely to have higher risk for developing hypertension younger than females [20], only male offspring were selected from each litter and used in subsequent experiments. Male offspring received either a control diet or HFD from post-weaning to 4 months of age. AICAR treatment was introduced by intraperitoneal injection at 50 mg/kg twice a day for 3 weeks throughout the pregnancy period (AICAR/P) or lactation period (AICAR/L) mixed in 0.9% sterile saline solution, and controls were treated with a 0.9% sterile saline solution vehicle. Another control group received AICAR treatment during the pregnancy period. The dose of AICAR used in pregnant rats was based on a previous study [21]. Male offspring were assigned to five groups (*n* = 7–8 for each group): control, HFD, AICAR/P, HFD + AICAR/L, and HFD + AICAR/P. Experimental protocol is illustrated in Figure 1. Only mother rats were given AICAR during pregnancy or lactation period. Male offspring was not treated with AICAR.

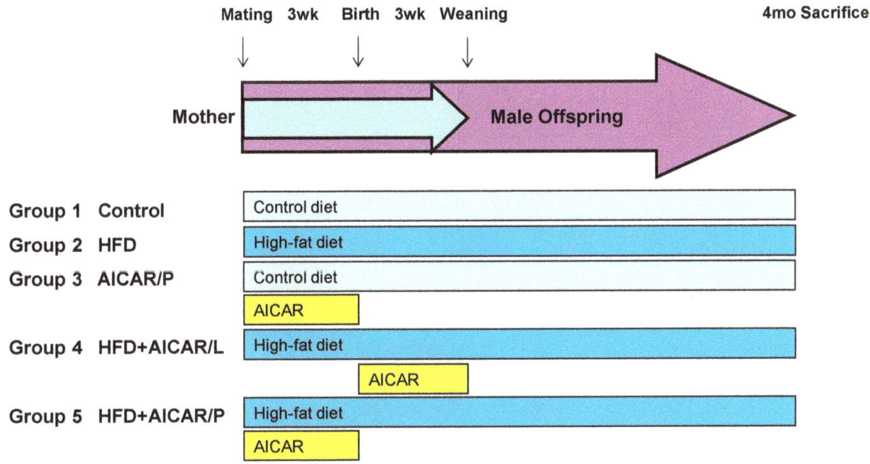

Figure 1. Animal protocol used in the present study.

BP was measured in conscious and previously trained rats by using an indirect tail-cuff method (BP-2000, Visitech Systems, Inc., Apex, NC, USA) at 4, 8, 12, and 16 weeks of age [7]. The rats were acclimated to restraint and tail-cuff inflation for 1 week prior to the measurement, to ensure accuracy and reproducibility. BP measurements were taken between 13.00 and 17.00 each day on a blinded basis by the same experienced research assistant. Rats were placed on specimen platform, and their tails were passed through tail cuffs and secured in place with tape. Following a 10 min warm-up period, ten preliminary cycles of tail-cuff inflation were performed to allow the rats to adjust to the inflating cuff. For each rat, five cycles were recorded at each time point. Average of values from three stable measurements was taken. Male offspring were killed at 16 weeks of age. An intraperitoneal injection of ketamine (50 mg/kg) and xylazine (10 mg/kg) were used to assess anesthesia. Rats were then euthanized by an intraperitoneal overdose of pentobarbital. Kidneys and heparinized blood samples were collected at the end of the study.

2.2. High-Performance Liquid Chromatography (HPLC)

The plasma levels of several components of the NO pathway, including L-citrulline (the precursor of L-arginine), L-arginine (the substrate for NO synthase), ADMA, and SDMA (an isomer of ADMA), were measured using HPLC with the *o*-phtalaldehyde-3-mercaptoprionic acid derivatization reagent [7]. Concentrations of 1–100 mM L-citrulline, 1–100 mM L-arginine, 0.5–5 mM ADMA, and 0.5–5 mM SDMA were used as standards. The recovery rate was approximately 95%.

2.3. Quantitative Real-Time Polymerase Chain Reaction (PCR)

RNA was extracted from the kidney cortex according to previously described methods [7]. Several genes related to the nutrient sensing signaling pathway were analyzed in this study, including SIRT-1 (*Sirt1*), SIRT-4 (*Sirt4*), AMPK-α2 (*Prkaa2*), -β2 (*Prkab2*), and -γ2 (*Prkag2*), PPAR-α (*Ppara*), -β (*Pparb*), and -γ (*Pparg*), and PGC-1α (encoded by *Ppargc1a*). The 18S rRNA gene (*Rn18s*) was used as a reference gene. Primer sequences are provided in Table 1. RNA expression levels were normalized to 18S rRNA levels and calculated according to the $\Delta\Delta$ comparative threshold (C_T) method. Values were then calculated relative to control to generate a $\Delta\Delta C_T$ value. The fold-increase of the experimental sample relative to the control was calculated using the formula $2^{-\Delta\Delta CT}$.

Table 1. Quantitative real-time polymerase chain reaction primers sequences.

Gene		Reverse
Sirt1	5 tggagcaggttgcaggaatcca 3	5 tggcttcatgatggcaagtggc 3
Sirt4	5 ccctttggaccatgaaaaga 3	5 cggatgaaatcaatgtgctg 3
Prkaa2	5 agctcgcagtggcttatcat 3	5 ggggctgtctgctatgagag 3
Prkab2	5 cagggccttatggtcaagaa 3	5 cagcgcatagagatggttca 3
Prkag2	5 gtgtgggagaagctctgagg 3	5 agaccacacccagaagatgc 3
Ppara	5 agaagttgcaggaggggatt 3	5 ttcttgatgacctgcacgag 3
Pparrb	5 gatcagcgtgcatgtgttct 3	5 cagcagtccgtctttgttga 3
Pparg	5 ctttatggagcctaagtttgagt 3	5 gttgtcttggatgtcctcg 3
Ppargc1a	5 cccattgagggctgtgatct 3	5 tcagtgaaatgccggagtca 3
Rn18s	5 gccgcggtaattccagctcca 3	5 cccgcccgctcccaagatc 3

2.4. Western Blot

Samples were subjected to electrophoresis, Western blot, and antibodies incubation using the methods published previously [7]. Briefly, 200 µg of kidney cortex were loaded on a 6–10% polyacrylamide gel and separated by electrophoresis (200 volts, 90 min). Following transfer to a nitrocellulose membrane (GE Healthcare Bio-Sciences Corp., Piscataway, NJ, USA), the membranes were incubated with Ponceau S red (PonS) stain solution (Sigma-Aldrich, St. Louis, MO, USA) for 10 min on the rocker to verify equal loading. After blocking with phosphate-buffered saline–Tween (PBS-T) containing 5% dry milk, the membranes were incubated with primary antibody. The blots were incubated overnight at 4 °C with a 1:1000 dilution of anti-phosphorylated AMPKα (Thr172) antibody (1:1000, #2535, Cell Signaling, Danvers, MA, USA). Following five washes with 0.1% Tween-Tris-buffered saline (TBS-T), the membranes were incubated for 1 h with horseradish peroxidase-labeled secondary antibody diluted 1:1000 in TBS-T. Bands were visualized using SuperSignal West Pico reagent (Pierce; Rockford, IL, USA) and quantified by densitometry as integrated optical density (IOD). IOD was then normalized to total protein PonS staining. Protein abundance was expressed as IOD/PonS.

2.5. Immunohistochemistry Staining

Paraffin-embedded kidney tissue sectioned at 3-µm thickness was deparaffinized in xylene and rehydrated in a graded ethanol series to phosphate-buffered saline. 8-Hydroxydeoxyguanosine (8-OHdG) is a DNA oxidation product that was measured to assess DNA damage. Nutrient sensing signal AMPKα2 and PGC-1α were also analyzed by immunohistochemistry. Following blocking with immunoblock (BIOTnA Biotech., Kaohsiung, Taiwan), the sections were incubated for 2 hr at room temperature with an anti-8-OHdG antibody (1:100, JaICA, Shizuoka, Japan), an anti-AMPKα2 antibody (1:400, Proteintech, Rosemont, IL, USA), and an anti-PGC-1α antibody (1:200, Abcam, Cambridge, MA, USA). Immunoreactivity was revealed using the polymer-horseradish peroxidase (HRP) labelling kit (BIOTnA Biotech). For the substrate-chromogen reaction, 3,3′-diaminobenzidine (DAB) was used. Identical staining protocol omitting incubation with primary antibody was employed to prepare samples that were used as negative controls. Quantitative analysis of positive cells per microscopic field (X400) in the renal sections was performed as we described previously [7].

2.6. Statistical Analysis

Values given in figures and tables represent mean ± standard error of mean (SEM). Statistical significance was determined using one-way ANOVA with a post-hoc Tukey test for multiple comparisons. In all cases, a *p*-value less than 0.05 was considered statistically significant. All analyses were performed using the Statistical Package for the Social Sciences software (SPSS, Chicago, IL, USA).

3. Results

HFD and AICAR treatment had no effect on the survival of male pups. Consumption of HF diet caused a greater body weight (BW) compared to controls, which was prevented by AICAR treatment in lactation (Table 2). The kidney weights and the ratios of kidney weight-to-body weight were lower in the HFD, HFD + AICAR/L, and HFD/AICAR/P groups compared to controls. AICAR treatment in lactation did not affect the kidney weight and the ratio of kidney weight-to-body weight vs. the control. At 16 weeks of age, a significant elevation of systolic BP (~20 mmHg) was noted in the HFD group compared with the other four groups.

Table 2. Measures of body weight, kidney weight, and blood pressure in 16-week-old male offspring exposed to high-fat diet (HFD) and 5-aminoimidazole-4-carboxamide riboside (AICAR) in pregnancy or lactation.

Groups	Control	HFD	AICAR/P	HFD + AICAR/L	HFD + AICAR/P
Number	7	8	8	8	8
BW (g)	610 ± 12	793 ± 17 [a]	606 ± 19	588 ± 26 [b]	644 ± 22
Left kidney weight (g)	2.36 ± 0.06	1.7 ± 0.06 [a]	2.02 ± 0.04	1.77 ± 0.07 [a]	1.5 ± 0.08 [a,c]
Left kidney weight/100 g BW	0.39 ± 0.01	0.31 ± 0.01 [a]	0.36 ± 0.01	0.31 ± 0.02 [a]	0.29 ± 0.01 [a,c]
Systolic blood pressure (mm Hg)	139 ± 1	164 ± 1 [a]	143 ± 1	146 ± 1 [b]	144 ± 1 [c]

HFD, high-fat diet; AICAR/P, AICAR treatment during pregnancy; HFD + AICAR/L, high-fat diet plus AICAR treatment during lactation; HFD + AICAR/P, high-fat diet plus AICAR treatment during pregnancy. BW, body weight; *n* = 7–8/group; [a] *p* < 0.05 vs. control; [b] *p* < 0.05 HFD vs. HFD + AICAR/L; [c] *p* < 0.05 HFD vs. HFD + AICAR/P.

As shown in Figure 2, systolic BP significantly increased in HFD group compared with that in control group from week 8 through 16. Compared to controls, AICAR treatment in pregnancy had no effect on systolic BP. While systolic BP was reduced by AICAR therapy in the HFD + AICAR/L and HFD + AICAR/P groups compared to that in the HDF group from 8 to 12 weeks of age. These data indicated that AICAR treatment either in pregnancy or lactation had similar protective effects on hypertension programmed by HFD.

We first evaluated the mRNA and protein level of AMPK (Figure 3). As a result, we demonstrated that HFD or AICAR in pregnancy had neglectable effect on renal mRNA expression of AMPK-α2, -β2, and -γ2. However, AICAR treatment in lactation significantly increased AMPK-α2, -β2, and -γ2 mRNA expression compared to the control as well as HFD group (Figure 3A). Also, AICAR treatment in pregnancy caused a higher AMPK-γ2 mRNA expression in the HFD + AICAR/P group than that in the controls. Additionally, HFD reduced phosphorylated AMPKα protein abundance compared to the controls (Figure 3B). Consistent with the change in mRNA level, AICAR treatment in lactation significantly increased the renal protein level of phosphorylated AMPKα in offspring kidneys.

We next analyzed AMPKα2 (Figure 3C) using the immunohistochemical examination of renal sections. Immunostaining of AMPKα2 in the glomeruli and renal tubules indicated intense staining in the control (145 ± 19 positive cells), AICAR/P group (135 ± 24 positive cells), and HFD + AICAR/L group (141 ± 26 positive cells), an intermediate level of staining in the HFD + AICAR/P group (95 ± 22 positive cells), and little staining in the HFD group (29 ± 11 positive cells) (Figure 3C).

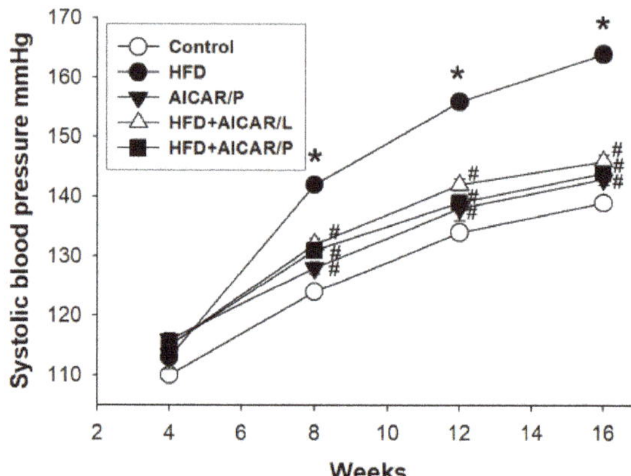

Figure 2. Effect of high-fat diet (HFD) and AICAR treatment in pregnancy (AICAR/P) or lactation (AICAR/L) on systolic blood pressure in male offspring. * $p < 0.05$ vs. control, # $p < 0.05$ vs. HFD.

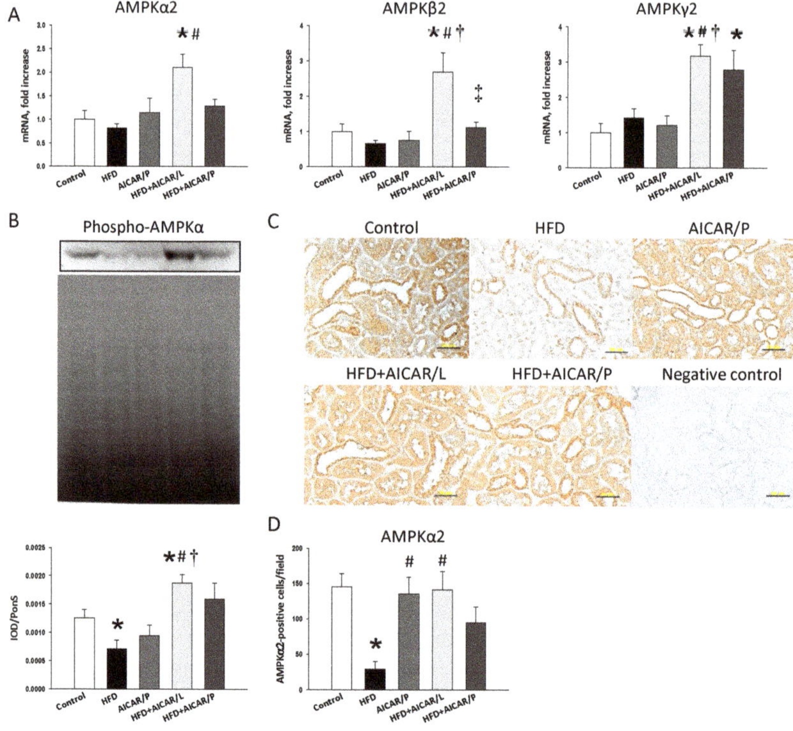

Figure 3. Effect of high-fat diet (HFD) and AICAR treatment in pregnancy (AICAR/P) or lactation (AICAR/L) on (**A**) mRNA expression of AMP-activated protein kinase (AMPK) α-, β-, and γ-subunits; (**B**) protein level of phosphorylated AMPKα (62 kDa) with represented blot and Ponceau S red (PonS) staining; (**C**) light microscopic findings of AMPKα2 immunostaining in the kidney cortex in 16-week-old male offspring; and (**D**) quantitative analysis of AMPKα2-positive cells per microscopic field (×400). Bar = 50 μm; * $p < 0.05$ vs. control, # $p < 0.05$ vs. HFD, † $p < 0.05$ vs. AICAR/P.

As AMPK is a key nutrient-sensing signal, we next analyzed the genes involved in the nutrient sensing pathway. As shown in Figure 4, HFD reduced renal mRNA expression of *Sirt1*. AICAR therapy in pregnancy significantly increased *Sirt1*, *Sirt4*, *Pparg* (encoding for PPARγ), and *Ppargc1a* (encoding for PGC-1α) in HFD + AICAR/P group compared to the other four groups. While AICAR treatment in pregnancy, unlike in lactation, had a negligible effect on nutrient-sensing signals. Also, we analyzed PGC-1α in the offspring kidneys by immunohistochemistry (Figure 5). Similar to phosphorylated AMPKα, AICAR treatment in lactation attenuated the reduction of PGC-1α expression caused by HFD (control: 221 ± 26 positive cells; HFD group: 75 ± 15 positive cells; HFD + AICAR/L group: 192 ± 23 positive cells). AICAR treatment in pregnancy also increased the immunostaining of PGC-1α (141 ± 24 positive cells) in the HFD + AIRCA/P group compared to that in the HFD group (Figure 5). Taken together, these findings indicated that HFD downregulated AMPK/SIRT1/PGC-1α pathway, which can be restored by AICAR treatment in lactation and to a lesser extent in pregnancy.

Because the interplay between ADMA-NO pathway and oxidative stress contributes to the pathogenesis of programmed hypertension [11,12], we further investigated whether AICAR treatment can mediate ADMA-NO pathway to prevent hypertension programmed by HFD (Table 3). Our results showed that plasma L-citrulline level (the precursor of L-arginine) did not differ among the five groups. HFD caused the decreases of plasma L-arginine level (the substrate for NO synthase) and L-arginine-to-ADMA ratio, an index of NO bioavailability [22]. AICAR therapy in pregnancy caused higher plasma L-arginine and ADMA levels in the AICAR/P groups than those in controls. AICAR therapy in lactation restored the increases of plasma SDMA levels induced by HFD. Additionally, AICAR therapy in pregnancy results in a higher plasma L-arginine level but a lower SDMA level in the HFD + AICAR/P groups than those in the HFD group.

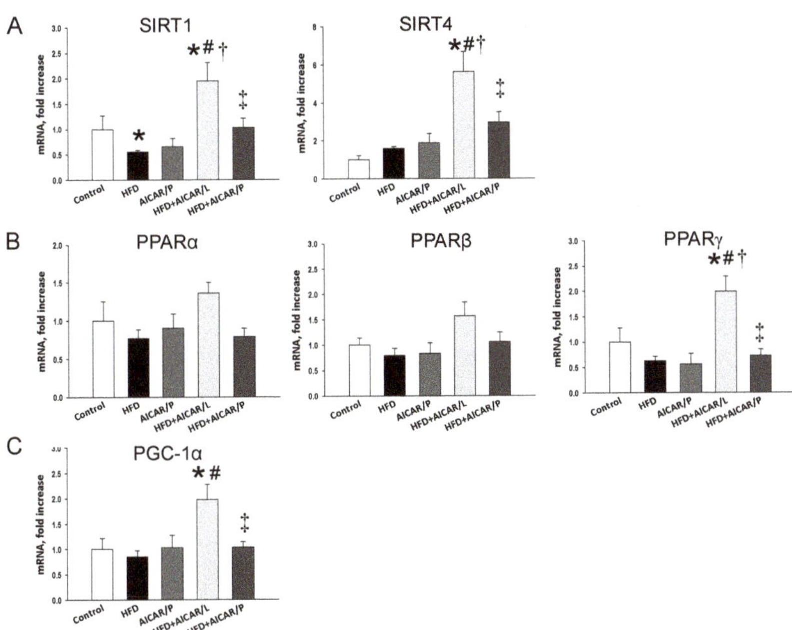

Figure 4. Effect of high-fat diet (HFD) and AICAR treatment in pregnancy (AICAR/P) or lactation (AICAR/L) on mRNA expression of (**A**) silent information regulator transcript 1 (SIRT1) and 4 (SIRT4); (**B**) peroxisome proliferator-activated receptor (PPAR) α-, β-, and γ-isoforms; and (**C**) PPARγ coactivator-1α (PGC-1α) in 16-week-old male offspring kidneys. * $p < 0.05$ vs. control, # $p < 0.05$ vs. HFD, † $p < 0.05$ vs. AICAR/P, ‡ $p < 0.05$ vs. HFD + AICAR/L.

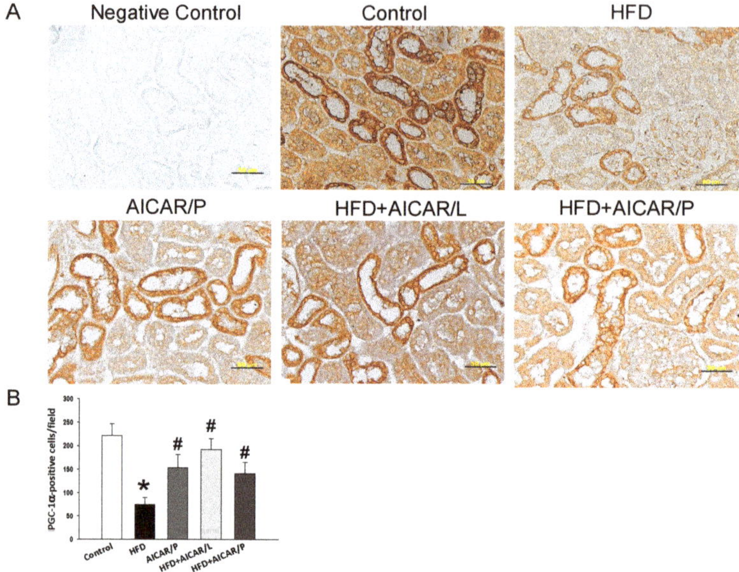

Figure 5. (**A**) Light microscopic findings of PPAR coactivator-1α (PGC-1α) immunostaining in the kidney cortex in 16-week-old male offspring. Bar = 50 μm; (**B**) quantitative analysis of PGC1α-positive cells per microscopic field (400×); * $p < 0.05$ vs. control, # $p < 0.05$ vs. HFD.

Table 3. Plasma ʟ-citrulline, ʟ-arginine, ADMA, and SDMA levels in 16-week-old male offspring exposed to high-fat diet (HFD) and AICAR in pregnancy or lactation.

Groups	Control	HFD	AICAR/P	HFD + AICAR/L	HFD + AICAR/P
L-citrulline	59.7 ± 4.5	53.7 ± 3.4	55.1 ± 4.6	57.1 ± 3.1	62.7 ± 3
L-arginine	141.4 ± 6.2	104.6 ± 4.4 [a]	171.2 ± 10.6 [a]	115.5 ± 3.3 [a]	140.2 ± 7.2 [c]
ADMA	1.35 ± 0.12	1.42 ± 0.09	1.73 ± 0.11 [a]	1.53 ± 0.06	1.19 ± 0.07
SDMA	0.57 ± 0.11	0.7 ± 0.03	0.7 ± 0.06	0.52 ± 0.02 [b]	0.49 ± 0.05 [c]
L-arginine-to-ADMA ratio	110 ± 12	75 ± 3 [a]	99 ± 3	76 ± 3 [a]	118 ± 4 [c]

ADMA, asymmetric dimethylarginine; SDMA, symmetric dimethylarginine; HFD, high-fat diet; AICAR/P, AICAR treatment during pregnancy; HFD + AICAR/L, high-fat diet plus AICAR treatment during lactation; HFD + AICAR/P, high-fat diet plus AICAR treatment during pregnancy; n = 7/group; [a] $p < 0.05$ vs. control; [b] $p < 0.05$ HFD vs. HFD + AICAR/L; [c] $p < 0.05$ HFD vs. HFD + AICAR/P.

We further assessed oxidative stress by immunostaining of 8-hydroxydeoxyguanosine (8-OHdG), a metabolite of oxidative damage to leukocyte DNA. Cytoplasmic and nuclear staining was present with little staining in the control group (5 ± 1 positive cells), a similar intermediate intensity in the AICAR/P (60 ± 3 positive cells), HFD + AICAR/L (55 ± 10 positive cells), and HFD + AICAR/P group (63 ± 9 positive cells), and intense staining in the HFD (155 ± 22 positive cells) (Figure 6). Additionally, HFD + AICAR/P group had a higher L-arginine level, and a lower ADMA level and L-arginine-to-ADMA ratio than those in the HFD + AICAR/L group. Taken together, our findings implied that AICAR therapy in pregnancy protects hypertension programmed by HFD is related to restoration of ADMA-NO pathway. Unlike its use in pregnancy, AICAR treatment in lactation had neglectable effect on ADMA-NO pathway.

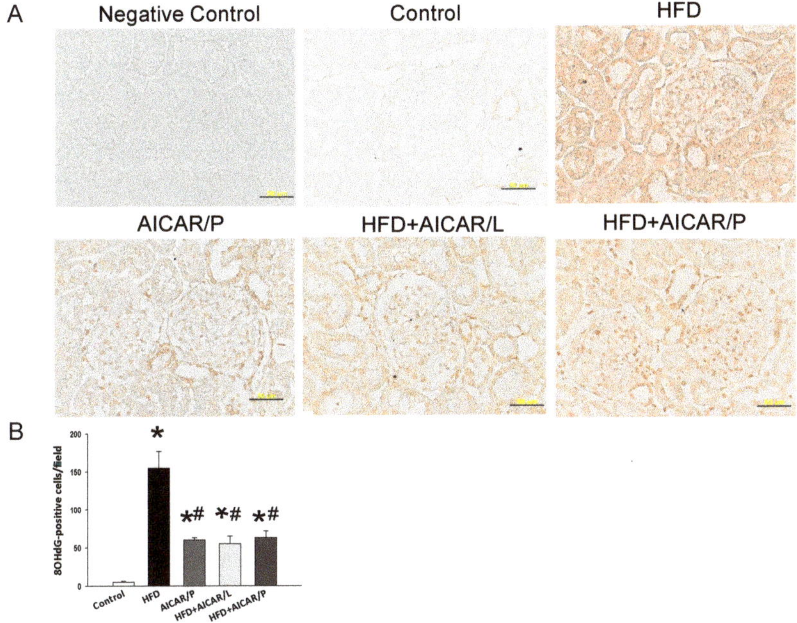

Figure 6. (**A**) Light microscopic findings of 8-hydroxydeoxyguanosine (8-OHdG) immunostaining in the kidney cortex in 16-week-old male offspring. Bar = 50 μm; (**B**) quantitative analysis of 8-OHdG-positive cells per microscopic field (×400); * $p < 0.05$ vs. control, # $p < 0.05$ vs. HFD.

4. Discussion

Our study describes, for the first time, direct AMPK activation using its activator AICAR protects male offspring against hypertension programmed by maternal plus post-weaning high saturated fat intake and puts special emphasis on the analysis of nutrient-sensing signals and oxidative stress. The major findings of this study were: (1) maternal plus post-weaning HFD resulted in hypertension and obesity in male adult offspring, which could be prevented by AICAR therapy used either during gestation or lactation; (2) HFD induced programmed hypertension is related to downregulation of AMPK/SIRT1/PGC-1α pathway; (3) therapeutic use of AICAR in lactation, to a greater extent than in pregnancy, activated AMPK signaling pathway; (4) the anti-hypertensive effect of AICAR therapy used in pregnancy and/or lactation is, at least in part, due to restoration of ADMA-NO pathway; and (5) both AICAR uses in pregnancy and lactation attenuated oxidative stress programmed by HFD in offspring kidneys.

Unlike other direct AMPK activators show isoform-specific activations, AICAR is a potent pan-activators for all 12 heterotrimeric AMPK complexes [23]. In line with previous studies using indirect AMPK activators [18,24–26], this is the first report of AICAR therapy activating AMPK signaling pathway to prevent hypertension of developmental origins [26]. We observed that the anti-hypertensive effect of AICAR either used during pregnancy or lactation was starting in week 8 and over time. Results from the present study support the argument that the reduction of BP is mainly due to reprogramming effect of AICAR instead of its acute hypotensive effect. Although AICAR has been shown good safety in a clinical trial to treat cancer patients [27], at high concentration it is toxic for mammalian cells [28]. As a result, we demonstrated that AICAR therapy in pregnancy or lactation showed similar offspring's BP. Of note is that AICAR use in lactation had a greater effect on activating AMPK signaling pathway than in pregnancy. As AICAR can induce AMPK activity in placenta [21], its transfer and metabolism by the placenta might explain the discrepancy. However, additional studies are required to elucidate the ideal dose and timing for AICAR use in programmed hypertension.

The results of this study showed that one positive effect of AICAR is relevant to activation of the AMPK/SIRT1/PGC-1α pathway. Early-life environmental insults can program AMPK and other nutrient-sensing signals to regulate PPARs and their target genes, hence provoking programmed hypertension [29]. Since that systemic BP was higher in AMPKα2 knockout mice than in wildtype mice [30], and that our previous studies showed AMPKα2 protein is related to programmed hypertension [25,31], we mainly focused on determining AMPKα2 protein level in the current study. Our previous study reported that resveratrol, a known natural activator of AMPK, prevents hypertension programmed by HFD associated with increased protein levels of SIRT1 and AMPKα2 [25]. Additionally, we recently found that resveratrol therapy activated AMPK/SIRT1/PGC-1α pathway and protected offspring against hypertension and oxidative stress in another hypertension model of programming [31].

Given that resveratrol has multifaceted biological functions, its reprogramming effects might not directly relate to AMPK activation. Results of the current study go beyond previous reports, showing that therapeutic use of direct AMPK activator AICAR in lactation activated AMPK/SIRT1/PGC-1α pathway and prevented the development of hypertension concurrently. Although AICAR use in pregnancy had a minor effect of mRNA expression of several nutrient-sensing signals and protein level of AMPKα2 compared to its use in lactation, both treatments displayed similar impacts on protein levels of PGC-1α. Future research should certainly further test whether other isoform-specific AMPK activators can serve as potential intervention to reverse the programming processes to prevent the developmental programming of hypertension and examine them in different kinds of programmed hypertension models.

Another beneficial effect of AICAR is via restoration of NO system. Our earlier report showed that metformin, an indirect AMPK activator, prevents the development of hypertension in spontaneously hypertensive rats via reducing ADMA and increasing NO production [32]. In the current study, AICAR therapy in pregnancy prevents hypertension that is associated with increased plasma levels of L-arginine and L-arginine-to-ADMA ratios, but decreased plasma levels of ADMA and SDMA. Since L-arginine-to-ADMA ratios represent NO bioavailability [22], both ADMA and SDMA are inhibitors of NO synthase [33], signals formed by NO pathway are supposed to cause vasodilatation. Therefore, our data are in accordance with the previous findings reported that AICAR ameliorates portal hypertension via preserving NO pathway [34]. We conducted the AICRA/P group to evaluate the programming effect of AICAR in control offspring. Although AICAR use in pregnancy had neglectable effects on BP and nutrient-sensing signal pathway in controls, our results showed higher plasma L-arginine and ADMA levels and greater intensity of 8-OHdG staining in the AICAR/P group compared to controls. Thus, whether AICAR use in pregnancy might induce other programming effects in control offspring deserves further clarification.

In this work, another preventive effect of AICAR against HFD-induced programmed hypertension might be, at least in part, due to reduction of oxidative stress. Oxidative stress is a well-known mechanism involved in the developmental programming of hypertension [12]. AMPK activation has been shown to suppress pro-oxidant enzymes and upregulate anti-oxidant enzymes, to reduce oxidative stress [35,36]. Our data demonstrated the presence of 8-OHdG staining, an oxidative stress damage marker, in the offspring exposed to HFD, which was attenuated by AICAR therapy. A similar pattern of results was obtained from AICAR therapy either used during pregnancy or lactation period. These findings support the notion that AMPK activation by AICAR in early life could prevent HFD-induced oxidative stress in adult offspring.

Our study has some limitations that are worth noting. Although current research is geared primarily to finding the beneficial effect of AICAR in the kidneys, its protective effect may come from other organs that regulate BP, such as the brain, the heart, and the vasculature. Although AMPK activation has been initially recognized as a dominant effect of AICAR, this compound also triggers

AMPK-independent effects [37]. Further studies should investigate organ- and isoform-specific effects of AICAR and other AMPK activators to clarify their relationships with programmed hypertension in different models of programmed hypertension. Second, we did not examine the different doses of AICAR, regardless of we did test AICAR in two therapeutic durations. Although therapeutic uses of AICAR in pregnancy or lactation exert similar BP-lowering effects, they had differential impacts on NO system and nutrient-sensing signals. This is an interesting topic for future work. We did not conduct a control group that received AICAR treatment during lactation because we have conducted the AICAR/P group to evaluate programming effects of AICAR on normal control offspring. Nevertheless, the programming effects of AICAR treatment during pregnancy or lactation might be different and that deserve further elucidation. Lastly, we did not evaluate sex difference in response to AICAR, as only male offspring were recruited in this study. The reason for this is due to the nature of programmed hypertension, hypertension occurred at an early age in males than females [20] and we measured BP in young adulthood.

5. Conclusions

In summary, the current findings cast a new light on therapeutic uses of AICAR in pregnancy or lactation to prevent hypertension programmed by HFD. Our results lend additional support to the notion that pharmacological AMPK activation can be a possible reprogramming strategy to improve the alarming scenario of hypertension and its related disorders. The ultimate challenge will be successful translation of the promising preclinical findings from animal studies into practical clinical applications.

Author Contributions: Conceptualization, Y.-L.T.; methodology, W.-L.T.; software, C.-N.H.; validation, Y.-L.T., W.-L.T., and C.-N.H.; formal analysis, W.-L.T. and C.-N.H.; investigation, Y.-L.T., W.-L.T., and C.-N.H.; data curation, Y.-L.T., W.-L.T., and C.-N.H.; writing—original draft preparation, Y.-L.T. and W.-L.T.; writing—review and editing, Y.-L.T., W.-L.T., and C.-N.H.; project administration, Y.-L.T.; funding acquisition, Y.-L.T. All authors have read and agreed to the published version of the manuscript.

Funding: This work was supported by grant CMRPG8H0831 from Chang Gung Memorial Hospital, Kaohsiung, Taiwan.

Acknowledgments: We would like to thank the Chang Gung Medical Foundation Kaohsiung Chang Gung Memorial Hospital Tissue Bank Core Lab (CLRPG8F1702) for excellent technical support.

Conflicts of Interest: The authors declare no conflict of interest.

References

1. Luyckx, V.A.; Bertram, J.F.; Brenner, B.M.; Fall, C.; Hoy, W.E.; Ozanne, S.E.; Vikse, B.E. Effect of fetal and child health on kidney development and long-term risk of hypertension and kidney disease. *Lancet* **2013**, *382*, 273–283. [CrossRef]
2. Barker, D.J.; Bagby, S.P.; Hanson, M.A. Mechanisms of disease: In utero programming in the pathogenesis of hypertension. *Nat. Clin. Pract. Nephrol.* **2006**, *2*, 700–707. [CrossRef] [PubMed]
3. Kett, M.M.; Denton, K.M. Renal programming: Cause for concern? *Am. J. Physiol. Regul. Integr. Comp. Physiol.* **2011**, *300*, R791–R803. [CrossRef]
4. Tain, Y.L.; Joles, J.A. Reprogramming: A preventive strategy in hypertension focusing on the kidney. *Int. J. Mol. Sci.* **2015**, *17*, 23. [CrossRef]
5. Buettner, R.; Schölmerich, J.; Bollheimer, L.C. High-fat diets: Modeling the metabolic disorders of human obesity in rodents. *Obesity* **2007**, *15*, 798–808. [CrossRef]
6. Torrens, C.; Ethirajan, P.; Bruce, K.D.; Cagampang, F.R.; Siow, R.C.; Hanson, M.A.; Byrne, C.D.; Mann, G.E.; Clough, G.F. Interaction between maternal and offspring diet to impair vascular function and oxidative balance in high fat fed male mice. *PLoS ONE* **2012**, *7*. [CrossRef]
7. Tain, Y.L.; Lin, Y.J.; Sheen, J.M.; Yu, H.R.; Tiao, M.M.; Chen, C.C.; Tsai, C.C.; Huang, L.T.; Hsu, C.N. High fat diets sex-specifically affect the renal transcriptome and program obesity, kidney injury, and hypertension in the offspring. *Nutrients* **2017**, *9*, 357. [CrossRef]
8. Williams, L.; Seki, Y.; Vuguin, P.M.; Charron, M.J. Animal models of in utero exposure to a high fat diet: A review. *Biochim. Biophys. Acta* **2014**, *1842*, 507–519. [CrossRef]

9. Khan, I.Y.; Taylor, P.D.; Dekou, V.; Seed, P.T.; Lakasing, L.; Graham, D.; Dominiczak, A.F.; Hanson, M.A.; Poston, L. Gender-linked hypertension in offspring of lard-fed pregnant rats. *Hypertension* **2003**, *41*, 168–175. [CrossRef]

10. Buettner, R.; Parhofer, K.G.; Woenckhaus, M.; Wrede, C.E.; Kunz-Schughart, L.A.; Schölmerich, J.; Bollheimer, L.C. Defining high-fat-diet rat models: Metabolic and molecular effects of different fat types. *J. Mol. Endocrinol.* **2006**, *36*, 485–501. [CrossRef] [PubMed]

11. Tain, Y.L.; Hsu, C.N. Targeting on asymmetric dimethylarginine related nitric oxide-reactive oxygen species imbalance to reprogram the development of hypertension. *Int. J. Mol. Sci.* **2016**, *17*, 2020. [CrossRef] [PubMed]

12. Tain, Y.L.; Hsu, C.N. Interplay between oxidative stress and nutrient sensing signaling in the developmental origins of cardiovascular disease. *Int. J. Mol. Sci.* **2017**, *18*, 841. [CrossRef]

13. Efeyan, A.; Comb, W.C.; Sabatini, D.M. Nutrient-sensing mechanisms and pathways. *Nature* **2015**, *517*, 302–310. [CrossRef] [PubMed]

14. Hardie, D.G.; Ross, F.A.; Hawley, S.A. AMPK: A nutrient and energy sensor that maintains energy homeostasis. *Nat. Rev. Mol. Cell Biol.* **2012**, *13*, 251–262. [CrossRef] [PubMed]

15. Rajani, R.; Pastor-Soler, N.M.; Hallows, K.R. Role of AMP-activated protein kinase in kidney tubular transport, metabolism, and disease. *Curr. Opin. Nephrol. Hypertens.* **2017**, *26*, 375–383. [CrossRef] [PubMed]

16. Cho, K.J.; Casteel, D.E.; Prakash, P.; Tan, L.; van der Hoeven, D.; Salim, A.A.; Kim, C.; Capon, R.J.; Lacey, E.; Cunha, S.R.; et al. AMPK and Endothelial Nitric Oxide Synthase Signaling Regulates K-Ras Plasma Membrane Interactions via Cyclic GMP-Dependent Protein Kinase 2. *Mol. Cell Biol.* **2016**, *36*, 3086–3099. [CrossRef] [PubMed]

17. Lindholm, C.R.; Ertel, R.L.; Bauwens, J.D.; Schmuck, E.G.; Mulligan, J.D.; Saupe, K.W. A high-fat diet decreases AMPK activity in multiple tissues in the absence of hyperglycemia or systemic inflammation in rats. *J. Physiol. Biochem.* **2013**, *69*, 165–175. [CrossRef]

18. Tain, Y.L.; Wu, K.L.H.; Lee, W.C.; Leu, S.; Chan, J.Y.H. Prenatal Metformin Therapy Attenuates Hypertension of Developmental Origin in Male Adult Offspring Exposed to Maternal High-Fructose and Post-Weaning High-Fat Diets. *Int. J. Mol. Sci.* **2018**, *19*, 1066. [CrossRef]

19. Ford, R.J.; Teschke, S.R.; Reid, E.B.; Durham, K.K.; Kroetsch, J.T.; Rush, J.W. AMP-activated protein kinase activator AICAR acutely lowers blood pressure and relaxes isolated resistance arteries of hypertensive rats. *J. Hypertens.* **2012**, *30*, 725–733. [CrossRef]

20. Grigore, D.; Ojeda, N.B.; Alexander, B.T. Sex differences in the fetal programming of hypertension. *Gend. Med.* **2008**, *5*, S121–S132. [CrossRef]

21. Banek, C.T.; Bauer, A.J.; Needham, K.M.; Dreyer, H.C.; Gilbert, J.S. AICAR administration ameliorates hypertension and angiogenic imbalance in a model of preeclampsia in the rat. *Am. J. Physiol. Heart Circ. Physiol.* **2013**, *304*, H1159–H1165. [CrossRef] [PubMed]

22. Bode-Böger, S.M.; Scalera, F.; Ignarro, L.J. The L-arginine paradox: Importance of the L-arginine/asymmetrical dimethylarginine ratio. *Pharmacol. Ther.* **2007**, *114*, 295–306. [CrossRef] [PubMed]

23. Steinberg, G.R.; Carling, D. AMP-activated protein kinase: The current landscape for drug development. *Nat. Rev. Drug Discov.* **2019**. [CrossRef] [PubMed]

24. Care, A.S.; Sung, M.M.; Panahi, S.; Gragasin, F.S.; Dyck, J.R.; Davidge, S.T.; Bourque, S.L. Perinatal Resveratrol Supplementation to Spontaneously Hypertensive Rat Dams Mitigates the Development of Hypertension in Adult Offspring. *Hypertension* **2016**, *67*, 1038–1044. [CrossRef]

25. Tain, Y.L.; Lin, Y.J.; Sheen, J.M.; Lin, I.C.; Yu, H.R.; Huang, L.T.; Hsu, C.N. Resveratrol prevents the combined maternal plus postweaning high-fat-diets-induced hypertension in male offspring. *J. Nutr. Biochem.* **2017**, *48*, 120–127. [CrossRef]

26. Tain, Y.L.; Hsu, C.N. AMP-Activated Protein Kinase as a Reprogramming Strategy for Hypertension and Kidney Disease of Developmental Origin. *Int. J. Mol. Sci.* **2018**, *19*, 1744. [CrossRef]

27. Van Den Neste, E.; Cazin, B.; Janssens, A.; González-Barca, E.; Terol, M.J.; Levy, V.; Pérez de Oteyza, J.; Zachee, P.; Saunders, A.; de Frias, M.; et al. Acadesine for patients with relapsed/refractory chronic lymphocytic leukemia (CLL): A multicenter phase I/II study. *Cancer Chemother. Pharmacol.* **2013**, *71*, 581–591. [CrossRef]

28. Ceschin, J.; Hürlimann, H.C.; Saint-Marc, C.; Albrecht, D.; Violo, T.; Moenner, M.; Daignan-Fornier, B.; Pinson, B. Disruption of Nucleotide Homeostasis by the Antiproliferative Drug 5-Aminoimidazole-4-carboxamide-1-β-d-ribofuranoside Monophosphate (AICAR). *J. Biol. Chem.* **2015**, *290*, 23947–23959. [CrossRef]

29. Tain, Y.L.; Hsu, C.N.; Chan, J.Y. PPARs Link Early Life Nutritional insults to later programmed hypertension and metabolic syndrome. *Int. J. Mol. Sci.* **2015**, *17*, 20. [CrossRef]

30. Wang, S.; Liang, B.; Viollet, B.; Zou, M.H. Inhibition of the AMP-activated protein kinase-α2 accentuates agonist-induced vascular smooth muscle contraction and high blood pressure in mice. *Hypertension* **2011**, *57*, 1010–1017. [CrossRef]

31. Chen, H.E.; Lin, Y.J.; Lin, I.C.; Yu, H.R.; Sheen, J.M.; Tsai, C.C.; Huang, L.T.; Tain, Y.L. Resveratrol prevents combined prenatal NG-Nitro-L-arginine-methyl ester (L-NAME) treatment plus postnatal high-fat diet induced programmed hypertension in adult rat offspring: Interplay between nutrient-sensing signals, oxidative stress and gut microbiota. *J. Nutr. Biochem.* **2019**, *70*, 28–37. [CrossRef] [PubMed]

32. Tsai, C.M.; Kuo, H.C.; Hsu, C.N.; Huang, L.T.; Tain, Y.L. Metformin reduces asymmetric dimethylarginine and prevents hypertension in spontaneously hypertensive rats. *Transl. Res.* **2014**, *164*, 452–459. [CrossRef] [PubMed]

33. Tain, Y.L.; Hsu, C.N. Toxic Dimethylarginines: Asymmetric Dimethylarginine (ADMA) and Symmetric Dimethylarginine (SDMA). *Toxins (Basel)* **2017**, *9*, 92. [CrossRef] [PubMed]

34. Hu, L.; Su, L.; Dong, Z.; Wu, Y.; Lv, Y.; George, J.; Wang, J. AMPK agonist AICAR ameliorates portal hypertension and liver cirrhosis via NO pathway in the BDL rat model. *J. Mol. Med. (Berl)* **2019**, *97*, 423–434. [CrossRef]

35. Trewin, A.J.; Berry, B.J.; Wojtovich, A.P. Exercise and Mitochondrial Dynamics: Keeping in Shape with ROS and AMPK. *Antioxidants* **2018**, *7*, 7. [CrossRef]

36. Song, P.; Zou, M.H. Regulation of NAD(P)H oxidases by AMPK in cardiovascular systems. *Free Radic. Biol. Med.* **2012**, *52*, 1607–1619. [CrossRef]

37. Philippe, C.; Pinson, B.; Dompierre, J.; Pantesco, V.; Viollet, B.; Daignan-Fornier, B.; Moenner, M. AICAR Antiproliferative Properties Involve the AMPK-Independent Activation of the Tumor Suppressors LATS 1 and 2. *Neoplasia* **2018**, *20*, 555–562. [CrossRef]

Article

The Effect of N-Acetylcysteine on Respiratory Enzymes, ADP/ATP Ratio, Glutathione Metabolism, and Nitrosative Stress in the Salivary Gland Mitochondria of Insulin Resistant Rats

Anna Zalewska [1,*], Izabela Szarmach [2], Małgorzata Żendzian-Piotrowska [3] and Mateusz Maciejczyk [3,*]

1 Experimental Dentistry Laboratory, Medical University, 15-222 Bialystok, Poland
2 Department of Orthodontics, Medical University, 15-222 Bialystok, Poland; orthod@umb.edu.pl
3 Department of Hygiene, Epidemiology and Ergonomics, Medical University, 15-222 Bialystok, Poland; mzpiotrowska@gmail.com
* Correspondence: azalewska426@gmail.com (A.Z.); mat.maciejczyk@gmail.com (M.M.)

Received: 20 January 2020; Accepted: 10 February 2020; Published: 12 February 2020

Abstract: This is the first study to assess the effect of N-acetylcysteine (NAC) on the mitochondrial respiratory system, as well as free radical production, glutathione metabolism, nitrosative stress, and apoptosis in the salivary gland mitochondria of rats with high-fat diet (HFD)-induced insulin resistance (IR). The study was conducted on male Wistar rats divided into four groups of 10 animals each: C (control, rats fed a standard diet containing 10.3% fat), C + NAC (rats fed a standard diet, receiving NAC intragastrically), HFD (rats fed a high-fat diet containing 59.8% fat), and HFD + NAC (rats fed HFD diet, receiving NAC intragastrically). We confirmed that 8 weeks of HFD induces systemic IR as well as disturbances in mitochondrial complexes of the parotid and submandibular glands of rats. NAC supplementation leads to a significant increase in the activity of complex I, II + III and cytochrome c oxidase (COX), and also reduces the ADP/ATP ratio compared to HFD rats. Furthermore, NAC reduces the hydrogen peroxide production/activity of pro-oxidant enzymes, increases the pool of mitochondrial glutathione, and prevents cytokine formation, apoptosis, and nitrosative damage to the mitochondria in both aforementioned salivary glands of HFD rats. To sum up, NAC supplementation enhances energy metabolism in the salivary glands of IR rats, and prevents inflammation, apoptosis, and nitrosative stress.

Keywords: apoptosis; inflammation; insulin resistance; NAC; salivary glands; mitochondrial activity; nitrosative stress

---

1. Introduction

N-acetyl-cysteine (NAC) is a derivative of a thiol-containing amino acid, which—directly or indirectly, by increasing the concentration of glutathione—demonstrates antioxidant properties. NAC is a cysteine donor for the synthesis of reduced glutathione (GSH). GSH is the most important intracellular antioxidant maintaining a reduced state of protein thiol groups, which is a prerequisite for sustaining the activity of these proteins. GSH can directly deactivate oxygen free radicals (ROS) or be used by glutathione peroxidase as a cofactor in the detoxification of hydrogen peroxide and peroxynitrite. NAC also stimulates the activity of glutathione reductase (GR), an enzyme that restores the reduced form of glutathione, using NADPH [1].

It has been evidenced that NAC reacts rapidly with highly reactive oxygen and nitrogen radicals. In pH 7 and at room temperature, the rate of HO$^\bullet$ elimination is $(1.36 \times 10^{10}\ \mathrm{M^{-1}.s^{-1}})$, the rate of NO$_2{}^\bullet$ nitrogen dioxide elimination is $(1.0 \times 10^7\ \mathrm{M^{-1}.s^{-1}})$, and the rate for carbonate radical CO$_3{}^{\bullet-}$ is $(1.0 \times 10^7$

$M^{-1}.s^{-1}$). Thus, NAC neutralizes free radicals produced by activated leukocytes, and is able to chelate transition metal ions, such as Cu^{2+} and Fe^{3+} as well as heavy metal ions, such as Cd^{2+}, Hg^{2+}, and Pb^{2+} by creating complexes with thiols, that are easily removable from the body [2]. Chelation of Cu and Fe ions prevents Fenton reactions and thus the production of one of the most dangerous free radicals: The hydroxyl radical. NAC inhibits the activation of nuclear factor kappa B and reduces subsequent expression of inflammatory cytokines [3]. It also lowers the degree of lipid peroxidation as well as the generation of superoxide anions by activating polymorphonuclear leukocytes independently of calcium [4]. Moreover, studies have demonstrated that NAC may affect the respiratory chain in the mitochondria, and inhibit the apoptotic pathway [5]. Experimental studies have indicated that long-term administration of NAC improves heart and brain mitochondrial activity [6] and prevents senile weakening of complexes I, IV, and V in hepatocyte mitochondria [7]. In vitro studies have shown that NAC supplementation has a protective effect on cytochrome c oxidase and complexes I, IV, and V, prevents drops in ATP concentration [8,9] and protects mitochondrial potential [10]. It was also demonstrated that NAC restores mitochondrial transport and improves calcium uptake in the cells of injured rat brains [11], which was attributed to the redox state of thiol groups in mitochondrial complexes.

It should be emphasized that excessive production of ROS, accompanied by a shortage of antioxidants, results in a condition defined as oxidative stress (OS). Oxidative stress leads to disturbances of cell metabolism and degradation of all cell components [12], resulting in cell death and dysfunctions of organs. Therefore, in recent years we have observed a growing interest in the therapeutic effect of NAC in a wide range of diseases in which OS plays a key role in initiation as well as progression.

In our previous studies we demonstrated that insulin resistance (IR) induced by a high-fat diet interferes with antioxidant systems of parotid and submandibular salivary glands, leading to the oxidation of DNA, proteins, and lipids. This was associated with morphological changes of salivary gland parenchyma, and it disturbed the qualitative and quantitative composition of saliva [13–15]. We also observed that IR induced by eight-week high-fat feeding impairs mitochondrial function in both salivary glands by enhancing ROS production and apoptosis, and demonstrated that the mitochondrial system and NOX both increase oxidative and nitrosative stress in both salivary glands [16]. We proved that NAC supplementation strengthens both enzymatic and non-enzymatic antioxidant mechanisms of parotid as well as submandibular glands, and prevents oxidative damage to and dysfunction of the parotid gland of rats with IR induced by high-fat diet [15]. Salivary redox imbalance increases oral cavity diseases associated with OS (tooth decay, gingivitis, periodontitis, oral mucosa ulceration, candidiasis, or burning mouth syndrome) [17–20]. Indeed, such lesions are found in a large percentage of people with obesity, insulin resistance, or diabetes mellitus [17,21–24]. Therefore, it is essential to maintain/restore effective functioning of antioxidant systems and thus maintain ROS concentration at a level enabling salivary redox balance.

However, so far there have been no studies evaluating the effect of NAC on mitochondrial oxidative/nitrosative stress and apoptosis, and no studies have assessed whether or not this substance can rescue mitochondrial function in the salivary glands in the rat model of IR.

Therefore, the aim of our study was to examine the effect of NAC supplementation on salivary glands: Their mitochondrial respiratory system and function, mitochondrial ROS production and glutathione metabolism, the activity of NOX and XO, as well as some parameters of nitrosative stress and apoptosis in the rat model of insulin resistance.

2. Materials and Methods

The study was approved by the Local Ethical Committee for Animal Experiments in Bialystok, No. 21/2017.

2.1. Animals

The experiment was performed on male Wistar rats with the initial body weight of approximately 50–72 g. Throughout the entire study, the rats were housed under standard laboratory animal housing conditions (21 °C, 12 h light/dark cycle), with free access to water and food [25].

After six days of adaptation to the conditions in the animal house, the rats were divided into four groups of 10 individuals each:

Group I—(C) control; rats receiving standard type LSM feed (Agropol Motycz Polska) containing 10.3% fat, 24.2% proteins and 65.5% carbohydrates, and 2 mL/kg body weight saline solution intragastrically (once a day, every day for eight weeks);

Group II—(C + NAC) rats receiving standard type LSM feed as well as N-acetylcysteine solution at a dose of 500 mg/kg body weight intragastrically (once a day, every day for eight weeks) at a volume of 2 mL/kg body weight (NAC, Sigma A9165) [26];

Group III—(HFD) rats fed high-fat diet (Research Diet, USA, catalog number D12492) containing 59.8% fat, 20.1% proteins, 20.1% carbohydrates [14] and 2 mL/kg body weight saline solution intragastrically (once a day, every day for eight weeks);

Group IV—(HFD + NAC) rats fed the abovementioned high-fat diet and, intragastrically (once a day, every day for eight weeks), N-acetylcysteine solution at a dose of 500 mg/kg body weight, at a volume of 2 mL/kg body weight.

Intragastric administration of saline and NAC solutions was performed at a fixed time during the day (between 8 and 9 a.m.) by two trained and authorized experts (A.Z., M.M.). Food intake was monitored once a week, and body weight was measured every two days, which allowed us to decide on an appropriate dose of NAC.

After eight weeks, upon an all-night refraining from any food intake, the concentration of glucose was determined, in phenobarbital anesthetic (80 mg $\times$ kg^{-1}, intraperitoneally), in blood collected from the caudal vein using a glucometer [27]. Then the rates of non-stimulated and stimulated saliva secretion were measured (pilocarpine hydrochloride at a dose of 5 mg/kg body weight, intraperitoneally; Sigma, Chemical Co., St Louis, MO, USA) in rats of all groups [15].

Next, blood was collected from the abdominal artery, and submandibular and parotid salivary glands were taken. The tissues were freeze-clamped with aluminum tongs, precooled in liquid nitrogen, and stored at −80 °C until the redox assays (but not longer than four weeks).

The plasma insulin level was determined by a commercial ELISA kit (Shibayagi Co., Gunma, Japan) according to the manufacturer's instructions. To confirm IR, insulin sensitivity was calculated by HOMA-IR index (homeostasis model assessment of insulin resistance = fasting insulin [U/mL] $\times$ fasting glucose [mM]/22.5) [28]. Plasma free fatty acids (FFA) were analyzed by gas chromatography (GC) [29].

2.2. Mitochondria Isolation

On the day of the biochemical part of the experiment, the collected salivary glands were slowly thawed at 4 °C, weighed and cut into small pieces with scissors. Thus prepared, glands were homogenized (1:10, *w/v*) with a Teflon-and-glass electric homogenizer in ice-cold mitochondrial isolation buffer containing 250 mM sucrose, 5 mM Tris-HCl and 2 mM ethylene glycol bis(2-aminoethyl) tetraacetic acid (EGTA), pH 7.4. To prevent protein degradation and proteolysis, protease inhibitors were added. Homogenate was centrifuged (500$\times$ *g*, 10 min, 4 °C) and the resulting supernatant was centrifuged twice at 8000$\times$ *g* for 10 min at 4 °C. The mitochondria pellet was resuspended in isolation buffer and used on the same day for assays.

In the mitochondrial fraction we did not detect cytoplasmic glyceraldehyde 3-phosphate dehydrogenase or the nuclear marker histone H3 (data not shown, western blot technique), which excludes extra-mitochondrial changes and proves purity of this fraction [30].

2.3. Mitochondrial Assays

The following were assayed in the mitochondrial fraction: The activity of mitochondrial complexes, ADP/ATP ratio, rate of hydrogen peroxide production, glutathione metabolism, pro-oxidative enzymes, nitrosative stress, and selected markers of inflammation and apoptosis. All the determinations were performed in duplicate samples and the results were standardized to 1 mg of total protein. The total protein content was measured using the bicinchoninic acid (BCA) method. A commercial kit (Thermo Scientific PIERCE BCA Protein Assay; Rockford, IL, USA) was used according to the manufacturer's instructions. The absorbance/fluorescence/chemiluminescence was measured with Infinite M200 PRO Multimode Microplate Reader from Tecan.

2.4. Activity of Mitochondrial Complexes

The activity of complex I (E.C. 1.6.5.3) was measured colorimetrically based on 2,6-dichloroindophenol reduction by electrons accepted from coenzyme Q_1, reduced after oxidation of NADH (reduced form of nicotinamide adenine dinucleotide) by complex I [31].

The activity of complex II (E.C. 1.3.5.1) and complex II + III (E.C. 1.10.2.2) was measured according to Rustin et al. [32]. The activity of succinate-ubiquinone reductase and succinate-cytochrome c reductase (respectively) were measured colorimetrically.

Cytochrome c oxidase (COX) activity was examined colorimetrically by measuring the oxidation of reduced cytochrome c at 550 nm wavelength [33].

2.5. ADP/ATP Ratio and Hydrogen Peroxide Production

The ADP/ATP ratio was measured using a bioluminescent method based on the conversion of ATP by luciferase. A commercial kit (ADP/ATP Ratio Assay Kit ab65313, Abcam, Burlingame, CA, USA) was used according to the manufacturer's instructions.

Hydrogen peroxide (H_2O_2) production was analyzed by measuring the increase in fluorescence at 530/590 nm wavelength due to the reaction of Amplex Red with H_2O_2 in the presence of horseradish peroxidase [34]. H_2O_2 formation rate was calculated using a standard curve of H_2O_2 stabilized solution.

2.6. Activity of Citrate Synthase

Citrate synthase (CS) activity was measured colorimetrically using the reaction with 5-thio-2-nitrobenzoic acid generated from 5,5′-dithiobis-2-nitrobenzoic acid during CS synthesis [35].

2.7. Reduced and Oxidized Glutathione and Redox Status

The level of total glutathione was measured based on an enzymatic reaction with 5,5′-dithiobis-(2-nitrobenzoic acid) (DTNB), NADPH and glutathione reductase (GR) [36].

Oxidized glutathione (GSSG) was estimated similarly to the assay performed for total glutathione; however, prior to the determination, the samples had been thawed and neutralized to pH 6–7 using 1 M chlorhydrol triethanolamine. Then, samples were incubated with 2-vinylpyridine [36].

The level of reduced glutathione (GSH) was calculated from the difference between the level of total glutathione and GSSG [36].

Redox (oxidation/reduction) status was calculated using the formula: $[GSH]^2/[GSSG]$.

2.8. Pro-Oxidant Enzymes

NADPH oxidase (NOX, E.C. 1.6.3.1) activity was measured by luminescence assay using lucigenin as an electron acceptor [37]. One unit of NOX activity was defined as the quantity of the enzyme required to release 1 nmol of superoxide radical for 1 min.

Xanthine oxidase (XO, E.C. 1.17.3.2.) activity was assessed based on uric acid production by measuring the increase in its optical density at 290 nm wavelength [38]. One unit of XO activity was defined as the amount of the enzyme required to release 1 μmol of uric acid for 1 min.

2.9. Inflammation and Apoptosis

Interleukin-1β (IL-1β) level was measured using the enzyme-linked immunosorbent assay (ELISA) method. A commercial kit (IL-1β ELISA kit, R&D Systems; Canada, Minneapolis, MN, USA) was used according to the manufacturer's instructions.

Caspase-3 (CAS-3, EC 3.4.22.56) activity was measured colorimetrically using Ac-Asp-Glu-Val-Asp -p-nitroanilide as a substrate [39]. The amount of p-nitroaniline released by CAS-3 was measured at 405 nm wavelength.

2.10. Nitrosative Stress

Nitric oxide (NO) level was measured colorimetrically using sulfanilamide and N-(1-naphthyl)-ethylenediamine dihydrochloride. The absorbance of the obtained product was measured at 490 nm wavelength [40].

Peroxynitrite level was measured colorimetrically based on peroxynitrite-mediated nitration leading to the formation of nitrophenol [41]. The absorbance of the obtained complex was measured at 320 nm wavelength.

Nitrotyrosine level was measured using the ELISA method. A commercial kit (Nitrotyrosine ELISA; Immundiagnostik AG, Bensheim, Germany) was used according to the manufacturer's instructions.

2.11. Statistical Analysis

The statistical analysis was performed using GraphPad Prism 8 for MacOS (GraphPad Software, La Jolla, CA, USA). Normality of the results distribution was determined using the Shapiro-Wilk test. The results were expressed as mean ± SD. One-way ANOVA, Tukey's multiple comparisons test, and Pearson correlation coefficient were used. Multiplicity adjusted p value was calculated. The statistical significance was defined as $p < 0.05$.

3. Results

3.1. General Characteristics

Eight weeks of high-fat feeding resulted in a significant increase in rat weight compared to the control rats fed normal chow ($p < 0.0001$). NAC administration prevented rat weight gain as body weight of HFD + NAC rats was significantly lower than of rats fed only high-fat diet ($p < 0.0001$). An eight-week high-fat diet resulted in a considerable increase in the level of glucose ($p < 0.0001$ and $p < 0.0001$, respectively), insulin ($p < 0.0001$ and $p < 0.0001$, respectively) and fatty acids ($p < 0.0001$ and $p < 0.0001$, respectively) compared to both the group fed standard diet and rats fed HFD + NAC solution. The HOMA-IR insulin sensitivity index was significantly higher in the high-fat diet group compared to the controls and rats fed HFD + NAC ($p = 0.0001$, $p = 0.0001$, respectively). The comparison of the described parameters between the control group and HFD + NAC group did not reveal any significant differences. Rats from HFD ($p = 0.0026$) and HFD + NAC ($p = 0.0083$) groups consumed considerably less feed compared to the control rats, with energy efficiency significantly higher in both groups fed high-fat diet (HFD, HFD + NAC) in relation to the controls ($p = 0.01$ and $p = 0.01$, respectively) (Table 1).

The secretion of non-stimulated saliva did not differ significantly between the studied groups, with HFD rats characterized by clearly decreased secretion of stimulated saliva compared to the control group ($p < 0.0001$). NAC supplementation prevented the reduction of stimulated saliva secretion in rats fed high-fat diet, and in this group the stimulated saliva secretion was significantly higher compared to HFD group ($p < 0.0001$) (Table 1).

Table 1. Effect of N-acetyl-cysteine (NAC) supplementation on body weight, plasma metabolic parameters, food intake, salivary flow rate and salivary gland weight.

	C	C + NAC	HFD	HFD + NAC
Final body weight (g)	280.0 ± 12.31	278.3 ± 26.62	380.9 ± 30.42 *	314.5 ± 32.78 *,#
Plasma glucose (mg/dL)	90.67 ± 6.19	91.77 ± 11.84	148.6 ± 7.99 *	98.50 ± 8.41 *,#
Plasma insulin (mIU/mL)	78.92 ± 8.46	76.86 ± 9.75	165.4 ± 12.79 *	84.49 ± 17.07 *,#
HOMA-IR	2.93 ± 1.36	2.56 ± 2.45	20.11 ± 1.96 *	2.89 ± 1.83 #
Plasma free fatty acids (μmol/L)	75.59 ± 10.99	69.80 ± 8.83	173.4 ± 11.99 *	86.38 ± 12.5 *,#
Food intake (mg/day)	20.98 ± 3.95	21.70 ± 6.93	11.12 ± 6.1 *	12.19 ± 5.68 *,#
Energy from chow (MJ/day)	0.19 ± 0.1	0.17 ± 0.1	0.31 ± 0.12 *	0.27 ± 0.1 *
Non-stimulated salivary flow (μL/min)	0.4229 ± 0.12	0.3813 ± 0.11	0.3707 ± 0.14	0.4195 ± 0.09
Stimulated salivary flow (μL/min)	110.2 ± 9.77	103.5 ± 10.76	70.08 ± 8.14 *	105.2 ± 12.05 *,#

C—control rats; C + NAC—control rats + N-acetylcysteine; HFD—rats fed high-fat diet; HFD + NAC—rats fed high-fat diet + N-acetylcysteine; PG—parotid glands; SMG—submandibular glands; HOMA-IR—homeostasis model assessment of insulin resistance; * $p < 0.05$ vs. control; # $p < 0.05$ vs. HFD.

3.2. Activity of Mitochondrial Complexes

An eight-week high-fat diet resulted in a considerable reduction of complex I activity in both parotid (−21%, $p < 0.0001$) and submandibular (−48%, $p < 0.0001$) glands compared to the control group. Chronic treatment with NAC significantly increased the activity of complex I in the parotid and submandibular glands of HFD + NAC rats in relation to the HFD group (+ 42%, $p < 0.0001$; +68%, $p < 0.0001$, respectively), to a level similar to the control rats fed normal chow (Figure 1).

High-fat diet and NAC supplementation did not affect the activity of complex II. In all the studied groups the activity of this complex remained at a similar level as in the control rats fed normal chow (Figure 1).

The activity of complex II + III in the parotid glands of rats fed a high-fat diet, although lower, did not differ from the control group. Chronic treatment with NAC significantly increased the activity of complex II + III in the parotid glands compared to the HFD group (+ 10%, $p < 0.0001$), but its activity in the HFD + NAC group did not differ significantly from the control group.

An eight-week high-fat diet resulted in a significant reduction in the activity of complex II + III in the submandibular glands compared to the control group (−12%, $p < 0.0001$). NAC supplementation prevented the reduction of complex II + III activity on the one hand, because the activity of this complex in the HFD + NAC group did not differ from the control group, but on the other hand it did not differ from the values obtained in HFD group (Figure 1).

COX activity in the parotid (+ 27%, $p < 0.0001$) and submandibular (+ 66%, $p < 0.0001$) glands of rats fed a high-fat diet for eight weeks was significantly elevated compared to the controls. NAC supplementation prevented an increase in COX activity in both salivary glands, as COX activity in the parotid and submandibular glands of HFD + NAC was significantly reduced compared to the HFD group (−40%, $p < 0.0001$, −23%, $p < 0.0001$, respectively) (Figure 1).

3.3. ADP/ATP Ratio and H_2O_2 Level

ADP/ATP ratio in the in the parotid (+ 20%, $p = 0.0019$) and submandibular (+ 24%, $p = 0.003$) glands of rats fed high-fat diet for eight weeks was significantly higher than in the group fed normal chow. NAC supplementation prevented ADP/ATP ratio from increasing in both salivary glands as ADP/ATP ratio in the parotid and submandibular glands of HFD + NAC rats was significantly reduced compared to the HFD group (−20%, $p = 0.005$, −24%, $p = 0.0034$, respectively) (Figure 2).

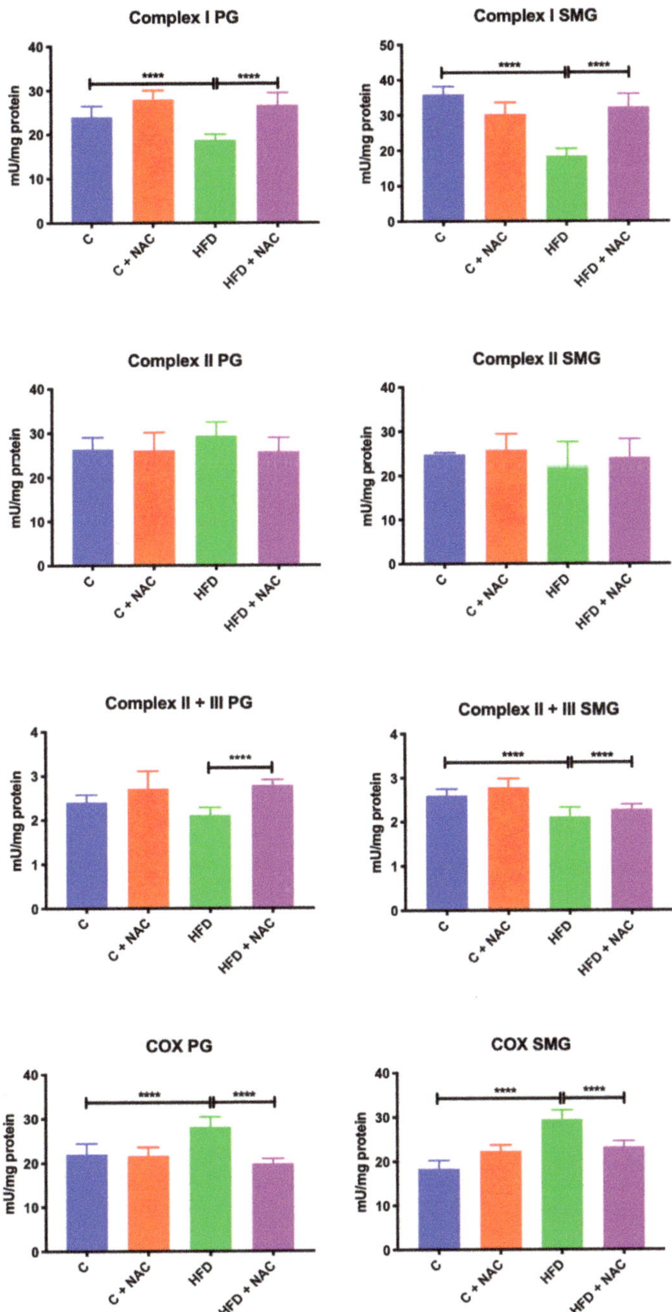

Figure 1. Effect of NAC supplementation on mitochondrial respiratory complexes in the parotid and submandibular glands of insulin resistant rats. C—control rats; C + NAC—control rats + N-acetylcysteine; HFD—rats fed high-fat diet; HFD + NAC—rats fed high-fat diet + N-acetylcysteine; PG—parotid glands; SMG—submandibular glands; COX—cytochrome c oxidase; **** $p < 0.0001$.

The level of H_2O_2 was significantly higher in the parotid and submandibular glands of rats fed with high-fat diet compared to the control rats (+ 48%, $p < 0.0001$, +14%, $p = 0.0017$, respectively). NAC supplementation prevented the increase in H_2O_2 production in both glands, so H_2O_2 concentration in the parotid and submandibular glands of HFD + NAC rats was lower compared to the HFD group, and at the same time it did not differ from the control group (-37%, $p < 0.0001$, -14%, $p = 0.0058$, respectively) (Figure 2).

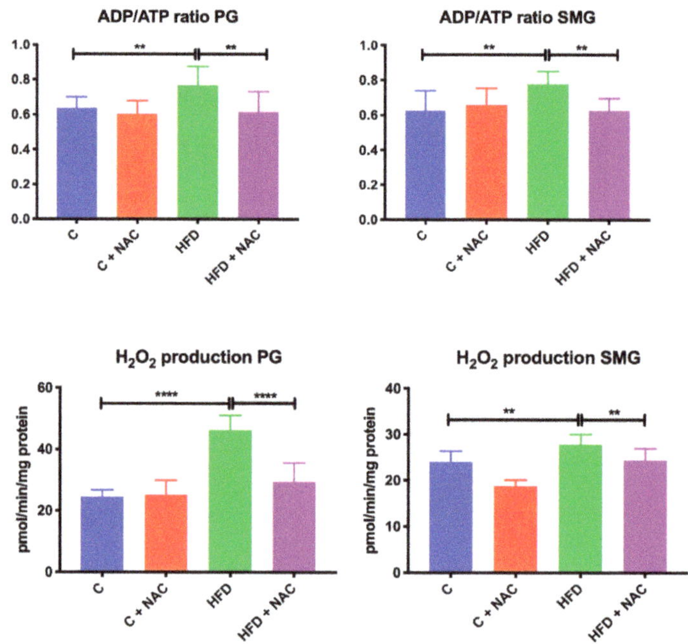

Figure 2. Effect of NAC supplementation on ADP/ATP ratio and hydrogen peroxide production in salivary gland mitochondria. C—control rats; C + NAC—control rats + N-acetylcysteine; HFD—rats fed high-fat diet; HFD + NAC—rats fed high-fat diet + N-acetylcysteine; PG—parotid glands; SMG—submandibular glands; H_2O_2—hydrogen peroxide. ** $p < 0.005$, **** $p < 0.0001$.

3.4. Activity of Mitochondrial CS

An eight-week high-fat diet resulted in a significant reduction of CS activity in both the parotid (-28%, $p < 0.0001$) and submandibular (-39%, $p < 0.0001$) glands compared to the control group. Chronic treatment with NAC considerably increased CS activity in the submandibular glands compared to the HFD group (+ 47%, $p < 0.0001$). In the parotid glands, NAC supplementation prevented the reduction of CS activity on the one hand—because the activity of this complex in the HFD + NAC group did not differ from the controls—but on the other hand CS activity did not differ from the values obtained in the HFD group (Figure 3).

3.5. GSH, GSSG and Redox Status

A high-fat diet consumed for eight weeks resulted in a significant reduction of GSH concentration and redox status in the mitochondria of both parotid (-50%, $p < 0.0001$, -57% $p < 0.0001$, respectively) and submandibular (-21%, $p = 0.0135$, -37%, $p = 0.04$, respectively) glands compared to the control group. In the mitochondria of HFD rats' parotid glands we also observed a significant increase in GSSG concentration compared to the controls (+ 37%, $p < 0.0001$). NAC supplementation resulted in increased GSH content and redox status in the mitochondria of parotid (+ 56%, $p = 0.0016$, +31%,

$p = 0.0077$, respectively) and submandibular (+ 26%, $p = 0.0194$, +51%, $p = 0.0039$, respectively) glands as well as decreased GSSG content (-17%, $p < 0.0001$) in the mitochondria of parotid glands compared to the HFD group and to the GSSG level of the control group (Figure 4).

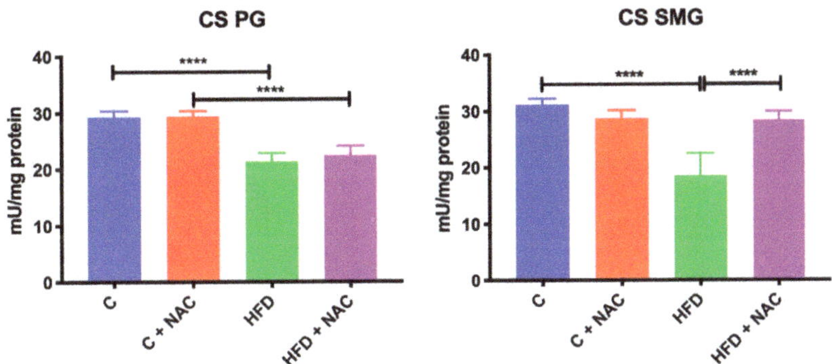

Figure 3. Effect of NAC supplementation on citrate synthase activity in the salivary gland mitochondria. C—control rats; C + NAC—control rats + N-acetylcysteine; HFD—rats fed high-fat diet; HFD + NAC—rats fed high-fat diet + N-acetylcysteine; PG—parotid glands; SMG—submandibular glands; CS—citrate synthase; **** $p < 0.0001$.

3.6. Activity of NOX and XO

A high-fat diet resulted in increased activity of NOX (+ 54%, $p < 0.0001$, +30%, $p < 0.0001$, respectively) and XO (+ 42%, $p < 0.0001$, +55%, $p < 0.0001$, respectively) in the mitochondria of both parotid and submandibular glands compared to the control group. NAC supplementation prevented an increase in NOX and XO activity in the mitochondria of both salivary glands as the activity of both NOX and XO in parotid (-50%, $p < 0.0001$, -30%, $p < 0.0001$, respectively) and submandibular (-27%, $p < 0.0001$, -22%, $p < 0.0001$, respectively) glands in the HFD + NAC group was significantly lower compared to HFD group (Figure 5).

3.7. Inflammation and Apoptosis

HFD rats were characterized by significantly increased IL-1β concentration and CAS-3 activity in the mitochondria of parotid (+ 80%, $p < 0.0001$, +56%, $p < 0.0001$, respectively) and submandibular (+ 172%, $p < 0.0001$; +94%, $p = 0.0004$; respectively) glands compared to the controls. NAC supplementation resulted in decreased IL-1β concentration and CAS-3 activity in the mitochondria of parotid (-33%, $p = 0.0007$; -40%, $p < 0.0001$, respectively) and submandibular (-33%, $p < 0.0001$; -22%, $p < 0.0001$; respectively) glands compared to HFD rats, in relation to the levels observed in the control group (Figure 6).

3.8. Nitrosative Stress

HFD rats showed significantly increased NO, peroxynitrite and nitrotyrosine concentration in the mitochondria of parotid (+ 65%, $p < 0.0001$; +63%, $p < 0.0001$, +12%, $p < 0.0001$; respectively) and submandibular (+ 62%, $p < 0.0001$; +86%, $p < 0.0001$; +23%, $p < 0.0001$; respectively) glands compared to the controls. NAC supplementation resulted in decreased NO, peroxynitrite, and nitrotyrosine concentration in the mitochondria of parotid (-48%, $p < 0.0001$; -42%, $p < 0.0001$, -19%, $p < 0.0001$; respectively) and submandibular (-50%, $p < 0.0001$; -67%, $p < 0.0001$; -15%, $p < 0.0001$; respectively) glands compared to HFD rats, in relation to the levels observed in the control group (Figure 7).

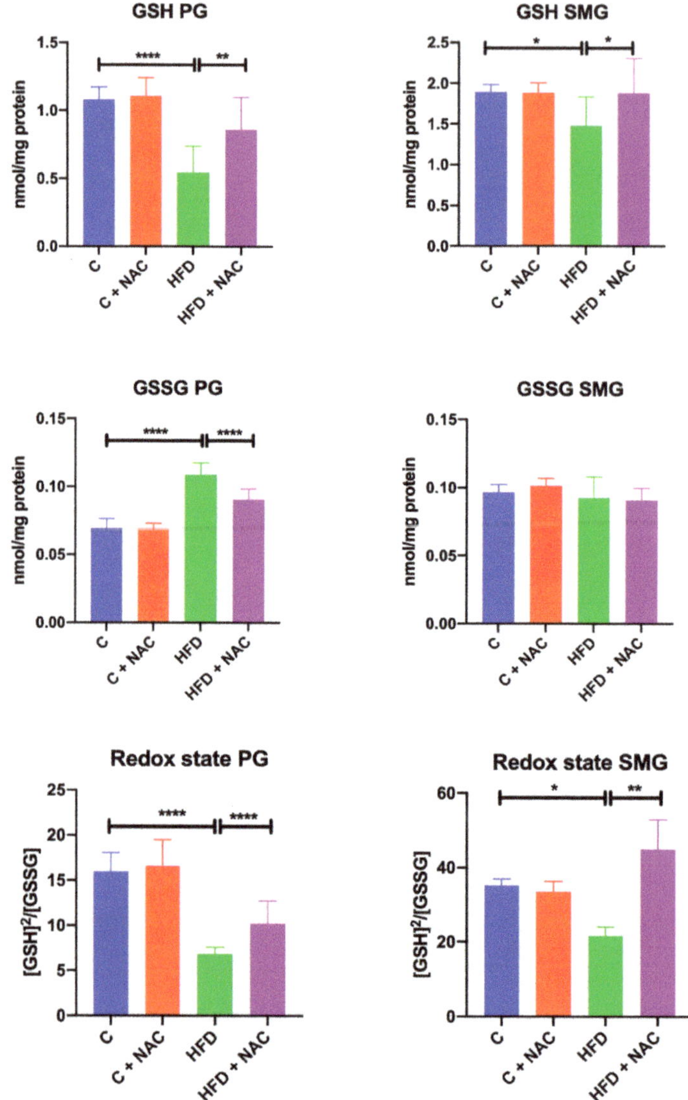

Figure 4. Effect of NAC supplementation on glutathione and redox status in the salivary gland mitochondria. C—control rats; C + NAC—control rats + N-acetylcysteine; HFD—rats fed high-fat diet; HFD + NAC—rats fed high-fat diet + N-acetylcysteine; GSH—reduced glutathione; GSSG—oxidized glutathione; * $p < 0.05$, ** $p < 0.005$, **** $p < 0.0001$.

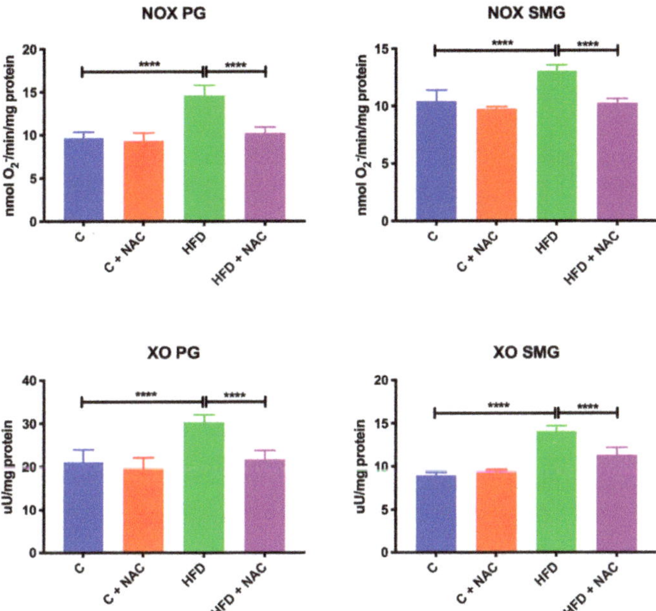

Figure 5. Effect of NAC supplementation on pro-oxidant enzymes in the salivary gland mitochondria. C—control rats; C + NAC—control rats + N-acetylcysteine; HFD—rats fed high-fat diet; HFD + NAC—rats fed high-fat diet + N-acetylcysteine; NOX—NADPH oxidase; XO—xanthine oxidase; **** $p < 0.0001$.

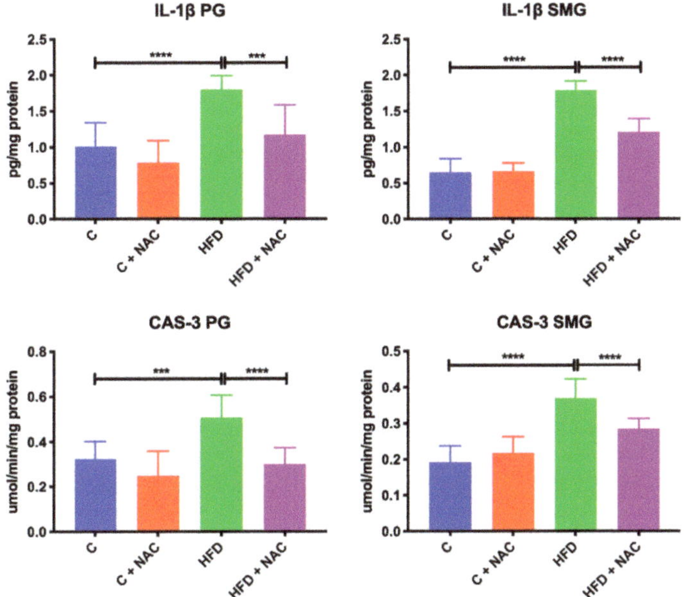

Figure 6. Effect of NAC supplementation on inflammation and apoptosis biomarkers in the salivary gland mitochondria. C—control rats; C + NAC—control rats + N-acetylcysteine; HFD—rats fed high-fat diet; HFD + NAC—rats fed high-fat diet + N-acetylcysteine; IL-1β—interleukin 1β; CAS-3—caspase 3; *** $p < 0.0005$, **** $p < 0.0001$.

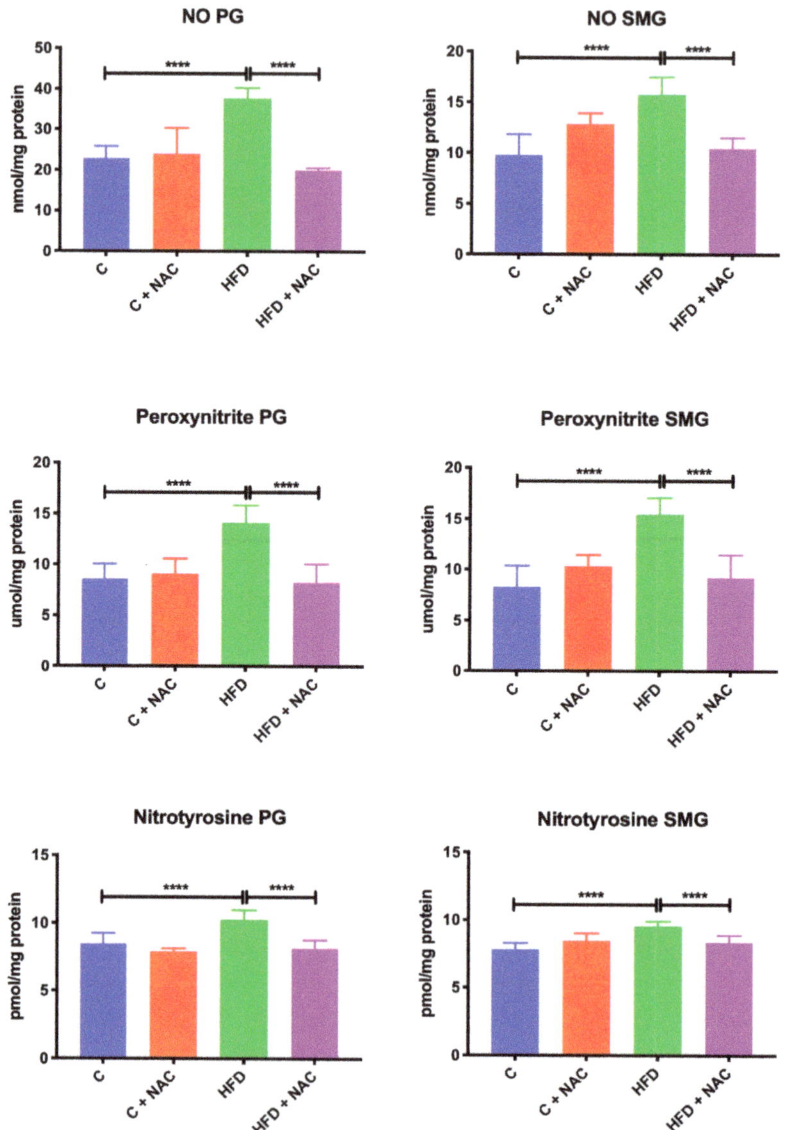

Figure 7. Effect of NAC supplementation on nitrosative stress biomarkers in the salivary gland mitochondria. C—control rats; C + NAC—control rats + N-acetylcysteine; HFD—rats fed high-fat diet; HFD + NAC—rats fed high-fat diet + N-acetylcysteine; NO—nitric oxide; **** $p < 0.0001$.

3.9. Correlations

In the parotid and submandibular glands of HFD + NAC rats we observed a negative correlation between GSH concentration and H_2O_2 production ($r = -0.72$, $p = 0.019$; $r = -0.902$, $p < 0.0001$, respectively) and between GSH and peroxynitrile ($r = -0.713$, $p = 0.021$; $r = -0.801$, $p = 0.005$, respectively). In the group of HFD + NAC rats we also demonstrated a positive correlation between CAS-3 activity and IL-2 concentration in submandibular glands ($r = 0.936$, $p < 0.0001$), and a positive correlation between CAS-3 and NOX activities in parotid glands ($r = 0.846$, $p = 0.002$).

4. Discussion

In our study we assessed the effect of NAC supplementation on salivary glands: Their mitochondrial respiratory system and function, mitochondrial ROS production and glutathione metabolism, the activity of NOX and XO, as well as some parameters of nitrosative stress and apoptosis in a rat model of insulin resistance.

HFD is a suitable model of human IR [13,14,16,42]. As expected, the implemented eight-week high-fat diet resulted in a significant increase in the body weight of rats, and decreased glucose tolerance assessed by a considerably higher concentration of blood glucose and insulin as well as HOMA-IR index compared to the control rats [43–45]. It was also observed that HFD induces OS [13–16] as well as mitochondrial dysfunction and cell death [16] in the salivary glands of rats and mice [30], which was confirmed in our study.

Chronic NAC treatment has been reported to be beneficial in IR and its complications [15,46–48]. We confirmed that NAC prevents hyperglycemia and hyperinsulinemia as well as impedes the development of IR. We also confirmed the preventing effect of NAC in increasing body weight seen in HFD rats compared to the HFD + NAC group. Despite the similar calorific value of the consumed food, HFD + NAC rats were characterized by lower final body weight compared to HFD rats, which most probably resulted from NAC-induced reduction in the intestinal absorption of the ingested chow [49]. Żukowski et al. [15] demonstrated that NAC supplementation protects parotid glands against oxidative stress, while the severity and extent of oxidative damage to submandibular glands of HFD + NAC rats was only reduced by NAC treatment. Our results confirmed the observations that NAC prevents parotid gland dysfunction, which was expressed as increased secretion of stimulated saliva in HFD + NAC rats.

The primary source of ROS formation in the body is the respiratory chain in mitochondria. There, electrons are transferred through the mitochondrial electron transport chain (mETC) to reduce molecular oxygen [50–52]. Interestingly, the uncontrolled production of ROS in mETC has been implicated in many pathological situations, and has also led to the dysfunction of the salivary glands in many systemic diseases [16,30,53]. Therefore, attempts should be taken to regulate/restore mitochondrial function to prevent oxidative damage to lipids, proteins, and DNA/RNA.

The effect of NAC treatment on mitochondrial: Respiratory complexes, H_2O_2 production and GSH, activity of NOX and XO as well as on nitrosative stress, inflammation, and apoptosis in a rat model of insulin resistance has not been described yet. Basically, we proved that NAC treatment provides beneficial effects on mitochondrial function. In fact, parotid and submandibular gland mitochondria of NAC-treated HFD rats showed a significant increase of complex I, II + III, and COX, as well as a decrease of ADP/ATP ratio compared to rats fed a high-fat diet. Thus, NAC reduces H_2O_2 production, increases the pool of mitochondrial GSH, and prevents inflammation, apoptosis, and nitrosative stress in the mitochondria of both salivary glands of HFD rats.

Mitochondrial complexes are of great importance as they play a key role in energy production. All respiratory chain enzymes are proteins with active thiol groups that can sense redox status in the cell, and simultaneously make these complexes very sensitive to oxidative modifications. The activity of respiratory chain enzymes is inhibited in the case of redox imbalance of sulfhydryl groups. This balance is disturbed by the deficiency of free sulfhydryl groups that are a ready source of reducing equivalents to quench radical species, which keeps thiol groups of the enzymes in a reduced state also by excess of free radicals [54]. NAC most likely prevents the reduction of mitochondrial complexes activity in HFD rats' salivary glands by protecting sulfhydryl groups from oxidation, as described for nerve cell mitochondria [55]. Moreover, Banaclocha et al. [9] argue that NAC thiol groups may participate in cofactor and substrate binding by forming covalent addiction products or charge transfer complexes. Thiol groups provided by NAC can also help preserve the tertiary structure of mitochondrial enzymes by serving as a donor for weak hydrogen bonds. The authors believe that these mechanisms can improve the affinity of the enzymes to substrates and the internal electron transfer rate, or change the affinity of the enzymes to oxygen.

Moreover, NAC has an indirect antioxidant effect by providing cysteine and boosting GSH synthesis. The observed increase in GSH concentration may also be related to the previously cited NAC-suppressed NF-kB (nuclear factor kappa B) activation and upregulation of the gene expression of this protein [56,57]. It should be noted that mitochondria do not synthesize their own GSH and therefore it must be transported from the cytosol through a multi-component system located in the internal mitochondrial membrane. The observed increase in GSH concentration in the mitochondria of both salivary glands of rats is most probably the result of increased cytosolic GSH concentration and a boosted rate of transport through mitochondrial membranes [58]. The main function of glutathione is to maintain a reduced state of thiol groups of proteins, and thus—similarly to NAC—it can determine efficient functioning of the respiratory chain and oxidative phosphorylation. Another mechanism through which GSH can prevent the dysfunction of respiratory chain enzymes is through direct interaction of this chain with ROS. GSH is used as a cofactor by glutathione peroxidase in the detoxification of H_2O_2 and peroxynitrite [59]. Therefore, it is not surprising that with increased GSH concentration we can observe the reduction of H_2O_2 and peroxynitrite concentrations in both salivary glands of HFD + NAC rats. It was demonstrated that H_2O_2 reacts easily with protein thiol groups as well as peroxynitrite with iron-sulfur cluster of mitochondrial electron transport enzymes, leading to their inactivation [59,60]. Similarly, NAC reacts rapidly with highly reactive oxygen and nitrogen radicals [61]; however, it is considered to be a weak antioxidant [62]. Moreover, the intracellular concentration of NAC, compared to the levels of GSH, ascorbate, and antioxidant enzymes, is too low for NAC to directly affect the antioxidant capacity [63].

It is noteworthy that there were no changes in GSH concentrations in the salivary glands of C + NAC rats, which is consistent with the previous observations [15,64]. It has been demonstrated that thiol supplementation, particularly when GSH content is undisturbed, results in an increased cysteine concentration without a concomitant rise in GSH level [64].

Treatment of rats with NAC resulted in a significant decrease in the ADP/ATP ratio compared to non-treated HFD rats, which was most probably connected with the maintenance of a reduced state of the said sulfhydryl groups of mitochondrial enzymes. It was shown that sulfhydryl groups of these enzymes are essential in the process of oxidative phosphorylation and in energetic metabolism [65]. It should be highlighted that a decreased ADP/ATP ratio can undoubtedly be influenced by the maintained (by NAC supplementation) activity of CS—an enzyme associated with tricarboxylic acid cycle, the efficient functioning of which is enabled by 12 ATP molecules.

Although continual energy production is one of the most important tasks of the mitochondria, these organelles are also involved in the initiation and execution of processes leading to cell death. It has been proven that the reduced activity of ETC (electron transport chain) complexes increases mitochondrial H_2O_2 production [66,67]. The observed increase in H_2O_2 concentration in the mitochondria of both salivary glands of HFD rats may also be caused by increased NADPH oxidase activity, particularly since mitochondria mainly contain the isoform NOX4 that predominantly produces H_2O_2 [68,69]. It should be noted that the increase in NOX activity may result from high concentrations of glucose [70] and free fatty acids [71], all of which were increased in the serum of HFD rats in our study. A certain amount of ROS is also provided by XO which catalyzes one- or two-electron molecular oxygen reduction with the formation of superoxides and H_2O_2 [60]. Excessive ROS production helped in inducing inflammation to a large extent. ROS may be implicated in the activation/intensification of numerous signaling pathways such as NF-κB, and cytokine release. Inflammatory cytokines promote cell damage, apoptosis, and consequently, organ damage, which we observed in the mitochondria of salivary glands of rats exposed to a high-fat diet. The observed anti-inflammatory effect of NAC was associated with the decreased activity of NOX and XO as well as lowered IL-2 concentration in both salivary glands of HFD + NAC rats. Further studies are required to explain whether the observed anti-inflammatory effect of NAC supplementation is associated with NAC-induced inhibition of proteasome activity followed by the downregulation of NF-κB activity, or whether this effect could be secondary to the systemic influence of NAC (preventing insulin resistance). The NAC-dependent regulation of mitochondrial ROS production and inflammation appears to be of key importance in reducing the rate of apoptosis in both salivary glands of rats fed a high-fat diet. In our experiment, NAC supplementation resulted in the decreased activity of caspase 3 in both salivary glands, with a simultaneous positive relationship between IL-2 concentration and caspase activity in submandibular glands, and NOX activity and caspase activity in the parotid glands of HFD + NAC rats.

We showed that HFD upregulated NO and peroxynitrite concentrations, as well as enhanced tyrosine nitration of mitochondrial proteins in the salivary glands of rats, which was previously described for the mitochondria of submandibular glands of HFD mice [30]. In the presence of NAC, NO production, and consequently, nitrosative stress were significantly reduced. It should be underlined that by decreasing NO levels in the salivary gland mitochondria, NAC may improve microvascular delivery of oxygen and salivary gland oxygenation and thus prevent changes in saliva secretion. It was demonstrated that hyperbaric oxygen treatment of previously irradiated head and neck cancer patients boosted non-stimulated and stimulated salivary flow rate to a level >0.2 mL/min and >0.7 mL/min, respectively [72].

One of the limitations of our experiment is the lack of histological examination. Thus, we can not show whether the diet in itself and/or NAC administration influence the total mitochondrial content or cause morphological changes in the salivary glands.

5. Conclusions

Dietary administration of NAC contributes to the preservation of mitochondrial enzymes in the salivary glands of HFD rats. NAC supplementation enhances energy metabolism in salivary glands of HFD rats by affecting respiratory chain enzymes and citrate synthase. Treatment with NAC prevents inflammation, apoptosis, and nitrosative stress in the salivary glands of HFD rats.

Author Contributions: Conceptualization, A.Z. and M.M.; Data curation, A.Z. and M.M.; Formal analysis, A.Z. and M.M.; Funding acquisition, A.Z. and M.M.; Investigation, A.Z. and M.M.; Methodology, A.Z., M.Ż.-P. and M.M.; Resources, A.Z., M.Ż.-P. and M.M.; Software, A.Z. and M.M.; Supervision, I.S. and M.Ż.-P.; Validation, A.Z. and M.M.; Visualization, A.Z. and M.M.; Writing—original draft, A.Z. and M.M.; Writing—review & editing, A.Z. and M.M. All authors have read and agreed to the published version of the manuscript.

Funding: This work was financed by the Medical University of Bialystok (grant numbers: SUB/1/DN/20/002/1209; SUB/1/DN/20/002/3330).

Conflicts of Interest: The authors declare no conflict of interest.

References

1. Lushchak, V.I. Glutathione homeostasis and functions: Potential targets for medical interventions. *J. Amino. Acids.* **2012**, *2012*, 736837. [CrossRef] [PubMed]
2. Ergin, B.; Guerci, P.; Zafrani, L.; Nocken, F.; Kandil, A.; Gurel-Gurevin, E.; Demirci-Tansel, C.; Ince, C. Effects of N-acetylcysteine (NAC) supplementation in resuscitation fluids on renal microcirculatory oxygenation, inflammation, and function in a rat model of endotoxemia. *Int. Care Med. Exp.* **2016**, *4*. [CrossRef]
3. Paterson, R.L.; Galley, H.F.; Webster, N.R. The effect of N-acetylcysteine on nuclear factor-kappa B activation, interleukin-6, interleukin-8, and intercellular adhesion molecule-1 expression in patients with sepsis. *Crit. Care Med.* **2003**, *31*, 2574–2578. [CrossRef] [PubMed]
4. Villagrasa, V.; Cortijo, J.; Marti-Cabrera, M.; Ortiz, J.L.; Berto, L.; Esteras, A.; Bruseghini, L.; Morcillo, E.J. Inhibitory effects of N-acetylcysteine on superoxide anion generation in human polymorphonuclear leukocytes. *J. Pharm. Pharmacol.* **1997**, *49*, 525–529. [CrossRef]
5. Shen, H.M.; Yang, C.F.; Ding, W.X.; Liu, J.; Ong, C.N. Superoxide radical-initiated apoptotic signalling pathway in selenite-treated HepG(2) cells: Mitochondria serve as the main target. *Free Radic. Biol. Med.* **2001**, *30*, 9–21. [CrossRef]

6. Cocco, T.; Sgobbo, P.; Clemente, M.; Lopriore, B.; Grattagliano, I.; Di Paola, M.; Villani, G. Tissue-specific changes of mitochondrial functions in aged rats: Effect of a long-term dietary treatment with N-acetylcysteine. *Free Radic. Biol. Med.* **2005**, *38*, 796–805. [CrossRef]

7. Miquel, J.; Ferrandiz, M.L.; De Juan, E.; Sevila, I.; Martinez, M. N-acetylcysteine protects against age-related decline of oxidative phosphorylation in liver mitochondria. *Eur. J. Pharmacol.* **1995**, *292*, 333–335. [CrossRef]

8. Martinez Banaclocha, M.; Martinez, N. N-acetylcysteine elicited increase in cytochrome c oxidase activity in mice synaptic mitochondria. *Brain Res.* **1999**, *842*, 249–251. [CrossRef]

9. Martinez Banaclocha, M. N-acetylcysteine elicited increase in complex I activity in synaptic mitochondria from aged mice: Implications for treatment of Parkinson's disease. *Brain Res.* **2000**, *859*, 173–175. [CrossRef]

10. Gonzalez, R.; Ferrin, G.; Hidalgo, A.B.; Ranchal, I.; Lopez-Cillero, P.; Santos-Gonzalez, M.; Lopez-Lluch, G.; Briceno, J.; Gomez, M.A.; Poyato, A.; et al. N-acetylcysteine, coenzyme Q10 and superoxide dismutase mimetic prevent mitochondrial cell dysfunction and cell death induced by d-galactosamine in primary culture of human hepatocytes. *Chem. Biol. Interact.* **2009**, *181*, 95–106. [CrossRef]

11. Xiong, Y.; Peterson, P.L.; Lee, C.P. Effect of N-acetylcysteine on mitochondrial function following traumatic brain injury in rats. *J. Neurotrauma* **1999**, *16*, 1067–1082. [CrossRef]

12. Lushchak, V.I. Classification of oxidative stress based on its intensity. *Exp. Clin. Sci.* **2014**, *13*, 922–937.

13. Kołodziej, U.; Maciejczyk, M.; Miąsko, A.; Matczuk, J.; Knaś, M.; Żukowski, P.; Żendzian-Piotrowska, M.; Borys, J.; Zalewska, A. Oxidative modification in the salivary glands of high fat-diet induced insulin resistant rats. *Front. Physiol.* **2017**. [CrossRef]

14. Zalewska, A.; Knaś, M.; Żendzian-Piotrowska, M.; Waszkiewicz, N.; Szulimowska, J.; Prokopiuk, S.; Waszkiel, D.; Car, H. Antioxidant profile of salivary glands in high fat diet- induced insulin resistance rats. *Oral. Dis.* **2014**, *20*, 560–566. [CrossRef]

15. Żukowski, P.; Maciejczyk, M.; Matczuk, J.; Kurek, K.; Waszkiel, D.; Żendzian-Piotrowska, M.; Zalewska, A. Effect of N-acetylcysteine on antioxidant defense, oxidative modification, and salivary gland function in a rat model of insulin resistance. *Oxid. Med. Cell Longev.* **2018**, *2018*. [CrossRef]

16. Zalewska, A.; Ziembicka, D.; Zendzian-Piotrowska, M.; Maciejczyk, M. The Impact of High-Fat Diet on Mitochondrial Function, Free Radical Production, and Nitrosative Stress in the Salivary Glands of Wistar Rats. *Oxid. Med. Cell Longev.* **2019**, *2019*, 2606120. [CrossRef]

17. Leite, R.S.; Marlow, N.M.; Fernandes, J.K. Oral health and type 2 diabetes. *Am. J. Med. Sci.* **2013**, *345*, 271–273. [CrossRef]

18. Modèer, T.; Blomberg, C.C.; Wondimu, B.; Julihn, A.; Marcus, C. Association between obesity, flow rate of whole saliva, and dental caries in adolescents. *Obesity* **2010**, *18*, 2367–2373. [CrossRef]

19. Su, H.; Velly, A.M.; Salah, M.H.; Benarroch, M.; Trifiro, M.; Schipper, H.M.; Gornitsky, M. Altered redox homeostasis in human diabetes saliva. *J. Oral. Pathol. Med.* **2012**, *41*, 235–241. [CrossRef]

20. Martinez, R.F.; Jaimes-Aveldañez, A.; Hernández-Pèrez, F.; Arenas, R.; Miguel, G.F. Oral Candida spp carriers: Its prevalence in patients with type 2 diabetes mellitus. *An Bras Dermatol* **2013**, *88*, 222–225. [CrossRef]

21. Dursun, E.; Akalin, F.A.; Genc, T.; Cinar, N.; Erel, O.; Yildiz, B.O. Oxidative stress and periodontal disease in obesity. *Medicine* **2016**, *95*, e3136. [CrossRef]

22. Zeigler, C.C.; Persson, G.R.; Wondimu, B.; Marcus, C.; Sobko, T.; Modèer, T. Microbiota in the oral subgingival biofilm is associated with obesity in adolescence. *Obesity* **2012**, *20*, 157–164. [CrossRef]

23. Carda, C.; Mosquera-Lloreda, N.; Salom, L.; Gomez de Ferraris, M.E.; Peydro, A. Structural and functional salivary disorders in type 2 diabetic patients. *Med. Oral. Pathol. Oral. Cir. Bucal.* **2006**, *11*, 309–313.

24. D'Aiuto, F.; Nibali, L.; Parkar, M.; Patel, K.; Suvan, J.; Donos, N. Oxidative stress, systemic inflammation, and severe periodontitis. *J. Dent. Res.* **2010**, *89*, 1241–1246. [CrossRef]

25. Knaś, M.; Maciejczyk, M.; Daniszewska, I.; Klimiuk, A.; Matczuk, J.; Kołodziej, U.; Waszkiel, D.; Ładny, J.R.; Żendzian-Piotrowska, M.; Zalewska, A. Oxidative Damage to the Salivary Glands of Rats with Streptozotocin-Induced Diabetes-Temporal Study: Oxidative Stress and Diabetic Salivary Glands. *J. Diabetes Res.* **2016**. [CrossRef]

26. Jaccob, A.A. Protective effect of N-acetylcysteine against ethanol-induced gastric ulcer: A pharmacological assessment in mice. *J Intercult Ethnopharmacol* **2015**, *4*, 90–95. [CrossRef]

27. Kononczuk, T.; Lukaszuk, B.; Miklosz, A.; Chabowski, A.; Zendzian-Piotrowska, M.; Kurek, K. Cerulein-Induced Acute Pancreatitis Affects Sphingomyelin Signaling Pathway in Rats. *Pancreas* **2018**, *47*, 898–903. [CrossRef]

28. Maciejczyk, M.; Zebrowska, E.; Zalewska, A.; Chabowski, A. Redox Balance, Antioxidant Defense, and Oxidative Damage in the Hypothalamus and Cerebral Cortex of Rats with High Fat Diet-Induced Insulin Resistance. *Oxid. Med. Cell. Longev.* **2018**, *2018*, 6940515. [CrossRef]

29. Bligh, E.G.; Dyer, W.J. A rapid method of total lipid extraction and purification. *Can. J. Biochem Physiol.* **1959**, *37*, 911–917. [CrossRef]

30. Zalewska, A.; Maciejczyk, M.; Szulimowska, J.; Imierska, M.; Błachnio-Zabielska, A. High-Fat Diet Affects Ceramide Content, Disturbs Mitochondrial Redox Balance, and Induces Apoptosis in the Submandibular Glands of Mice. *Biomolecules* **2019**, *9*, 877. [CrossRef]

31. Janssen, A.J.M.; Trijbels, F.J.M.; Sengers, R.C.A.; Smeitink, J.A.M.; Van Den Heuvel, L.P.; Wintjes, L.T.M.; Stoltenborg-Hogenkamp, B.J.M.; Rodenburg, R.J.T. Spectrophotometric assay for complex I of the respiratory chain in tissue samples and cultured fibroblasts. *Clin. Chem.* **2007**, *53*, 729–734. [CrossRef]

32. Rustin, P.; Chretien, D.; Bourgeron, T.; Gérard, B.; Rötig, A.; Saudubray, J.M.; Munnich, A. Biochemical and molecular investigations in respiratory chain deficiencies. *Clin. Chim. Acta* **1994**. [CrossRef]

33. Wharton, D.C.; Tzagoloff, A. Cytochrome oxidase from beef heart mitochondria. *Methods Enzymol.* **1967**. [CrossRef]

34. Muller, F.L.; Liu, Y.; Van Remmen, H. Complex III releases superoxide to both sides of the inner mitochondrial membrane. *J. Biol. Chem.* **2004**. [CrossRef]

35. Srere, P.A. Citrate synthase. [EC 4.1.3.7 Citrate oxaloacetate-lyase (CoA-acetylating)]. *Methods Enzymol.* **1969**. [CrossRef]

36. Griffith, O.W. Determination of glutathione and glutathione disulfide using glutathione reductase and 2-vinylpyridine. *Anal. Biochem.* **1980**, *106*, 207–212. [CrossRef]

37. Griendling, K.K.; Minieri, C.A.; Ollerenshaw, J.D.; Alexander, R.W. Angiotensin II stimulates NADH and NADPH oxidase activity in cultured vascular smooth muscle cells. *Circ. Res.* **1994**, *74*, 1141–1148. [CrossRef]

38. Prajda, N.; Weber, G. Malignant transformation-linked imbalance: Decreased xanthine oxidase activity in hepatomas. *FEBS Lett.* **1975**, *59*, 245–249. [CrossRef]

39. Meki, A.R.; Esmail, E.-D.; Hussein, A.A.; Hassanein, H.M. Caspase-3 and Heat Shock Protein-70 in Rat Liver Treated with Aflatoxin B1: Effect of Melatonin. *Toxicon* **2004**, 93–100. [CrossRef]

40. Grisham, M.B.; Johnson, G.G.; Lancaster, J.R. Quantitation of nitrate and nitrite in extracellular fluids. *Methods Enzymol.* **1996**, *268*, 237–246.

41. Beckman, J.S.; Ischiropoulos, H.; Zhu, L.; van der Woerd, M.; Smith, C.; Chen, J.; Harrison, J.; Martin, J.C.; Tsai, M. Kinetics of superoxide dismutase- and iron-catalyzed nitration of phenolics by peroxynitrite. *Arch. Biochem. Biophys.* **1992**, *298*, 438–445. [CrossRef]

42. Kurek, K.; Miklosz, A.; Lukaszuk, B.; Chabowski, A.; Gorski, J.; Zendzian-Piotrowska, M. Inhibition of Ceramide De Novo Synthesis Ameliorates Diet Induced Skeletal Muscles Insulin Resistance. *J. Diabetes Res.* **2015**, *2015*, 154762. [CrossRef]

43. Ebertz, C.E.; Bonfleur, M.L.; Bertasso, I.M.; Mendes, M.C.; Lubaczeuski, C.; Araujo, A.C.; Paes, A.M.; Amorim, E.M.P.; Balbo, S.L. Duodenal jejunal bypass attenuates non-alcoholic fatty liver disease in western diet-obese rats. *Acta Cir. Bras.* **2014**, *29*, 609–614. [CrossRef]

44. Gan, K.X.; Wang, C.; Chen, J.H.; Zhu, C.J.; Song, G.Y. Mitofusin-2 ameliorates high-fat diet- induced insulin resistance in liver of rats. *World J. Gastroenterol.* **2013**, *19*, 1572–1581. [CrossRef]

45. Qu, D.M.; Song, G.Y.; Gao, Y.; Wang, J.; Hu, S.G.; Han, M. Expression pattern of PGC-1 alpha and Mfn2 in insulin resistance state after recovery in rat. *Jichu Yixue Yu Linchuang* **2008**, *28*, 133–137.

46. Ceriello, A.; Testa, R. Antioxidant anti-inflammatory treatment in type 2 diabetes. *Diabetes Care* **2009**, *32*, 232–236. [CrossRef]

47. Haber, C.A.; Lam, T.K.; Yu, Z.; Gupta, N.; Goh, T.; Bogdanovic, E.; Giacca, A.; Fantus, I.G. N-acetylcysteine and taurine prevent hyperglycemia-induced insulin resistance in vivo: Possible role of oxidative stress. *Am. J. Physiol. Endocrinol. Metab.* **2003**, *285*, E744–E753. [CrossRef]

48. Song, D.; Hutchings, S.; Pang, C.C.Y. Chronic N-acetylcysteine prevents fructose-induced insulin resistance and hypertension in rats. *Eur. J. Pharmacol.* **2005**, *508*, 205–210. [CrossRef]

49. Diniz, Y.S.; Rocha, K.K.; Souza, G.A.; Galhardi, C.M.; Ebaid, G.M.X.; Rodrigues, H.G.; Novelli Filho, J.V.B.; Cicogna, A.C.; Novelli, E.L.B. Effects of N-acetylcysteine on sucrose-rich diet-induced hyperglycaemia, dyslipidemia and oxidative stress in rats. *Eur. J. Pharmacol.* **2006**, *543*, 151–157. [CrossRef]

50. Murphy, M.P. How mitochondria produce reactive oxygen species. *Biochem. J.* **2009**, *417*, 1–13. [CrossRef]
51. Turrens, J.F. Mitochondrial formation of reactive oxygen species. *J. Physiol.* **2003**, *552*, 335–344. [CrossRef] [PubMed]
52. Cadenas, E.; Davies, K.J. Mitochondrial free radical generation, oxidative stress, and aging. *Free Radic. Biol. Med.* **2000**, *29*, 222–230. [CrossRef]
53. Xiang, R.L.; Huang, Y.; Zhang, Y.; Cong, X.; Zhang, Z.J.; Wu, L.L.; Yu, G.Y. Type 2 diabetes-induced hyposalivation of the submandibular gland through PINK1/Parkin-mediated mitophagy. *J. Cell Physiol.* **2020**, *235*, 232–244. [CrossRef]
54. Zhang, Y.; Marcillat, O.; Giulivi, C.; Ernster, L.; Davies, K.J. The oxidative inactivation of mitochondrial electron transport chain components and ATPase. *J. Biol. Chem.* **1990**, *265*, 16330–16336. [PubMed]
55. Martinez, M.; Martinez, N.; Hernandez, A.I.; Ferrandiz, M.L. Hypothesis: Can N-acetylcysteine be beneficial in Parkinson's disease? *Life Sci.* **1999**, *64*, 1253–1257. [CrossRef]
56. Oh, S.H.; Lim, S.C. A rapid and transient ROS generation by cadmium tiggers apoptosis via caspase-dependent pathway in HepG2 cells and this is inhibited through N-acetylcysteine-mediated catalase upregulation. *Toxicol. Appl. Pharmacol.* **2006**, *212*, 212–223. [CrossRef]
57. Zaragoza, A.; Diez-Fernandez, C.; Alvarez, A.M.; Andres, D.; Cascales, M. Effect of N-acetylcysteine and deferoxamine on endogenous antioxidant defense system gene expression in a rat hepatocyte model of cocaine cytotoxicity. *Biochim. Biophys. Acta Mol. Cell. Res.* **2000**, *1496*, 183–195. [CrossRef]
58. Martensson, J.; Lai, J.C.; Meister, A. High-affinity transport of glutathione is part of a multicomponent system essential for mitochondrial function. *Proc. Natl. Acad. Sci. USA* **1990**, *87*, 7185–7189. [CrossRef]
59. Selles, B.; Hugo, M.; Trujillo, M.; Srivastava, V.; Wingsle, G.; Jacquot, J.P.; Radi, R.; Rouhier, N. Hydroperoxide and peroxynitrite reductase activity of poplar thioredoxin-dependent glutathione peroxidase 5: Kinetics, catalytic mechanism and oxidative inactivation. *Biochem. J.* **2012**, *442*, 369–380. [CrossRef]
60. Hopkins, R.Z.; Li, R.Y. *Essential of Free Radical Biology and Medicine*; Cell Med Press AIMSCI, Inc.: Raleigh, NC, USA, 2017.
61. Samuni, Y.; Goldstein, S.; Dean, O.M.; Berk, M. The chemistry and biological activities of N-acetylcysteine. *Biochim. Biophys. Acta* **2013**, *1830*, 4117–4129. [CrossRef]
62. Winterbourn, C.C.; Metodiewa, D. Reactivity of biologically important thiol compounds with superoxide and hydrogen peroxide. *Free Radic. Biol. Med.* **1999**, *27*, 322–328. [CrossRef]
63. Russell, S.L.; Reisine, S. Investigation of xerostomia in patients with rheumatoid arthritis. *JADA* **1998**, *129*, 733–739. [CrossRef] [PubMed]
64. Burgunder, J.M.; Varriale, A.; Lauterburg, B.H. Effect of N-acetylcysteine on plasma cysteine and glutathione following paracetamol administration. *Eur. J. Clin. Pharmacol.* **1989**, *36*, 127–131. [CrossRef] [PubMed]
65. Haugaard, N.; Lee, N.H.; Kostrzewa, R.; Horn, R.S.; Haugaard, E.S. The role of sulfhydryl groups in oxidative phosphorylation and ion transport by rat liver mitochrondia. *Biochim. Biophys. Acta* **1969**, *172*, 198–204. [CrossRef]
66. Garcia-Ruiz, I.; Solis-Munoz, P.; Fernandez-Moreira, D.; Grau, M.; Colina, F.; Munoz-Yague, T.; Solis-Herruzo, J.A. High-fat diet decreases activity of the oxidative phosphorylation complexes and causes nonalcoholic steatohepatitis in mice. *Dis. Models Mech.* **2014**, *7*, 1287–1296. [CrossRef]
67. Valenti, D.; Manente, G.A.; Moro, L.; Marra, E.; Vacca, R.A. Deficit of complex I activity in human skin fibroblasts with chromosome 21 trisomy and overproduction of reactive oxygen species by mitochondria: Involvement of the cAMP/PKA signalling pathway. *Biochem. J.* **2011**, *435*, 679–688. [CrossRef]
68. Block, K.; Gorin, Y.; Abboud, H.E. Subcellular localization of Nox4 and regulation in diabetes. *Proc. Natl. Acad. Sci. USA* **2009**, *106*, 14385–14390. [CrossRef]
69. Takac, I.; Schroder, K.; Zhang, L.; Lardy, B.; Anilkumar, N.; Lambeth, J.D.; Shah, A.M.; Morel, F.; Brandes, R.P. The E-loop is involved in hydrogen peroxide formation by the NADPH oxidase Nox4. *J. Biol. Chem.* **2011**, *286*, 13304–13313. [CrossRef]
70. Brownlee, M. Biochemistry and molecular cell biology of diabetic complication. *Nature* **2001**, *414*, 813–820. [CrossRef]

71. Hatanaka, E.; Dermargos, A.; Hirata, A.E.; Vinolo, M.A.; Carpinelli, A.R.; Newsholme, P.; Armelin, H.A.; Curi, R. Oleic, linoleic and linolenic acids increase ros production by fibroblasts via NADPH oxidase activation. *PLoS ONE* **2013**, *8*, e58626. [CrossRef]
72. Forner, L.; Hyldegaard, O.; von Brockdorff, A.S.; Specht, L.; Andersen, E.; Jansen, E.C.; Hillerup, S.; Nauntofte, B.; Jensen, S.B. Does hyperbaric oxygen treatment have the potential to increase salivary flow rate and reduce xerostomia in previously irradiated head and neck cancer patients? A pilot study. *Oral Oncol.* **2011**, *47*, 546–551. [CrossRef] [PubMed]

Review

Preventive Role of L-Carnitine and Balanced Diet in Alzheimer's Disease

Alina Kepka [1,*], **Agnieszka Ochocinska** [1,*], **Małgorzata Borzym-Kluczyk** [2], **Ewa Skorupa** [1], **Beata Stasiewicz-Jarocka** [3], **Sylwia Chojnowska** [4] and **Napoleon Waszkiewicz** [5]

[1] Department of Biochemistry, Radioimmunology and Experimental Medicine, The Children's Memorial Health Institute, 04-730 Warsaw, Poland; e.skorupa@ipczd.pl

[2] Department of Pharmaceutical Biochemistry, Medical University of Bialystok, 15-089 Bialystok, Poland; malgorzata.borzym-kluczyk@umb.edu.pl

[3] Department of Medical Genetics, Medical University of Bialystok, 15-089 Bialystok, Poland; beata.stasiewicz.jarocka@gmail.com

[4] Faculty of Health Sciences, Lomza State University of Applied Sciences, 18-400 Lomza, Poland; sylwiacho3@gmail.com

[5] Department of Psychiatry, Medical University of Bialystok, 15-089 Bialystok, Poland; napwas@wp.pl

* Correspondence: a.kepka@ipczd.pl (A.K.); a.ochocinska@ipczd.pl (A.O.); Tel.: +48-22-815-73-01 (A.K.); +48-22-815-73-01 (A.O.)

Received: 3 June 2020; Accepted: 29 June 2020; Published: 3 July 2020

Abstract: The prevention or alleviation of neurodegenerative diseases, including Alzheimer's disease (AD), is a challenge for contemporary health services. The aim of this study was to review the literature on the prevention or alleviation of AD by introducing an appropriate carnitine-rich diet, dietary carnitine supplements and the MIND (Mediterranean-DASH Intervention for Neurodegenerative Delay) diet, which contains elements of the Mediterranean diet and the Dietary Approaches to Stop Hypertension (DASH) diet. L-carnitine (LC) plays a crucial role in the energetic metabolism of the cell. A properly balanced diet contains a substantial amount of LC as well as essential amino acids and microelements taking part in endogenous carnitine synthesis. In healthy people, carnitine biosynthesis is sufficient to prevent the symptoms of carnitine deficiency. In persons with dysfunction of mitochondria, e.g., with AD connected with extensive degeneration of the brain structures, there are often serious disturbances in the functioning of the whole organism. The Mediterranean diet is characterized by a high consumption of fruits and vegetables, cereals, nuts, olive oil, and seeds as the major source of fats, moderate consumption of fish and poultry, low to moderate consumption of dairy products and alcohol, and low intake of red and processed meat. The introduction of foodstuffs rich in carnitine and the MIND diet or carnitine supplementation of the AD patients may improve their functioning in everyday life.

Keywords: Alzheimer's disease; L-carnitine; carnitine supplementation; Mediterranean diet; MIND diet

---

1. Introduction

Recent progress in the medical sciences have significantly prolonged human life and the incidence of old age diseases. Old age diseases are a rapidly growing cause of disability and/or death [1]. There was proof that some diet supplements may have a positive therapeutic effect on patients with neurodegenerative diseases, particularly in AD [2]. Recently, the attention of physicians as well as their patients is increasingly directed to dietary supplements, which have become a more and more attractive option in neurodegenerative disease treatment [3].

Neurodegenerative diseases, including AD, are characterized by a loss of neurons and synapses of the brain cortex and some subcortical regions, resulting in the atrophy and degeneration of the involved regions in the temporal and parietal lobes as well as parts of the frontal lobe and part of the callosal gyrus [4,5]. The extracellular deposition of amyloid-β, the intracellular deposition of protein tau (τ) and microglia activation are causes of AD's pathological symptoms [6]. Recently, it has been reported that an essential role in the pathogenesis of AD may be also played by other proteins such as α-synuclein and the protein TDP-43 [7]. It has been confirmed that there are genetic predispositions to the onset of AD. People with the defective e4 variant of the apolipoprotein E gene, which is involved in amyloid-β degradation, may already have symptoms of AD in the early period of life [8,9]. The accumulation of β-amyloid plates in neurons decreases the efficiency of electron transport in the respiratory chain, leading to a decrease in ATP production, the induction of oxidative stress and the disturbance of Ca^{++} homeostasis. On the other hand, the accumulation of the protein τ inside neurons blocks the intracellular transport of proteins, nutrients and neurotrophins. The primary symptoms of AD present as disturbances of recent memory, the concentration of attention and orientation. The leading symptoms of AD are aphasia (speech disturbance), apraxia (inability of movement despite non-handicapped motoric functions), agnosia (lack of sensory ability to recognize objects despite correct sensory function), and disturbances in sleep and wakefulness. The deepening symptoms in AD patients, connected with disturbances in simple and complex everyday life activities, make independent existence impossible. Ophthalmic symptoms—e.g., the impairment of chromatic vision, scotoma, the reduction of contrast sensitivity, disturbances in the mobility of the eyeballs, and retina degeneration—also occur in AD. The deposition of β-amyloid concrements in the eyes of AD patients were demonstrated through eye ground imaging [10].

Acetylcholine (Ach) plays a crucial role in the cognitive functions of the brain [5]. A decrease in the activity of the mechanisms responsible for the biosynthesis and degradation of Ach, such as choline acetyltransferase, acetylcholine esterase, transport system with high affinity to choline (HACU) and follicular acetylcholine transporter (VAChT), was found in patients with AD [5]. The impairment of the transmission of nervous signals caused by a decrease in the density of muscarinic and nicotinic receptors, a decrease in the intraneural concentration of Ach and Ach accumulation in synaptic vesicles was demonstrated in AD [7,11,12]. Numerous articles document disturbances in the biosynthesis and metabolism of carnitine in AD patients [13–15]. Low concentrations of free carnitine, acetyl-L-carnitine (ALC) and other acylcarnitines were found in plasma and tissues in many studies on AD [13,16]. A progressive decrease in ALC and other acyl-carnitine serum levels in healthy subjects (HS) through to subjective memory complaints (SMC) and mild cognitive impairment (MCI) up to Alzheimer's disease (AD) were reported. ALC significantly decreased on average by 21% in SMC, 27% in MCI and 36% in AD as compared to in HS [16]. The deficit of ALC suggests a perturbed transport of fatty acids into the mitochondria for beta-oxidation, as well as suppressed energy metabolism. The results of transcriptomic studies showing a significant decrease in the activity of the carnitine shuttle in AD patients are consistent with the above-described hypothesis. The deficit of the carnitine shuttle might contribute to the mitochondrial dysfunctions supposed to be responsible for many neurodegenerative diseases including AD. The decreased serum levels of some acyl-carnitines found in MCI subjects might indirectly signal an impending progression of dementia and might be used as biomarkers of phenotype conversion from MCI to AD [16]. The results described above suggest that serum ALC and other acyl-L-carnitine levels decrease along a continuum from HS to SMC and MCI subjects, up to patients with AD [13,14,16]. It was demonstrated that ALC facilitates cholinergic neurotransmission directly or by providing an acetyl group that may be used for acetylcholine synthesis [17,18]. It was reported that ALC, by the stimulation of the synthesis of nerve growth factor receptors in the hippocampus and basal forebrain, prevents the loss of muscarinic receptors as well as nerve growth factor (NGF) and directly or indirectly modulates N-methyl-D-aspartate receptor (NMDA). The excessive stimulation of NMDA receptors by glutamic acid may induce the uncontrolled influx of Ca^{++} into cells, causing damage to and the death of neurons [19]. In addition, it was proven that carnitine significantly increases dopamine levels in the cortex, hippocampus and striatum of the rat brain [20]. In animal experiments, it was found that ALC supplementation prevents the hyperphosphorylation of the protein τ induced by homocysteine (a new marker of AD) and inhibits the phosphorylation of β-amyloid [21]. It has been shown that ALC and L-carnitine reduce apoptosis through the mitochondrial pathway [18]. Suchy et al. [22] reported that carnitine supplementation reduces damage to the murine brain caused by free radicals and improves cognitive performance. A beneficial effect of ALC on cognition and behavior in aging and AD subjects was reported [23]. The possible mechanisms of ALC action in AD may also involve facilitating the rebuilding of cell membranes, as well as improving synaptic function, enhancing cholinergic activity, restoring the brain energy supply, protecting against toxins, and exerting neurotrophic effects via stimulating NGF and the acetylation of proteins [24].

The introduction of an appropriate balanced diet, besides carnitine supplementation (especially ALC), is an important factor that can delay or prevent the onset of AD. Many studies have suggested that the diet MIND (Mediterranean-DASH Intervention for Neurodegenerative Delay)—composed of the DASH (Dietary Approaches to Stop Hypertension) diet and the Mediterranean diet, considered one of the healthiest diets on the earth—appears to be the most effective in preventing neurodegeneration, especially Alzheimer's disease. [25]. Adherence to the Mediterranean diet may not only reduce the risk of AD but also diminish pre-dementia syndromes and their progression to overt dementia. Based on the current evidence, there are no definitive dietary recommendations for the prevention of AD. However, the following dietary advice for lowering the risk of AD and inhibiting cognitive decline, as well as decreasing all-cause mortality in AD patients, is suggested: a high level of consumption of fats from fish, vegetable oils, non-starchy vegetables, and low glycemic index fruit; a diet low in foods with added sugars; and a moderate wine or beer intake should be encouraged [26].

2. AD Epidemiology

Contemporary society aging is connected with the occurrence of neurodegenerative disorders leading to dementia, creating serious problems in complex medical and social care. It was estimated that in 2006, 7.3 million (mln) Europeans (12.5 per 1000 residents) in 27 member states, between 30 and 99 years of age, suffered from different kinds of dementia. Among Europeans, dementia affected women (4.9 mln) more frequently, than men (2.4 mln). Together with an increase in the average life span, especially in developed countries, the frequency of dementia has dramatically increased. The number of people with dementia worldwide is expected to double every 20 years, reaching almost 75 million by 2030 [27]. Most recent regional estimates of the age-standardized prevalence of dementia in people aged 60 and over fall between 5.6 and 7.6%. Although dementia was traditionally viewed as more prevalent in developed countries, these estimates range from 4.6% in Central Europe to 8.7% in North Africa and the Middle East [28]. Currently, there are over 70 million people in the world's oldest group, over 80 years. It is estimated that by 2050, the population over 80 will increase several times. According to data from the World Health Organization (WHO), in 2030, there will be 65 million; in 2040, there will be 80 million; and in 2050, there will be 115 million dementia patients [29,30]. The seven places with the largest numbers of people with dementia in 2001 were as follows: China (5.0 mln), the European Union (5.0 mln), the USA (2.9 mln), India (1.5 mln), Japan (1.1 mln), Russia (1.1 mln) and Indonesia (1.0 mln) [30]. Currently, in the developed countries, the population of people over 65 constitutes 14% (about 170 mln), while in the developing countries, it is about 5% (about 248 mln). According to the expected demographic changes, it is estimated that in 2030, people over 65 will constitute 23% (275 mln) in developed countries and 10% (680 mln people) in developing countries. It is estimated that AD may affect about 6% of the people of the world over 65 and 40% of people over 85 [31]. It is estimated that by 2050, there will be in the world two billion people aged 60 years or over, of which 131 million will be affected by dementia, but depression will be the second cause of disability worldwide by 2020. Preventing or delaying the onset of dementia and depression should therefore be a world public health priority [29,30,32]. Today, in the world, almost 44 million people suffer from AD, and each year, 4.5 million new cases of dementia are diagnosed. In the world, the risk of dementia doubles every 5 years, with AD responsible for 2/3 of all cases of dementia. In the USA, AD is placed in third position after neoplasms and cardiovascular diseases concerning social security costs [31].

For Poland's demographics, prognoses as Poland's population gets old are also not optimistic, and it has been estimated that in the next 25 years, the average age of men will increase by 7.2 years and that of women, by 4.5 years. It should be mentioned that in 2010, the expected length of life in Poland averaged 75.9 years (men, 71.9 years; women, 80.1 years). According to the official Polish data from 1999, population studies were conducted in two centers, among the population of the Warsaw district of Mokotów and in Świebodzin (rural and small town communities). The epidemiological data from Świebodzin revealed that the prevalence of AD in the population over 65 was 3.5% and that of vascular dementia, 3.6%. In the second study (Mokotów, Warsaw district, Poland), dementia was found in the population from 65 to 84 years of age to be at the level of 5.7% (including AD at 2.3% and vascular dementia at 2.7%) [33]. Population data published in 2007 by Bdzan et al. [34] regarding the studied rural populations (Pruszcz Gdański, Trąbki Wielkie and Pszczółki, Poland) showed that the prevalence of dementia was estimated at 6.7%. The prevalence of dementia in total rural populations was 3.0% for men and 8.8% for women. The prevalence of Alzheimer's disease was 1.1% for men and 4.0% for women, and that of vascular dementia, 1.9% and 3.5%, respectively. Bdzan et al.'s [34] study presents data showing that rural populations had a higher incidence of dementia disorders than that previously reported in other Polish regions and that women had a higher dementia risk than men. It should be stressed that in 2005–2010, the annual increase in the number of dementia patients remained levels not exceeding 2%, but in 2010–2015, the annual increase in the number of dementia patients reached 3–4%. The Polish National Statistics Agency predicts an increase in the number of people reaching retirement age to 23.8% of the Polish population (4.8 million) by 2030 [35]. According to the Polish National Statistics Agency report of 2016, presently in Poland, people above 65 constitute 14.7% of the population, and increases to 24.5% by 2035 and to above 30% by 2050 are predicted. In Poland, AD presently affects more than 300,000 people, mostly elderly people, and it is estimated that the number of AD subjects will triple by 2050, reaching almost a million [36].

The above figures show the importance of the problem facing health services not only in Poland but also over the whole world [37]. It should be taken into consideration that the further aging of the Polish population will increase the numbers of people that take advantage of medical and social care, are disabled and suffer from non-infectious chronic diseases. In the existing situation, the Polish system of health care—similarly to the systems in other countries—must take up the challenge of creating an efficient system of care for people suffering from dementia.

3. Prevention of Alzheimer's Disease—Mediterranean and MIND Diets

In contemporary medicine, the prevention of neurodegenerative diseases is one of the main objectives for many branches of health services. The appropriate types and quality of food may be important factors for preventing and supporting the treatment of AD. The early prevention of AD starts by reducing risk factors: reducing the smoking of cigarettes, preventing hypertension, insuring optimal concentrations of homocysteine, preventing type 2 diabetes, combating insulin resistance and obesity, limiting stress, avoiding toxins, and mental and physical training are important components of AD prevention. Appropriate nutrition is an essential and modifiable factor that plays a key role in preventing and/or delaying the onset of dementia, including AD. It was reported that a reduction in amount of fried meat and an increase in the amounts of other foods (such as fish, cheese, vegetables and vegetable oil) in the diet significantly reduced the incidence of AD [38]. Diets rich in advanced glycation end products (AGEs), which arise during long-lasting food thermal processing (heating, frying and irradiation), significantly accelerate the development of AD [39]. High concentrations of AGEs are contained in (a) products containing sugar (candy, cookies, chocolate biscuits, cakes, fizzy drinks, pastries and sauces); (b) processed meats such as sausages and conserved and preserved meat; (c) processed dairy products; (d) food containing trans fats such as margarine and cream; and (e) highly fried products such as fried potatoes, crisped cakes etc. The lowest amounts of AGEs are found in fresh fruits, vegetables, seafood, products with short thermal processing, and raw and non-processed food. A correctly composed diet rich in polyunsaturated (from the families of omega-3 and omega-6) and monounsaturated fatty acids as well as antioxidative vitamins (E, C and β-carotene) reduces the risk of AD development. It was reported that a decrease in the impairment of cognitive function and dementia risk were connected with a decrease in the consumption of milk and dairy products. The consumption of full-fat dairy products may be connected with worsening cognitive function in older people. It was reported that moderate alcohol consumption may be connected with a decrease in the risk of dementia due to AD [40,41]; however, recently, the Lancet published a statement concerning alcohol, that there is "no safe limit, even one drink a day, increases risk of chronic non infectious diseases" [42].

A large study demonstrated the association of consuming a Mediterranean diet with a decrease in the incidence of AD, which creates hope for using the Mediterranean diet as a modifiable risk factor in protection against AD [43,44]. A Mediterranean diet rich in vegetables with low starch contents, fruit with low glycemic indices, cereal products, legumes, plant oils (olive, colza, linen and sunflower) and fish (especially sea fish: halibut, herring, mackerel and sardines) and a diet containing moderate amounts of meat and dairy products positively affect health conditions and may decrease the risk of the development of many diseases including neurodegenerative diseases [25,45].

It was reported that people adhering to a Mediterranean diet have a 28% decreased risk of cognitive disturbances and 48% decreased risk of AD in comparison to people who do not consume a Mediterranean diet [46]. It seems that a complex nutritional strategy initiated in the early stages of cognitive impairment is the most pragmatic approach for controlling the progress of AD [47]. Presently, for the elderly and people at risk of AD, nutrition based on the MIND (Mediterranean-DASH Intervention for Neurodegenerative Delay) diet—composed of DASH (Dietary Approaches to Stop Hypertension) diet and the Mediterranean diet (MD), considered as the healthiest diet on earth, constituting a careful nutritional program—is recommended [48]. In our opinion, the MIND diet should be a very important element of many disease prevention strategies, including those for dementia and AD (Table 1).

Mediterranean and Asian diets are currently considered the healthiest. Good eating habits are effective in combating the risk of age-related diseases, especially cardiovascular and neurodegenerative diseases. Foods and drinks of plant origin such as green tea, extra-virgin olive oil, red wine, spices, berries and aromatic herbs have beneficial effects in the prevention of amyloid diseases, e.g., Alzheimer's disease, Parkinson's disease and prion diseases. Mediterranean and Asian diets are becoming more and more attractive for the prevention and treatment of neurodegenerative diseases, as it has been proven that they can inhibit the production of amyloidogenic peptides, increase the activity of antioxidant enzymes, activate autophagy and reduce inflammation [49] (Table 1). Currently, some foods or food groups traditionally considered harmful such as eggs and red meat have been partially rehabilitated, but there is still a negative correlation of cognitive functions with saturated fatty acids. Protective effects against cognitive decline of elevated fish consumption and a high intake of monounsaturated fatty acids and polyunsaturated fatty acids (PUFA), particularly n-3 PUFA, was confirmed [40]. Cheese and yoghurt should be moderately consumed, while meat should be rarely consumed. Wine or beer (mostly non-alcoholic beer) during main meals are also one of the components of the Mediterranean diet. Beer consumption, and its content of bioavailable silicon, reduces the accumulation of aluminum in the body and brain tissue and lipid peroxidation, and protects the brain against neurotoxic effects by regulating antioxidant enzymes [26].

Table 1. Type and frequency of consumption of foods in the Mediterranean-DASH Intervention for Neurodegenerative Delay (MIND) diet pattern with a role in Alzheimer's prevention [26,48], modified.

No.	Type of Food	Components	Intake Categories	Characteristics
1.	green, leafy vegetables	kale, spinach, kohlrabi, different varieties of lettuce, cooked greens and salads	≥6 servings/week	vegetables rich in vitamins C and A
2.	all other vegetables	celery, cabbage, beets, cucumbers, cauliflower, zucchini, tomatoes, leeks, garlic and onion	1 serving/day	to choose non-starchy vegetables with a lot of nutrients and a low number of calories
3.	berries	strawberries, blueberries, raspberries, blackberries	≥2 servings/week	a source of antioxidants
4.	nuts and almonds	pineapple, pistachios, macadamia, pecans, peanuts and Brazilian walnut	≥5 servings/week	are a source of unsaturated fatty acids and antioxidants; contain vitamins E, B_1 and PP; and reduce the level of "bad" cholesterol
5.	whole grains	oatmeal, quinoa, brown rice, whole-wheat pasta and 100% whole-wheat bread	≥3 servings/day	a source of fiber, folic acid, vitamin B_3, iron, zinc, magnesium and phosphorus
6.	fish	salmon, sardines, trout, tuna and mackerel	≥1 serving/week	high amounts of omega-3 fatty acids
7.	poultry	chicken or turkey	≥2 servings/week	fried chicken is not encouraged on the MIND diet
8.	beans	lentils, soybeans, string beans, broad beans, green peas, chickpeas and white beans	≥4 servings/week	a source of fiber, protein, vitamins and minerals
9.	olive oil	cold pressed oils		a source of vitamins A, D, E, and K; polyunsaturated fatty acids; use fat for long frying at a high smoke point
10.	milk, dairy products	low-fat: milk, cheese, yoghurt, buttermilk, kefir and cottage cheese	≥2 glasses/day; 280–400 g semi-skimmed cheese or 1 slice (30 g) of yellow cheese.	a source of protein and minerals: calcium, potassium, phosphorus, magnesium, zinc, manganese, iron; high in vitamins: B_2, B_{12}, A, D, E, and K and probiotics
11.	wine	red and white	≤1 glass serving/day	both red and white wine may benefit the brain; red wine is recommended because a lot of research has focused on the red wine compound resveratrol
	beer	non-alcoholic beer	regular beer consumption is not recommended for some risk-group populations (pregnant, children, people affected by liver diseases)	source: carbohydrates, protein/amino acids (proline, glutamic and aspartic acid, glycine, alanine), minerals (fluoride, potassium, phosphorus, calcium, sodium, magnesium, silicon), vitamins (B1–B6, folic acid), and other compounds, such as polyphenols

4. Physiological Properties of L-Carnitine

L-carnitine (2-hydroxy-4-trimethylammonium butyrate) (LC) plays many important roles in the intracellular functions of the body. The most important role is played by obtaining cellular energy from fatty acids (FAs) in the mitochondrial matrix and maintaining mitochondria coenzyme A (CoA) homeostasis during mitochondrial FA oxidation [50]. Maintaining the mitochondrial homeostasis of CoA is an extremely important function of carnitine in supporting optimal cellular CoA and acyl-CoA concentrations. CoA is necessary for the activation and oxidation of the FAs from adipose tissue in the mitochondria for ATP synthesis. FA oxidation reduces glucose oxidation in the tissues where glucose is not an essential fuel, and amino acid catabolism for gluconeogenesis and energy production. FAs are a very efficient source of human energy, as the yield of energy from the total oxidation of FAs is 37.7 kJ/g as compared to 16.7 kJ/g from protein or carbohydrates. Another role of carnitine in human metabolism is participation in the processes of the detoxication of toxic exogenous compounds (e.g., some xenobiotics, including ampicillin, valproic acid and salicylic acid), which are excreted by the kidneys in combination with carnitine [51]. The next important role of carnitine is its contribution to the catabolism of branched-chain ketoacids derived from branched-chain amino acids (valine, leucine and isoleucine). L-carnitine also inhibits free radical production and demonstrates antioxidant action [51].

The richest source of L-carnitine is red meat consumed by adults, and milk consumed by infants and children [52], whereas plants contain only trace of carnitine [53] (Table 2). The standard human diet covers about 3/4 of the requirements for L-carnitine, and the remaining 1/4 is synthesized in the human body from lysine and methionine with the participation of ascorbic acid, niacin, piridoxin and Fe^{2+} [54] in the liver, kidneys, brain [50] and placenta [55]. Nutritional carnitine is actively (sodium dependent) and passively transported from the intestinal contents into enterocytes, with 54–86% bioavailability, depending on the amount of carnitine in a meal. The bioavailability of carnitine from dietary supplements (0.5–6.0 g) is much lower than that from the diet and reaches only 14–18%. It should be mentioned that organisms maintain carnitine homeostasis, and along with an increase in carnitine consumption, there is decrease in carnitine absorption [56]. Consumed but not absorbed carnitine is degraded, mainly by intestinal microorganisms to trimethylamine (TMA), which can reach the liver, where it is transformed into the toxic trimethylamine N-oxide (TMNO), excreted in the urine. To date, TMA and TMNO have mostly been treated as nontoxic substances, but more recently, they have been treated as potentially carcinogenic agents, because of the possibility of their transformation into N-nitrosodimethylamine (NDMA) [57]. However, no evidence of clinical implications of TMAO in the central nervous system has been documented as yet, but TMAO is present at detectable levels in the cerebro-spinal fluid (CSF). In a small tested groups of subjects, TMAO levels in the CSF were apparently unrelated to the diagnosis of neurological disorders such as AD [58].

In the human intestine, carnitine is acetylated to biologically active acetyl-L-carnitine (γ-trimethyl-β-acetylbutyre-betaine) (ALC). Since ALC is transferred through intestinal serous membranes better than non-acetylated L-carnitine, the intracellular acetylation of carnitine may facilitate its diffusion across the serosal membrane [56]. LC and its short chain fatty esters do not combine with plasma proteins, though blood cells contain LC, but the speed of the transport of LC between erythrocytes and the plasma is very slow [59]. In the circulation, about 75% of LC occurs in a free state, 15% occurs as ALC, and the remaining 10% occurs as esters of carnitine with other acids (e.g., propionyl-L-carnitine) [60]. In human tissues, carnitine is localized mainly in skeletal and cardiac muscles (98%), whereas in the liver, kidneys and brain resides only about 1.5% [52,61]. In the adult brain, about 80% of carnitine exists as free carnitine (FC); 10–15%, as ALC; and less than 10%, as long chain acylcarnitines [62]. The plasma of healthy people on a normal diet contains only small amounts of LC (about 0.5–1% of total body L-carnitine) [63], amounting to about 40–60 µM in plasma [59]. There are at least three pharmacokinetic compartments for LC. The first consists of extracellular fluid (including plasma), constituting the central compartment, reflecting the initial carnitine content; the second compartment is relatively small, with fast turnover, involving the liver, kidneys and other organs; the third and the greatest compartment involves the skeletal and heart muscles. The turnover times (or average times for which carnitine resides in compartments) amount to about 1, 12 and 191 h for compartments one, two and three, respectively, and the total carnitine turnover time for the whole human organism amounts to about 66 days. The velocity constant for the transport of carnitine from muscle to plasma is about 0.0005 h^{-1}, which means that the half-time for carnitine transportation from muscle to plasma is about 5–6 days. Muscle carnitine content is rather stable, because the transport of carnitine to and from muscle cells is a very slow process, lasting weeks and even months [59]. Since the taking of carnitine from the muscles is a long process in carnitine deficiency, it is advisable to apply carnitine supplementation to quickly restore appropriate concentrations of carnitine in the first (extracellular fluids) and second compartments (internal organs excluding muscles) with short turnover times. It was reported that in humans, after the per os administration of carnitine at doses of 30–100 mg/kg, carnitine's peak plasma concentrations were 27–91 μM after 3 h, and the carnitine concentrations returned to basal levels by 24 h. After the intravenous injection of single doses (0.5 g) of LC, the plasma concentrations of ALC and LC returned to their initial values by 12 h [58].

In the homeostasis of carnitine and its ester, the main role in the maintenance of appropriate concentrations is played by the kidneys [64]. Carnitine is not metabolized in humans but undergoes filtration in the glomeruli and is almost totally (98–99%) reabsorbed in the renal canaliculi. Sodium-dependent carnitine organic cation transporter (OCTN2) plays a key role in renal carnitine reabsorption [65]. The renal threshold for free carnitine is in the range of its physiological plasma concentration.

4.1. Role of L-Carnitine and Acetyl-L-Carnitine in Human Brain Metabolism

L-carnitine is actively transported to the brain through the blood–brain barrier by the organic cation transporter OCTN2 and accumulates in neural cells especially as acetylcarnitine. LC, besides its important role in the metabolism of lipids, is also a potent antioxidant (free radical scavenger) and thus protects brain tissues against oxidative damage [66]. Mitochondria, as the main source of cell energy as well as ROS and antioxidants, play a key role in the production of ATP, regulating apoptosis and detoxication, maintaining membrane potentials and the distributions of ions in appropriate compartments of the cell, etc. Maintaining mitochondrial homeostasis is crucial for the development and proper action of neurons. Dietary supplements applied for maintaining mitochondrial homeostasis include L-carnitine; coenzyme Q_{10}; mitoquinone mesylate; and other mitochondrion-targeted antioxidants such as N-acetylcysteine; vitamins C, E, K_1 and B; sodium pyruvate; and lipoic acid [67].

Even though the brain's basic energetic substrate is glucose, LC is important in brain energetic lipid metabolism, taking part in the transport of the long chain fatty acids from the cytoplasm to mitochondria [52] and acetyl groups from mitochondria to the cytoplasm [62]. Maintaining a suitable relationship between cellular acetyl-CoA and CoA, L-carnitine ensures correct energetic cell metabolism [68]. It should be mentioned that connecting acyl and acetyl groups to LC increases LC's hydrophobicity, which facilitates LC's crossing of the blood–brain barrier. After the penetration of brain mitochondria with long chain fatty acids, LC reacts with free coenzyme A (CoASH)—a reaction catalyzed by carnitine palmitoyltransferase II (CPT II)—releasing LC [50] (Figure 1).

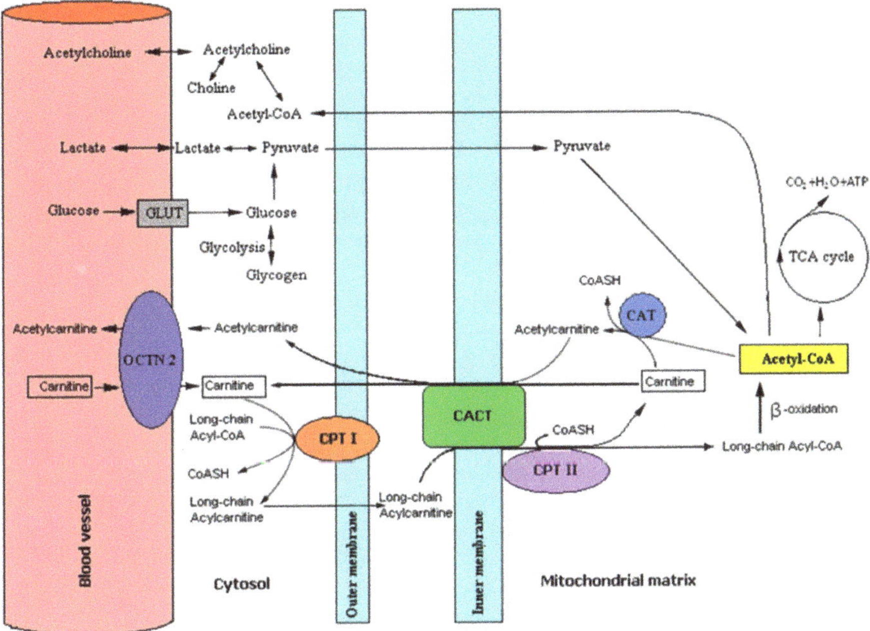

Figure 1. The role of carnitine and other substances in the brain's energy supply. Abbreviations: GLUT, glucose transporters; OCTN2, sodium-dependent carnitine organic cation transporter; CPT I, carnitine palmitoyltransferase I; CPT II, carnitine palmitoyltransferase II; CACT, carnitine acylcarnitine translocase; CAT, carnitine acetyltransferase; CoASH, coenzyme A; Acetyl-CoA, acetyl coenzyme A; TCA cycle, tricarboxylic acid cycle.

Therefore, LC improves energetic brain homeostasis by supplying acyl groups to the mitochondria of the brain cells. ALC is an important factor expanding the brain's sources of energy; therefore, treatment with ALC is responsible for a reduction in brain glycolytic flow and the enhancement of the utilization of alternative energy sources, such as fatty acids or ketone bodies [69], that may reduce brain glucose utilization [70]. ALC can provide an acetyl moiety that can be oxidized for energy production; used as a precursor for acetylcholine; or incorporated into glutamate, glutamine and γ-aminobutyric acid, as well as into lipids for myelination and cell growth [71]. ALC also has many other functions in the body. ALC's ability to freely pass the brain–blood barrier could help brain fatty acid transport for oxidation in mitochondria that improves brain energy metabolism. ALC supplementation positively influences the activity of enzymes in the tricarboxylic acid cycle, the electron transport chain and amino acid metabolism [72]. ALC acts neuroprotectively by improving the energetic function of brain mitochondria, the elimination of oxidative products, the stabilization of the cell membranes and neurotrophic factor production (stimulates protein and phospholipid biosynthesis), and it exerts anti-apoptotic functions, as well as modulating the expression of genes coding proteins, and protects neural cells from excitotoxicity [73]. Additionally, ALC provides acetyl groups for the synthesis of acetylcholine [17] and acetylation of nuclear histones [74]. ALC counteracts stressogenic agents by maintaining proper plasma concentrations of β-endorphin and cortisol [75]. ALC stimulates α-secretase activity and physiological amyloid precursor protein (APP) metabolism. In particular, ALC favors the delivery of disintegrin and metalloproteinase domain-containing protein 10 (ADAM10), the most accredited α-secretase, to the post-synaptic compartment and, consequently, positively modulates ADAM10's enzymatic activity toward APP [76]. Palmitoylcarnitine (ester of carnitine with palmitic acid) can stimulate the expression of GAP-43 (also named B-50, neuromodulin, F1, pp45), a protein involved in neural development, neuroplasticity and neurotransmission [77]. ALC increases the concentration of the brain-derived neurotrophic factor (BDNF), that is lowered in AD patients [78,79]. ALC increases astrocytes' glutathione concentrations (an important cellular antioxidant), which are lowered with age [80]. ALC is recommended for delaying the onset and development of Alzheimer's and Parkinson's diseases, and alleviating the symptoms of senile depression and memory disturbances connected with age; it also improves studying and memory [23,81,82]. It was confirmed that the multidirectional positive role of ALC in the dopaminergic system [20] depended on the slowing down of the progressive deterioration of dopaminergic receptors with a simultaneous increase in the concentrations of dopamine (a neurotransmitter responsible for mood, processes of thinking, the coordination of movement and resistance to stress) in neurons [83]. It was reported that the LC supplementation of experimental animals significantly increased their levels of neurotransmitters such as noradrenaline, adrenaline and serotonin, especially in brain regions rich in cholinergic neurons, i.e., the brain cortex, hippocampus and striatum [20,84]. ALC improves dopamine metabolism, prevents degenerative changes in dopamine-producing neurons and reduces the age-dependent process of the destruction of receptors that bind dopamine [85]. Cristofano et al. [16] showed a progressive decrease in ALC and other acyl-carnitines' serum levels in people changing from normal to AD and concluded that the decreased serum concentrations of ALC and hence its disturbed functions may predispose to AD and contribute to neurodegeneration. Clinical studies in humans demonstrated positive effects of ALC on brain function, cognition and memory that led to the suggestion that ALC may slow or reverse mild cognitive impairment and the progression of dementia in Alzheimer's disease [58].

ALC supplementation is recommended for improving brain and nervous system action, memory, the speed of learning and memorization, the level of brain energy, psychological conditions and mood, and the effects of therapies for brain neurodegenerative disorders and peripheral neuropathies [66,81].

4.2. Recommendations for L-Carnitine Content in the Diet

LC is an essential nutritional component delivered in food produced from animals, because LC endogenic synthesis is insufficient to cover metabolic needs. Primary carnitine deficiency is rare, but secondary carnitine deficiency is more frequent, being associated with several inborn errors of metabolism and acquired medical or iatrogenic conditions, for example, in patients under valproate and zidovudine treatment. Other chronic conditions such as diabetes mellitus, heart failure and Alzheimer's disease in connection with diseases creating increased catabolism may cause secondary carnitine deficiency [86].

Presently, there are no published recommended carnitine reference values. In the majority of cases, the estimated average daily carnitine requirements for an adult person amount to 20–200 mg, which is covered by diet and endogenous synthesis. Meat, fish and dairy products provide at least 80% of the required LC [87]. Based on rational dietary rules supporting good health, it is important to introduce carnitine-rich food [61,88] (Table 2), including carnitine supplementation. However, it should be taken into consideration that the bioavailability of LC from food is about four times higher than that from dietary supplements. Additionally, it should be taken into consideration that a high-fat, low-carbohydrate diet might be capable of boosting the endogenous synthesis of carnitine and its metabolites [89].

Table 2. The amount of L-carnitine in product groups [61,88] modified.

Type of Food	Total L-Carnitine Content
Ruminant meat	**(mg/100 g)**
kangaroo meat	637
horseflesh	423
beef	98.2–139
beef steak	232
beef kidneys	31.0
beef liver	15.6
lamb	106–113
goat meat	95.0–99.0
pork	20.0–30.0
pork liver	10.7
Poultry, bird meat	**(mg/100 g)**
duck	73.0
pigeon	52.8
turkey	51.0
chicken	34.0
quail	29.1
pheasant	13.5
Fish	**(mg/100 g)**
salmon	5.96
zebrafish	2.80–8.95
yellow catfish	5.93
Milk	**(mg/100 mL)**
sheep	10.2–12.7
goat	4.50–7.50
cow	7.80–9.60
Milk products	**(mg/100 g)**
yoghurt	40.0
buttermilk	38.0
cottage cheese	22.5–26.6
sour cream	19.7
coffee cream	16.6
cheese	14.0–28.0
Mushrooms	**(mg/100 g)**
Pleureotus ostreatus—oyster mushrooms	53.0
champignon	29.8
Cantharellus cibarius—chanterelle	13.3
other mushrooms	1.00–6.00
Vegetables	**(mg/100 g)**
cucumber	4.45
cauliflower	3.26
carrot	3.73
maize	0.68
peas	0.60
Fruits	**(mg/100 g)**
avocado	1.72
guava	0.82
bananas	0.39
apples	0.29
orange	0.22

4.3. Supplementation with L-Carnitine and Its Derivatives

In the prevention and treatment of patients with Alzheimer's disease, supplementation with carnitine is essential for complementing intracellular and extracellular carnitine resources. In addition, LC supplementation is intended to facilitate the elimination of toxic metabolites that may interfere with mitochondrial homeostasis, thereby interfering with the production of cellular energy, further leading to increased ROS production in many neurodegenerative diseases.

LC in the form of powder, fluid, tablets or capsules has been approved by the American Food and Drug Administration (FDA) for the treatment of primary and secondary carnitine deficiency. Experimental data obtained in in vitro and in vivo research did not demonstrate toxicity of LC. No side effects (including allergic reactions) were observed after the oral administration of LC in humans. However, some people using LC showed symptoms of alimentary tract intolerance (periodical nausea, diarrhea and tummy ache). People using large doses of LC may emit a fish body odor caused by the trimethylamine produced in the gut from LC by intestinal bacteria [56]. There were no published cases of LC intoxication. For the LC supplementation of the healthy adults, there were administered 250 mg to 2.0 g (highest safe dose) of LC daily, in several doses [90]. Daily LC doses greater than 2.0 g appeared to offer no advantage, since the gut mucosal absorption of carnitine appears to be saturated at about a 2.0 g daily dose [91]. A meta-analysis of 21 double-blind, randomized, placebo-controlled studies lasting from three months to one year showed that ALC either improved cognitive deficits or delayed the progression of cognitive decline. Improved cognitive function and delayed progression of cognitive decline were both statistically and clinically significant, with the magnitude of the effects increasing over time. Most studies used daily doses of LC of 1.5–2.0 g, which were well tolerated [92]. The treatment of ALC with doses of 2.25–3.0 g/day in patients with mild (initial) dementia caused by AD and vascular dementia (VD) led to a significant clinical improvement in patients with AD compared to in VD patients and placebo-treated patients [93]. In another study, 11 patients suffering from senile dementia of the Alzheimer's type were treated intravenously with ALC at 30 mg/kg for 10 days, and 1.5 g/day per os for 50 days, in three daily doses. It was concluded that the intravenous and oral administration of multiple doses of ALC increases ALC plasma and CSF concentrations in patients suffering from AD, which suggests that ALC easily crosses the blood–brain barrier [94]. The bioavailability of LC in food supplements depends on the applied doses [59,89]. As a general guideline, the average therapeutic ALC dose is 1.0 g, given two to three times daily for a total of 2.0–3.0 g. No advantage appears to exist in giving an oral dose greater than 2.0 g of ALC at one time, since absorption studies indicate the saturation of GI receptors (receptors coupled with proteins G) at this dose [89]. The reported side effects of LC (especially at high doses) include agitation, headaches, diarrhea, nausea, vomiting, anorexia and abdominal discomfort, mostly of mild or moderate severity [95].

4.4. Choosing Proper Form of Carnitine Supplements

LC and its derivatives have been proposed as drugs or as adjuncts to conventional medicine for many conditions, including stable angina, intermittent claudication, diabetic neuropathy, kidney disease and dialysis, hyperthyroidism, male infertility, erectile dysfunction, chronic fatigue syndrome, AD and memory impairment [95]. Many specimens containing LC differing only in form (powder, liquid, tablets or capsules) are available on the market. Free or in combination with organic acids, e.g., citric, fumaric or orotic acids, LC given as a dietary supplement is perfectly bioavailable. A combination of LC and arginine facilitates the release of ammonia as an end product of protein and amino acid metabolism. It was reported that for epileptics, exceptionally beneficial is a combination of LC with taurine because taurine acts as a modulator of membrane excitability in the central nervous system by inhibiting the release of other neurotransmitters and decreasing the mitochondrial release of calcium [96]. When administering organic salts of LC, it should be taken into consideration that a given free carnitine content corresponds to a higher weight of drug [59,89]. However, healthy people should avoid the oral intake of LC at amounts higher than 1.0–2.0 g/day in 3–4 doses [97]. The bioavailability of LC from foods is 54–87% of the LC content and is dependent on the amount of

LC in the meal. The absorption of LC from dietary supplements (0.5–6.0 g/day) is primarily passive, and the bioavailability is 14–18% of LC in the dose. LC that is unabsorbed in the gastrointestinal tract is mostly degraded by microorganisms in the large intestine [58].

4.4.1. Pure L-Carnitine (LC)

A prevalent and the most economical form of L-carnitine supplementation is orally administered LC powder or solution. LC powder or solution is recommended for people taking care of their appearance and pursuing a reduction in or maintaining the proper level of their body weight. LC is also recommended for supporting the circulatory system by reducing the risk of ischemic heart disease and other circulatory disorders [98]. LC, as an organic osmoprotectant, has been proven to have protective roles against the production of proinflammatory mediators and apoptosis in primary human corneal epithelial cells exposed to hyperosmotic media, as well as in dry-eye patients [99,100]. It was reported that LC supplementation improved the depression state in patients undergoing hemodialysis [101]. A deficiency of carnitines in terminally ill HIV/AIDS patients requires supplementation. Significant reductions in serum lactate after LC supplementation may have clinical significance in patients taking certain antiretroviral drugs [102]. L-carnitine plays an important role in energy metabolism. Skeletal muscles store about 95% of the total of 20 g carnitine contained in the adult human body, but high-intensity physical exercise decreases the muscle's carnitine content. It has been proven that L-carnitine supplementation may enhance athletic performance when coupled with physical exercise itself [103].

4.4.2. Acetyl L-Carnitine (ALC)

L-carnitine acetylation increases L-carnitine's hydrophobicity, which permits ALC's crossing of the blood–brain barrier. ALC shows neuroprotective action on nervous cells, supports energetic metabolism and the regeneration of nerve cell structures, and alleviates mitochondrial dysfunction and apoptosis [104], improving memory and creativity. Several studies have suggested a beneficial effect of ALC on cognition and behavior in aging and AD subjects [23]. In AD patients, ALC improves clinical and cognitive functions in the short and medium term (3 and 6 months) in varied doses (1.5–3.0 g/day). Additionally, with 12 month treatment at a dose of 2.0 g/day, ALC slows down the deterioration of cognitive function in AD patients [19]. Other experimental data confirm the effectiveness of ALC supplementation in protecting against brain damage, e.g., the application of 100 mg/kg body weight of ALC reduced the volume of brain injury and improved victims' behavior after traumatic brain injury. Additionally, ALC reduces oxidative stress and improves the function of mitochondrial membranes provoked by neurotoxic glutamate action [105]. The efficacy, safety and tolerability of ALC were studied during a double-blind, placebo-controlled, 12-week trial in patients with initial dementia caused by AD and vascular dementia (VD). The trial ended with the conclusion that ALC (carnicetine) can be recommended at doses of 2.25–3.0 g/day for the treatment of the early stages of AD and VD. ALC was well-tolerated [93]. Recently, the unique pharmacological properties of ALC have been confirmed, which allow us to look at this molecule as a representative of the next generation of antidepressants with a safe profile, especially for older people [106].

4.4.3. Propionyl L-Carnitine (PLCAR)

Propionyl-L-carnitine (PLCAR), or L-carnitine esterified with propionic acid, is more stable and bioavailable than free carnitine, with significantly stronger action than free L-carnitine, especially in the circulatory system and cardiac muscle. PLCAR was recommended for the treatment of diseases of the peripheral arteries and other disturbances of the cardiovascular system [98,107]. According to some data, PLCAR increases the concentration of L-carnitine in muscles independently of insulin levels, which is beneficial during the application of low carbohydrate diets. It was proven that PLCAR prevents the peripheral neuropathy connected with diabetes or toxic chemotherapy, improving nervous conductivity and blood flow in peripheral nerves [81].

4.4.4. Acetyl L-Carnitine Arginate (ALCA)

Damaged mitochondria are associated with decreased ATP production and increased reactive oxygen species production, both of which characterize AD patients. Arginine is a substrate for nitric oxide (NO) synthesis that improves the extension of blood vessel walls and improves blood circulation in the organism [108]. Acetyl L-Carnitine Arginate is an assistant to mitochondrial function, that helps to promote cell growth and cellular differentiation and may even play a role in the slowing the aging process. ALCA helps to boost mitochondrial energy production and promotes targeted benefits for the brain, heart and central nervous system. Supplementation with ALCA results in an increase in resting nitrate/nitrite levels in pre-diabetics, without any statistically significant changes in other substances connected with metabolic or oxidative stress (malondialdehyde, xanthine oxidase activity and hydrogen peroxide) [109]. Alpha-lipoic acid (ALA) (an eight-carbon saturated fatty acid)—a compound with strong antioxidative action, required by the pyruvate dehydrogenase complex for starting the tricarboxylic acids cycle—is frequently added to ALC. An important function of ALA in organisms is a redirection of anaerobic to aerobic metabolism in cells, preventing the acidification of the organism and enabling the generation of much more energy than can be generated by anaerobic metabolism [110]. It was reported that old rats that were fed with ALC and ALA significantly reduced their numbers of severely damaged mitochondria and increased the numbers of intact mitochondria in the hippocampus. The above results suggest that feeding ALC with ALA may also ameliorate age-associated mitochondrial ultrastructural decay in humans [110].

4.4.5. Glycine-Propionyl- L-Carnitine (GPLC)

Glycine improves the gut absorption and transfer through the intestinal walls of L-carnitine and facilitates the reaching of the circulation by L-carnitine. The combination of L-carnitine with glycine and propionic acid (GPLC) significantly improves the absorption and utilization of L-carnitine by cells. GPLC supports the production of nitric oxide (NO), an important substance facilitating blood circulation during physical training [111]. By stimulating NO synthesis, GPLC induced the distension of blood vessels, lowering pressure on the blood vessel walls. GPLC facilitates the faster and more efficient transport of fats to muscle cells (an important source of energy during prolonged physical exercise, when glycogen stores are exhausted) and facilitates the elimination of metabolic waste products. GPLC exhibits strong antioxidative properties that protect cells against the action of free radicals of oxygen and nitrogen. It was reported that GPLC significantly increases glutathione levels and decreases the levels of markers reflecting an increased speed of protein and lipid oxidation in humans [112]. GPLC increases the effectivity of citric acid cycle and inhibits lactate synthesis, prolonging the time of effective physical activity [113]. Some studies showed that GPLC significantly blocked D-galactosamine-induced pro-inflammatory cytokine (TNF-α and IL-6) production and, at the same time, inhibited the expression of α-smooth muscle actin, collagen-I and transforming growth factor-β. It has been demonstrated that GPLC has hepatoprotective effects against fulminant hepatic failure and chronic liver injury induced by D-galactosamine. Blommer et al. [114] determined the effect of GPLC on oxidative stress biomarkers at rest and on reactive hyperemia during the exercise of trained men. They found a decrease in lipid peroxidation with the oral intake of GPLC at rest, and in previously sedentary subjects. Whereas short-term ischemia–reperfusion in trained men results in a modest and transient increase in blood oxidative stress biomarkers, oral GPLC supplementation during short-term ischemia–reperfusion in trained men does not attenuate the increase in oxidative stress biomarkers.

4.4.6. L-Carnitine-L-Tartate (LCLT)

L-carnitine-L-tartate (LCLT) contains L-tartate coupled with L-carnitine. L-tartate intensifies L-carnitine's action by decreasing glucose absorption from the gastro-intestinal tract and decreasing the deposition of spare fats. LCLT is recommended for persons desiring to reduce fat tissue and improve efficiency and muscle force during training. Some research has shown that LCLT supplementation beneficially affects markers of hypoxic stress following resistance exercise. Muscle oxygenation was reduced by LCLT in the trial upper arm occlusion and following each set of resistance exercise. Despite reduced oxygenation, plasma malondialdehyde, a marker of membrane damage, was attenuated during the LCLT trial. The hypoxic stress was attenuated with LCLT supplementation [115]. The use of LCLT relieved the damage caused by metabolic stress and the hypoxic chain of events leading to muscle damage after exercise (reduced post-exercise serum levels of hypoxanthine, xanthine oxidase and myoglobin and perceived muscle soreness) [116]. In addition, a positive effect of LCLT on the endocrine system has been demonstrated. The supplementation of LCLT increased androgen receptor content in the muscle, which may result in increased testosterone uptake and thus enhanced luteinizing hormone secretion via feedback mechanisms, which may promote recovery post resistance exercise [117]. Chronic LCLT supplementation increased carbohydrate oxidation during exercise [118]. The influence of LCLT on markers of purine catabolism (hypoxanthine, xanthine oxidase and serum uric acid), circulating cytosolic proteins (myoglobin, fatty acid-binding protein and creatine kinase), free radical formation, and muscle tissue disruption after squat exercise was examined. Exercise-induced increases in plasma malondialdehyde (a lipid peroxidation product) returned to resting values sooner with LCLT supplementation than with a placebo. The above data indicate that LCLT supplementation is effective in assisting recovery from high-repetition squat exercise [119].

5. Conclusions

From the point of view of rational nutrition standards and the MIND diet favoring good health, it is beneficial to introduce nutritional products rich in L-carnitine and its derivatives or the supplementation of the diet with L-carnitine, and especially ALC, for the prevention and/or alleviation of dementia and other AD symptoms.

Despite the controversy concerning the consumption of red meat—based on the belief that a high consumption of red meat increases the risk of cancer, particularly colon cancer—a properly balanced diet should contain animal food. Excessive restrictions or the extreme avoidance of eating meat or dairy products in the human diet eliminates many bioactive substances necessary for the correct development and functioning of the organism. Therefore, it is necessary to implement diets with food products rich in carnitine and its derivatives.

It should be stressed that correct nutrition is an important element of lifestyle that may be an important factor for healthy, slow, favorable aging and delaying the development of neurodegenerative diseases including dementia and AD. We should keep in mind that a correct diet rich in vegetables with low starch content, fruit with low glycemic indices, cereals, legumes, vegetable oils and sea fish with reasonable amounts of meat and dairy products is essential for our health and lowers the risk of dementia and AD.

Author Contributions: Conception and design of the study: A.K.; acquisition of data: A.O., E.S. and B.S.-J.; analysis and interpretation of data: M.B.-K. and S.C.; writing—original draft preparation: A.K. and N.W.; review and/or revision of the manuscript: A.K., N.W. and A.O.; study supervision: A.K. and N.W.; acceptance of the final version: all authors. All authors have read and agreed to the published version of the manuscript.

Funding: This research received no external funding.

Acknowledgments: The authors would like to thank Max Opęchowski, an American Student of the English Division of the Białystok Medical University, for the critical reading of the manuscript.

Conflicts of Interest: The authors declare no conflict of interest.

References

1. Strafella, C.; Caputo, V.; Galota, M.R.; Zampatti, S.; Marella, G.; Mauriello, S.; Cascella, R.; Giardina, E. Application of precision medicine in neurodegenerative diseases. *Front. Neurol.* **2018**, *9*, 701:1–701:9. [CrossRef]

2. Mancuso, C.; Bates, T.E.; Butterfield, D.A.; Calafato, S.; Cornelius, C.; De Lorenzo, A.; Dinkova Kostova, A.T.; Calabrese, V. Natural antioxidants in Alzheimer's disease. *Expert Opin. Investig. Drugs* **2007**, *16*, 1921–1931. [CrossRef] [PubMed]

3. Winiarska-Mieczan, A.; Baranowska-Wójcik, E.; Kwiecień, M.; Grela, E.R.; Szwajgier, D.; Kwiatkowska, K.; Kiczorowska, B. The role of dietary antioxidants in the pathogenesis of neurodegenerative diseases and their impact on cerebral oxidoreductive balance. *Nutrients* **2020**, *12*, 435. [CrossRef] [PubMed]

4. Wenk, G.L. Neuropathologic changes in Alzheimer's disease. *J. Clin. Psychiatr.* **2003**, *64*, 7–10.

5. Schliebs, R.; Arendt, T. The cholinergic system in aging and neuronal degeneration. *Behav. Brain Res.* **2011**, *221*, 555–563. [CrossRef]

6. Ouerfurth, H.W.; LaFerla, F.M. Alzheimer's disease. *N. Engl. J. Med.* **2010**, *362*, 329–344. [CrossRef]

7. Kaźmierczak, A.; Adamczyk, A.; Benigna-Strosznajder, J. The role of extracellular α-synuclein in molecular mechanisms of cell heath. *Post. Hig. Med. Dosw.* **2013**, *67*, 1047–1057. [CrossRef]

8. Dean, D.C., 3rd; Jerskey, B.A.; Chen, K.; Protas, H.; Thiyyagura, P.; Roontiva, A.; O'Muircheartaigh, J.; Dirks, H.; Waskiewicz, N.; Lehman, K.; et al. Brain differences in infants at differential genetic risk for late-onset Alzheimer disease: A cross-sectional imaging study. *JAMA Neurol.* **2014**, *71*, 11–22. [CrossRef]

9. Ng, S.; Lin, C.C.; Hwang, Y.H.; Hsieh, W.S.; Liao, H.F.; Chen, P.C. Mercury, APOE, and children's neurodevelopment. *Neurotoxicology* **2013**, *37*, 85–92. [CrossRef]

10. Dehghani, C.; Frost, S.; Jayasena, R.; Masters, C.L.; Kanagasingam, Y. Ocular biomarkers of Alzheimer's disease: The role of anterior eye and potential future directions. *Investig. Ophthalmol. Vis. Sci.* **2018**, *59*, 3554–3563. [CrossRef]

11. Nunes-Tavares, N.; Santos, L.E.; Stutz, B.; Brito-Moreira, J.; Klein, W.L.; Ferreira, S.T.; de Mello, F.G. Inhibition of choline acetyltransferase as a mechanism for cholinergic dysfunction induced by amyloid-β peptide oligomers. *J. Biol. Chem.* **2012**, *287*, 19377–19385. [CrossRef] [PubMed]

12. Szutowicz, A.; Bielarczyk, H.; Jankowska-Kulawy, A.; Pawełczyk, T.; Ronowska, A. Acetyl-CoA the key factor for survival or death of cholinergic neurons in course of neurodegenerative diseases. *Neurochem. Res.* **2013**, *38*, 1523–1542. [CrossRef]

13. Lodeiro, M.; Ibáñez, C.; Cifuentes, A.; Simó, C.; Cedazo-Mínguez, Á. Decreased cerebrospinal fluid levels of L-carnitine in non-apolipoprotein E4 carriers at early stages of Alzheimer's disease. *J. Alzheimers Dis.* **2014**, *41*, 223–232. [CrossRef] [PubMed]

14. Pan, X.; Nasaruddin, M.B.; Elliott, C.T.; McGuinness, B.; Passmore, A.P.; Kehoe, P.G.; Hölscher, C.; McClean, P.L.; Graham, S.F.; Green, B.D. Alzheimer's disease-like pathology has transient effects on the brain and blood metabolome. *Neurobiol. Aging* **2016**, *38*, 151–163. [CrossRef]

15. Thomas, S.C.; Alhasawi, A.; Appanna, V.P.; Auger, C.; Appanna, V.D. Brain metabolism and Alzheimer's disease: The prospect of a metabolite-based therapy. *J. Nutr. Health Aging* **2015**, *19*, 58–63. [CrossRef] [PubMed]

16. Cristofano, A.; Sapere, N.; La Marca, G.; Angiolillo, A.; Vitale, M.; Corbi, G.; Scapagnini, G.; Intrieri, M.; Russo, C.; Corso, G.; et al. Serum levels of acyl-carnitines along the continuum from normal to Alzheimer's dementia. *PLoS ONE* **2016**, *11*, e0155694. [CrossRef]

17. White, H.L.; Scates, P.W. Acetyl-L-carnitine as a precursor of acetylcholine. *Neurochem. Res.* **1990**, *15*, 597–601. [CrossRef]

18. Traina, G. The neurobiology of acetyl-L-carnitine. *Front Biosci. (Landmark Ed)* **2016**, *21*, 1314–1329. [CrossRef]

19. Calvani, M.; Carta, A.; Caruso, G.; Benedetti, N.; Iannuccelli, M. Action of acetyl-L-carnitine in neurodegeneration and Alzheimer's disease. *Ann. N. Y. Acad. Sci.* **1992**, *663*, 483–486. [CrossRef]

20. Juliet, P.A.; Balasubramaniam, D.; Balasubramaniam, N.; Panneerselvam, C. Carnitine: A neuromodulator in aged rats. *J. Gerontol. A- Biol. Sci. Med. Sci.* **2003**, *58*, 970–974. [CrossRef]

21. Zhou, P.; Chen, Z.; Zhao, N.; Liu, D.; Guo, Z.Y.; Tan, L.; Hu, J.; Wang, Q.; Wang, J.Z.; Zhu, L.Q. Acetyl-L-carnitine attenuates homocysteine-induced Alzheimer-like histopathological and behavioral abnormalities. *Rejuvenation Res.* **2011**, *14*, 669–679. [CrossRef] [PubMed]

22. Suchy, J.; Chan, A.; Shea, T.B. Dietary supplementation with a combination of alpha-lipoic acid, acetyl-L-carnitine, glycerophosphocholine, docosahexaenoic acid, and phosphatidylserine reduces oxidative damage to murine brain and improves cognitive performance. *Nutr. Res.* **2009**, *29*, 70–74. [CrossRef] [PubMed]

23. Pettegrew, J.W.; Klunk, W.E.; Panchalingam, K.; Kanfer, J.N.; McClure, R.J. Clinical and neurochemical effects of acetyl-L-carnitine in Alzheimer's disease. *Neurobiol. Aging* **1995**, *16*, 1–4. [CrossRef]

24. Onofrj, M.; Ciccocioppo, F.; Varanese, S.; di Muzio, A.; Calvani, M.; Chiechio, S.; Osio, M.; Thomas, A. Acetyl-L-carnitine: From a biological curiosity to a drug for the peripheral nervous system and beyond. *Expert Rev. Neurother.* **2013**, *13*, 925–936. [CrossRef]

25. Grodzicki, W.; Dziendzikowska, K. The Role of selected Bioactive Compounds in the Prevention of Alzheimer's Disease. *Antioxidants* **2020**, *9*, 229. [CrossRef]

26. Sánchez-Muniz, F.J.; Macho-González, A.; Garcimartín, A.; Santos-López, J.A.; Benedí, J.; Bastida, S.; González-Muñoz, M.J. The nutritional components of beer and its relationship with neurodegeneration and Alzheimer's disease. *Nutrients* **2019**, *11*, 1558. [CrossRef]

27. De Boer, B.; Hamers, J.P.H.; Zwakhalen, S.M.G.; Tan, F.E.S.; Verbeek, H. Quality of care and quality of life of people with dementia living at green care farms: A cross-sectional study. *BMC Geriatr.* **2017**, *17*, 155. [CrossRef]

28. Gonçalves-Pereira, M.; Cardoso, A.; Verdelho, A.; Alves da Silva, J.; Caldas de Almeida, M.; Fernandes, A.; Raminhos, C.; Ferri, C.P.; Prina, A.M.; Prince, M.; et al. The prevalence of dementia in a Portuguese community sample: A 10/66 Dementia Research Group study. *BMC Geriatr.* **2017**, *17*, 261. [CrossRef]

29. World Health Organization. *Neurological Disorders: Public Health Challenges*; World Health Organization: Geneva, Switzerland, 2006; pp. 204–207.

30. Ferri, C.P.; Prince, M.; Brayne, C.; Brodaty, H.; Fratiglioni, L.; Ganguli, M.; Hall, K.; Hasegawa, K.; Hendrie, H.; Huang, Y.; et al. Global prevalence of dementia: A Delphi consensus study. *Lancet* **2005**, *366*, 2112–2117. [CrossRef]

31. Fillit, H.M. The pharmacoeconomics of Alzheimer's disease. *Am. J. Manag. Care* **2000**, *6*, S1139–S1144.

32. Moore, K.; Hughes, C.F.; Ward, M.; Hoey, L.; McNulty, H. Diet, nutrition and the ageing brain: Current evidence and new directions. *Proc. Nutr. Soc.* **2018**, *77*, 152–163. [CrossRef] [PubMed]

33. Gabryelewicz, T. Prevalence of dementia syndromes among the residents of the Mokotow district of Warsaw, aged 65–84 (in Polish). *Psychiatr. Pol.* **1999**, *3*, 353–366.

34. Bdzan, L.B.; Turczyński, J.; Szabert, K. Prevalence of dementia in a rural population. *Psychiatr. Pol.* **2007**, *41*, 181–188. [PubMed]

35. Principal Polish Statistical Office. *Demographic Yearbook of Poland*; Principal Polish Statistical Office: Warsaw, Poland, 2004; p. 207. (In Polish)

36. Szczudlik, A.; Barcikowska-Kotowicz, M.; Gabryelewicz, T.; Opala, G.; Parnowski, T.; Kuźnicki, J.; Rossa, A.; Sadowska, A. *The Situation of People with Alzheimer's Disease in Poland*; RPO Report; Szczudlik, A., Ed.; RPO: Warsaw, Poland, 2016. (In Polish)

37. Shah, H.; Albanese, E.; Duggan, C.; Rudan, I.; Langa, K.M.; Carrillo, M.C.; Chan, K.Y.; Joanette, Y.; Prince, M.; Rossor, M.; et al. Research priorities to reduce the global burden of dementia by 2025. *Lancet Neurol.* **2016**, *15*, 1285–1294. [CrossRef]

38. Perrone, L.; Grant, W.B. Observational and ecological studies of dietary advanced glycation end products in national diets and Alzheimer's disease incidence and prevalence. *J. Alzheimers Dis.* **2015**, *45*, 965–979. [CrossRef]

39. Lubitz, I.; Ricny, J.; Atrakchi-Baranes, D.; Shemesh, C.; Kravitz, E.; Liraz-Zaltsman, S.; Maksin-Matveev, A.; Cooper, I.; Leibowitz, A.; Uribarri, J.; et al. High dietary advanced glycation end products are associated with poorer spatial learning and accelerated Aβ deposition in an Alzheimer mouse model. *Aging Cell* **2016**, *15*, 309–316. [CrossRef]

40. Solfrizzi, V.; Panza, F.; Frisardi, V.; Seripa, D.; Logroscino, G.; Imbimbo, B.P.; Pilotto, A. Diet and Alzheimer's disease risk factors or prevention: The current evidence. *Expert Rev. Neurother.* **2011**, *11*, 677–708. [CrossRef]

41. Anastasiou, C.A.; Yannakoulia, M.; Kosmidis, M.H.; Dardiotis, E.; Hadjigeorgiou, G.M.; Sakka, P.; Arampatzi, X.; Bougea, A.; Labropoulos, I.; Scarmeas, N. Mediterranean diet and cognitive health: Initial results from the hellenic longitudinal investigation of ageing and diet. *PLoS ONE* **2017**, *12*, e0182048. [CrossRef]

42. GBD 2016 DALYs and HALE Collaborators. Global, regional, and national disability-adjusted life-years (DALYs) for 333 diseases and injuries and healthy life expectancy (HALE) for 195 countries and territories, 1990-2016: A systematic analysis for the Global Burden of Disease Study 2016. *Lancet* **2017**, *390*, 1260–1344. [CrossRef]

43. Yusufov, M.; Weyandt, L.L.; Piryatinsky, I. Alzheimer's disease and diet: A systematic review. *Int. J. Neurosci.* **2017**, *127*, 161–175. [CrossRef]

44. Singh, B.; Parsaik, A.K.; Mielke, M.M.; Erwin, P.J.; Knopman, D.S.; Petersen, R.C.; Roberts, R.O. Association of mediterranean diet with mild cognitive impairment and Alzheimer's disease: A systematic review and meta-analysis. *J. Alzheimers Dis.* **2014**, *39*, 271–822. [CrossRef]

45. Dochniak, M.; Ekiert, K. Nutrition in prevention and treatment of Alzheimer's and Parkinson's diseases. *Nurs. Public Health* **2015**, *5*, 199–208. (In Polish)

46. Parnowski, T. *Alzheimer's Disease*; Wydawnictwo Lekarskie PZWL: Warsaw, Poland, 2010. (In Polish)

47. Abbatecola, A.M.; Russo, M.; Barbieri, M. Dietary patterns and cognition in older persons. *Curr. Opin. Clin. Nutr. Metab. Care* **2018**, *21*, 10–13. [CrossRef] [PubMed]

48. Morris, M.C.; Tangney, C.C.; Wang, Y.; Sacks, F.M.; Bennett, D.A.; Aggarwal, N.T. MIND diet associated with reduced incidence of Alzheimer's disease. *Alzheimers Dement.* **2015**, *11*, 1007–1014. [CrossRef]

49. Stefani, M.; Rigacci, S. Beneficial properties of natural phenols: Highlight on protection against pathological conditions associated with amyloid aggregation. *Biofactors* **2014**, *40*, 482–493. [CrossRef] [PubMed]

50. Kępka, A.; Szajda, S.D.; Waszkiewicz, N.; Płudowski, P.; Chojnowska, S.; Rudy, M.; Szulc, A.; Ładny, J.R.; Zwierz, K. Carnitine: Function, metabolism and value in hepatic failure during chronic alcohol intoxication. *Post. Hig. Med. Dosw.* **2011**, *65*, 645–653. [CrossRef] [PubMed]

51. Czeczot, H.; Ścibior, D. Role of L-carnitine in metabolism, nutrition and therapy. *Post. Hig. Med. Dosw.* **2005**, *59*, 9–19.

52. Kępka, A.; Chojnowska, S.; Okungbowa, O.E.; Zwierz, K. The role of carnitine in the perinatal period. *Dev. Period Med.* **2014**, *18*, 417–425.

53. Bourdin, B.; Adenier, H.; Perrin, Y. Carnitine is associated with fatty acid metabolism in plants. *Plant. Physiol. Biochem.* **2007**, *45*, 926–931. [CrossRef]

54. Hurot, J.M.; Cucherat, M.; Haugh, M.; Fouque, D. Effects of L-carnitine supplementation in maintenance hemodialysis patients: A systematic review. *J. Am. Soc. Nephrol.* **2002**, *13*, 708–714.

55. Oey, N.A.; van Vlies, N.; Wijburg, F.A.; Wanders, R.J.; Attie-Bitach, T.; Vaz, F.M. L-carnitine is synthesized in the human fetal-placental unit: Potential roles in placental and fetal metabolism. *Placenta* **2006**, *27*, 841–846. [CrossRef]

56. Rebouche, C.J. Kinetics, pharmacokinetics, and regulation of L-carnitine and acetyl-L-carnitine metabolism. *Ann. N. Y. Acad. Sci.* **2004**, *1033*, 30–41. [CrossRef] [PubMed]

57. Bain, M.A.; Fornasini, G.; Evans, A.M. Trimethylamine: Metabolic, pharmacokinetic and safety aspects. *Curr. Drug Metab.* **2005**, *6*, 227–240. [CrossRef] [PubMed]

58. Del Rio, D.; Zimetti, F.; Caffarra, P.; Tassotti, M.; Bernini, F.; Brighenti, F.; Zini, A.; Zanotti, I. The gut microbial metabolite trimethylamine-N-oxide is present in human cerebrospinal fluid. *Nutrients* **2017**, *9*, 1053. [CrossRef] [PubMed]

59. Evans, A.M.; Fornasini, G. Pharmacokinetics of L-carnitine. *Clin. Pharmacokinet.* **2003**, *42*, 941–967. [CrossRef]

60. Lavon, L. Perturbation of serum carnitine levels in human adults by chronic renal disease and dialysis therapy. *Am. J. Clin. Nutr.* **1981**, *34*, 1314–1320.

61. Bodkowski, R.; Patkowska-Sokoła, B.; Nowakowski, P.; Jamroz, D.; Janczak, M. Products of animal origin—The most important L-carnitine source in human diet (in Polish). *Prz. Hod.* **2011**, *10*, 22–25.

62. Jones, L.L.; McDonald, D.A.; Borum, P.R. Acylcarnitines: Role in brain. *Prog. Lipid Res.* **2010**, *49*, 61–75. [CrossRef]

63. Calvani, M.; Benatti, P.; Mancinelli, A.; D'Iddio, S.; Giordano, V.; Koverech, A.; Amato, A.; Brass, E.P. Carnitine replacement in end-stage renal disease and hemodialysis. *Ann. N. Y. Acad. Sci.* **2004**, *1033*, 52–66. [CrossRef]

64. Jacob, C.; Belleville, F. L-carnitine: Metabolism, function and value in pathology. *Pathol. Biol. (Paris)* **1992**, *40*, 910–919.

65. Hedayati, S.S. Dialysis-related carnitine disorder. *Semin. Dial.* **2006**, *19*, 323–328. [CrossRef] [PubMed]

66. Ribas, G.S.; Vargas, C.R.; Wajner, M. L-carnitine supplementation as a potential antioxidant therapy for inherited neurometabolic disorders. *Gene* **2014**, *533*, 469–476. [CrossRef] [PubMed]
67. Valero, T. Mitochondrial biogenesis: Pharmacological approaches. *Curr. Pharm. Des.* **2014**, *20*, 5507–5509. [CrossRef] [PubMed]
68. Zammit, V.A.; Ramsay, R.R.; Bonomini, M.; Arduini, A. Carnitine, mitochondrial function and therapy. *Adv. Drug Deliv. Rev.* **2009**, *61*, 1353–1362. [CrossRef] [PubMed]
69. Aureli, T.; Miccheli, A.; Ricciolini, R.; Di Cocco, M.E.; Ramacci, M.T.; Angelucci, L.; Ghirardi, O.; Conti, F. Aging brain: Effect of acetyl-L-carnitine treatment on rat brain energy and phospholipid metabolism. A study by 31P and 1H NMR spectroscopy. *Brain Res.* **1990**, *526*, 108–112. [CrossRef]
70. Aureli, T.; Di Cocco, M.E.; Puccetti, C.; Ricciolini, R.; Scalibastri, M.; Miccheli, A.; Manetti, C.; Conti, F. Acetyl-L-carnitine modulates glucose metabolism and stimulates glycogen synthesis in rat brain. *Brain Res.* **1998**, *796*, 75–81. [CrossRef]
71. Ferreira, G.C.; McKenna, M.C. L-Carnitine and acetyl-L-carnitine roles and neuroprotection in developing brain. *Neurochem. Res.* **2017**, *42*, 1661–1675. [CrossRef]
72. Gorini, A.; D'Angelo, A.; Villa, R.F. Action of L-acetylcarnitine on different cerebral mitochondrial populations from cerebral cortex. *Neurochem. Res.* **1998**, *23*, 1485–1491. [CrossRef]
73. Calabrese, V.; Giuffrida Stella, A.M.; Calvani, M.; Butterfield, D.A. Acetylcarnitine and cellular stress response: Roles in nutritional redox homeostasis and regulation of longevity genes. *J. Nutr. Biochem.* **2006**, *17*, 73–88. [CrossRef]
74. Madiraju, P.; Pande, S.V.; Prentki, M.; Madiraju, S.R. Mitochondrial acetylcarnitine provides acetyl groups for nuclear histone acetylation. *Epigenetics* **2009**, *4*, 399–403. [CrossRef]
75. Martignoni, E.; Facchinetti, F.; Sances, G.; Petraglia, F.; Nappi, G.; Genazzani, A.R. Acetyl-L-carnitine acutely administered raises beta-endorphin and cortisol plasma levels in humans. *Clin. Neuropharmacol.* **1988**, *11*, 472–477. [CrossRef] [PubMed]
76. Marszałek, M. Alzheimer's disease against peptides products of enzymatic cleavage of APP protein. Forming and variety of fibrillating peptides—Some aspects (in Polish). *Post. Hig. Med. Dosw.* **2016**, *70*, 787–796. [CrossRef] [PubMed]
77. Nałęcz, K.A.; Miecz, D.; Berezowski, V.; Cecchelli, R. Carnitine: Transport and physiological functions in the brain. *Mol. Aspects Med.* **2004**, *25*, 551–567. [CrossRef] [PubMed]
78. Respondek, M.; Buszman, E. Regulation of neurogenesis: Factors affecting of new neurons formation in adult mammals brain (in Polish). *Post. Hig. Med. Dosw.* **2015**, *69*, 1451–1461.
79. Kazak, F.; Yarim, G.F. Neuroprotective effects of acetyl-l-carnitine on lipopolysaccharide-induced neuroinflammation in mice: Involvement of brain-derived neurotrophic factor. *Neurosci Lett.* **2017**, *658*, 32–36. [CrossRef]
80. Abdul, H.M.; Calabrese, V.; Calvani, M.; Butterfield, D.A. Acetyl-L-carnitine-induced up-regulation of heat shock proteins protects cortical neurons against amyloid-beta peptide 1-42-mediated oxidative stress and neurotoxicity: Implications for Alzheimer's disease. *J. Neurosci. Res.* **2006**, *82*, 398–408. [CrossRef]
81. Head, K.A. Peripheral neuropathy: Pathogenic mechanisms and alternative therapies. *Altern. Med. Rev.* **2006**, *11*, 294–329.
82. Hudson, S.; Tabet, N. Acetyl-L-carnitine for dementia. *Cochrane Database Syst. Rev.* **2003**, *2*, CD003158:1–CD003158:38. [CrossRef]
83. Scheggi, S.; Rauggi, R.; Nanni, G.; Tagliamonte, A.; Gambarana, C. Repeated acetyl-l-carnitine administration increases phospho-Thr34 DARPP-32 levels and antagonizes cocaine-induced increase in Cdk5 and phospho-Thr75 DARPP-32 levels in rat striatum. *Eur. J. Neurosci.* **2004**, *19*, 1609–1620. [CrossRef]
84. Smeland, O.B.; Meisingset, T.W.; Borges, K.; Sonnewald, U. Chronic acetyl-L-carnitine alters brain energy metabolism and increases noradrenaline and serotonin content in healthy mice. *Neurochem. Int.* **2012**, *61*, 100–107. [CrossRef]
85. Robinson, B.L.; Dumas, M.; Cuevas, E.; Gu, Q.; Paule, M.G.; Ali, S.F.; Kanungo, J. Distinct effects of ketamine and acetyl L-carnitine on the dopamine system in zebrafish. *Neurotoxicol. Teratol.* **2016**, *54*, 52–60. [CrossRef] [PubMed]
86. Evangeliou, A.; Vlassopoulos, D. Carnitine metabolism and deficit-when supplementation is necessary? *Curr. Pharm. Biotechnol.* **2003**, *4*, 211–219. [CrossRef]

87. Rigault, C.; Mazué, F.; Bernard, A.; Demarquoy, J.; Le Borgne, F. Changes in l-carnitine content of fish and meat during domestic cooking. *Meat Sci.* **2008**, *78*, 331–335. [CrossRef] [PubMed]

88. Rospond, B.; Chłopicka, J. The biological function of L-carnitine and its content in the particular food examples (in Polish). *Prz. Lek.* **2013**, *70*, 85–91.

89. Kelly, G.S. L-Carnitine: Therapeutic applications of a conditionally essential amino acid. *Altern. Med. Rev.* **1998**, *3*, 345–360.

90. Pękala, J.; Patkowska-Sokoła, B.; Bodkowski, R.; Jamroz, D.; Nowakowski, P.; Lochyński, S.; Librowski, T. L-carnitine-metabolic functions and meaning in humans life. *Curr. Drug Metab.* **2011**, *12*, 667–678. [CrossRef] [PubMed]

91. Chen, N.; Yang, M.; Zhou, M.; Xiao, J.; Guo, J.; He, L. L-carnitine for cognitive enhancement in people without cognitive impairment. *Cochrane Database Syst. Rev.* **2017**, *3*, CD009374:1–CD009374:28. [CrossRef] [PubMed]

92. Wollen, K.A. Alzheimer's disease: The pros and cons of pharmaceutical, nutritional, botanical, and stimulatory therapies, with a discussion of treatment strategies from the perspective of patients and practitioners. *Altern. Med. Rev.* **2010**, *15*, 223–244. [PubMed]

93. Gavrilova, S.I.; Kalyn, I.B.; Kolykhalov, I.V.; Roshchina, I.F.; Selezneva, N.D. Acetyl-L-carnitine (carnicetine) in the treatment of early stages of Alzheimer's disease and vascular dementia. *Zh. Nevrol. Psikhiatr. Im. S. S. Korsakova* **2011**, *111*, 16–22.

94. Parnetti, L.; Gaiti, A.; Mecocci, P.; Cadini, D.; Senin, U. Pharmacokinetics of i.v. and oral acetyl-L-carnitine in a multiple dose regimen in patients with senile dementia of Alzheimer type. *Eur. J. Clin. Pharmacol.* **1992**, *42*, 89–93. [CrossRef]

95. Wang, S.M.; Han, C.; Lee, S.J.; Patkar, A.A.; Masand, P.S.; Pae, C.U. A review of current evidence for acetyl-l-carnitine in the treatment of depression. *J. Psychiatr. Res.* **2014**, *53*, 30–37. [CrossRef] [PubMed]

96. Gaby, A.R. Natural approaches to epilepsy. *Altern. Med. Rev.* **2007**, *12*, 9–24. [PubMed]

97. Bain, M.A.; Milne, R.W.; Evans, A.M. Disposition and metabolite kinetics of oral L-carnitine in humans. *J. Clin. Pharmacol.* **2006**, *46*, 1163–1170. [CrossRef]

98. Ferrari, R.; Merli, E.; Cicchitelli, G.; Mele, D.; Fucili, A.; Ceconi, C. Therapeutic effects of L-carnitine and propionyl-L-carnitine on cardiovascular diseases: A review. *Ann. N. Y. Acad. Sci.* **2004**, *1033*, 79–91. [CrossRef] [PubMed]

99. Hua, X.; Deng, R.; Li, J.; Chi, W.; Su, Z.; Lin, J.; Pflugfelder, S.C.; Li, D.Q. Protective effects of L-carnitine against oxidative injury by hyperosmolarity in human corneal epithelial cells. *Invest. Ophthalmol. Vis. Sci.* **2015**, *56*, 5503–5511. [CrossRef] [PubMed]

100. Seen, S.; Tong, L. Dry eye disease and oxidative stress. *Acta Ophthalmol.* **2017**, *96*, e412–e420. [CrossRef] [PubMed]

101. Tashiro, K.; Kaida, Y.; Yamagishi, S.I.; Tanaka, H.; Yokoro, M.; Yano, J.; Sakai, K.; Kurokawa, Y.; Taguchi, K.; Nakayama, Y.; et al. L-Carnitine supplementation improves self-rating depression scale scores in uremic male patients undergoing hemodialysis. *Lett. Drug Des. Discov.* **2017**, *14*, 737–742. [CrossRef] [PubMed]

102. Cruciani, R.A.; Revuelta, M.; Dvorkin, E.; Homel, P.; Lesage, P.; Esteban-Cruciani, N. L-carnitine supplementation in patients with HIV/AIDS and fatigue: A double-blind, placebo-controlled pilot study. *HIV AIDS (Auckl)* **2015**, *7*, 65–73. [CrossRef] [PubMed]

103. Benvenga, S. Effects of L-carnitine on thyroid hormone metabolism and on physical exercise tolerance. *Horm. Metab. Res.* **2005**, *37*, 566–571. [CrossRef]

104. Zhang, Z.Y.; Fan, Z.K.; Cao, Y.; Jia, Z.Q.; Li, G.; Zhi, X.D.; Yu, D.S.; Lv, G. Acetyl-L-carnitine ameliorates mitochondrial damage and apoptosis following spinal cord injury in rats. *Neurosci. Lett.* **2015**, *604*, 18–23. [CrossRef]

105. Bigford, G.E.; Del Rossi, G. Supplemental substances derived from foods as adjunctive therapeutic agents for treatment of neurodegenerative diseases and disorders. *Adv. Nutr.* **2014**, *5*, 394–403. [CrossRef]

106. Chiechio, S.; Canonico, P.L.; Grilli, M. L-Acetylcarnitine: A mechanistically distinctive and potentially rapid-acting antidepressant drug. *Int. J. Mol. Sci.* **2017**, *19*, 11. [CrossRef] [PubMed]

107. Evans, A.M.; Mancinelli, A.; Longo, A. Excretion and metabolism of propionyl-L-carnitine in the isolated perfused rat kidney. *J. Pharmacol. Exp. Ther.* **1997**, *281*, 1071–1076. [PubMed]

108. Ścibior, D.; Czeczot, H. Arginine-metabolism and functions in the human organism. *Post. Hig. Med. Dosw.* **2004**, *58*, 321–332.

109. Bloomer, R.J.; Fisher-Wellman, K.H.; Tucker, P.S. Effect of oral acetyl L-carnitine arginate on resting and postprandial blood biomarkers in pre-diabetics. *Nutr. Metab. (Lond.)* **2009**, *6*, 25:1–25:11. [CrossRef]
110. Aliev, G.; Liu, J.; Shenk, J.C.; Fischbach, K.; Pacheco, G.J.; Chen, S.G.; Obrenovich, M.E.; Ward, W.F.; Richardson, A.G.; Smith, M.A.; et al. Neuronal mitochondrial amelioration by feeding acetyl-L-carnitine and lipoic acid to aged rats. *J. Cell. Mol. Med.* **2009**, *13*, 320–333. [CrossRef]
111. Bescós, R.; Sureda, A.; Tur, J.A.; Pons, A. The effect of nitric-oxide-related supplements on human performance. *Sports Med.* **2012**, *42*, 99–117. [CrossRef]
112. Ganai, A.A.; Jahan, S.; Ahad, A.; Abdin, M.Z.; Farooqi, H. Glycine propionyl L-carnitine attenuates D-galactosamine induced fulminant hepatic failure in Wistar rats. *Chem. Biol. Interact.* **2014**, *214*, 33–40. [CrossRef]
113. Jacobs, P.L.; Goldstein, E.R. Long-term glycine propionyl-l-carnitine supplemention and paradoxical effects on repeated anaerobic sprint performance. *J. Int Soc. Sports Nutr.* **2010**, *7*, 35:1–35:8. [CrossRef]
114. Bloomer, R.J.; Smith, W.A.; Fisher-Wellman, K.H. Oxidative stress in response to forearm ischemia-reperfusion with and without carnitine administration. *Int. J. Vitam. Nutr. Res.* **2010**, *80*, 12–23. [CrossRef]
115. Spiering, B.A.; Kraemer, W.J.; Hatfield, D.L.; Vingren, J.L.; Fragala, M.S.; Ho, J.Y.; Thomas, G.A.; Häkkinen, K.; Volek, J.S. Effects of L-carnitine L-tartrate supplementation on muscle oxygenation responses to resistance exercise. *J. Strength. Cond. Res.* **2008**, *22*, 1130 1135. [CrossRef] [PubMed]
116. Spiering, B.A.; Kraemer, W.J.; Vingren, J.L.; Hatfield, D.L.; Fragala, M.S.; Ho, J.Y.; Maresh, C.M.; Anderson, J.M.; Volek, J.S. Responses of criterion variables to different supplemental doses of L-carnitine L-tartrate. *J. Strength. Cond. Res.* **2007**, *21*, 259–264. [CrossRef]
117. Kraemer, W.J.; Spiering, B.A.; Volek, J.S.; Ratamess, N.A.; Sharman, M.J.; Rubin, M.R.; French, D.N.; Silvestre, R.; Hatfield, D.L.; Van Heest, J.L.; et al. Androgenic responses to resistance exercise: Effects of feeding and L-carnitine. *Med. Sci. Sports Exerc.* **2006**, *38*, 1288–1296. [CrossRef] [PubMed]
118. Abramowicz, W.N.; Galloway, S.D. Effects of acute versus chronic L-carnitine L-tartrate supplementation on metabolic responses to steady state exercise in males and females. *Int. J. Sport Nutr. Exerc. Metab.* **2005**, *15*, 386–400. [CrossRef]
119. Volek, J.S.; Kraemer, W.J.; Rubin, M.R.; Gómez, A.L.; Ratamess, N.A.; Gaynor, P. L-Carnitine L-tartrate supplementation favorably affects markers of recovery from exercise stress. *Am. J. Physiol. Endocrinol. Metab.* **2002**, *282*, E474–E482. [CrossRef] [PubMed]

Article

Antioxidant Activity with Increased Endogenous Levels of Vitamin C, E and A Following Dietary Supplementation with a Combination of Glutathione and Resveratrol Precursors

Priscilla Biswas [1,*,†], Cinzia Dellanoce [2,†], Alessandra Vezzoli [2], Simona Mrakic-Sposta [2], Mauro Malnati [3], Alberto Beretta [1] and Roberto Accinni [1]

[1] SoLongevity Research, 20121 Milan, Italy; alberto.beretta@solongevity.com (A.B.); roberto.accinni@gmail.com (R.A.)
[2] Institute of Clinical Physiology, National Council of Research (IFC-CNR), ASST Grande Ospedale Metropolitano Niguarda, 20162 Milan, Italy; cinzia.dellanoce@ospedaleniguarda.it (C.D.); alessandra.vezzoli@cnr.it (A.V.); simona.mrakicsposta@cnr.it (S.M.-S.)
[3] Unit of Viral Evolution and Transmission, IRCCS Ospedale San Raffaele, 20132 Milan, Italy; mauro.malnati@hsr.it
*   Correspondence: priscilla.biswas@hsr.it; Tel.: +39-02-26434903
†   These authors contributed equally to this work.

Received: 2 September 2020; Accepted: 20 October 2020; Published: 22 October 2020

Abstract: The effects of two different dietary supplements on the redox status of healthy human participants were evaluated. The first supplement (GluS, Glutathione Synthesis) contains the precursors for the endogenous synthesis of glutathione and the second (GluReS, Glutathione and Resveratrol Synthesis) contains in addition polydatin, a precursor of resveratrol. To assess the influence of GluS and GluReS on the redox status, ten thiol species and three vitamins were measured before (t0) and after 8 weeks (t1) of dietary supplementation. An inflammatory marker, neopterin, was also assessed at the same time points. Both supplements were highly effective in improving the redox status by significantly increasing the reduced-glutathione (GSH) content and other reduced thiol species while significantly decreasing the oxidized species. The positive outcome of the redox status was most significant in the GluRes treatment group which also experienced a significant reduction in neopterin levels. Of note, the endogenous levels of vitamins C, E and A were significantly increased in both treatment groups, with best results in the GluReS group. While both dietary supplements significantly contributed to recognized antioxidant and anti-inflammatory outcomes, the effects of GluReS, the combination of glutathione and resveratrol precursors, were more pronounced. Thus, dietary supplementation with GluReS may represent a valuable strategy for maintaining a competent immune status and a healthy lifespan.

Keywords: glutathione; polydatin; precursors; vitamins; supplements; antioxidant; anti-inflammatory; sirtuins; aging; lifespan

---

1. Introduction

Oxidation [1] and inflammation [2] are key physiologic processes that are at the basis of several chronic diseases. Aging [3] and external factors contribute to oxidative stress, which is an exacerbation of oxidative reactions that cause damage to DNA [4], lipids [5] and proteins/enzymes [6]. A link between oxidation and inflammation exists, with antioxidants having been shown to help to prevent excessive inflammatory responses [7].

Several antioxidant factors [8] are continuously activated to delay the age-related pro-oxidizing shift in the redox state. Glutathione represents one of the most important and ubiquitous among those factors [9,10]. Glutathione is a thiol tripeptide formed by glutamate (Glu), cysteine (Cys) and glycine (Gly) which is synthesized in the cytosol of all mammalian cells [11] where it reaches a very high concentration, in the mM range [10] and is actively transported in the inter-membrane space of mitochondria [12]. Most of glutathione synthesis occurs in the liver where it is subsequently secreted and transported to other organs, to assist in maintaining inter-organ glutathione homeostasis. Glutathione is also the principal mediator of detoxification reactions for protection from xenobiotic insults (drugs, toxins, etc.) [10].

The levels of glutathione decline with age [13–15] and their replenishment through dietary supplementation represents a potential strategy to achieve a healthy aging [16]. However, the majority of exogenously ingested glutathione is destroyed within the gastrointestinal tract. To overcome this limitation, different approaches have been attempted, including delivering glutathione by liposomes [17]. Thus far, the most effective approach has been shown to be the induction of endogenous synthesis of glutathione through dietary supplementation of its precursors [18].

This study was designed to test, in healthy human participants, the antioxidant and anti-inflammatory efficacy of two dietary supplements containing either the glutathione precursors glutamine, α-ketoglutarate and alanine [19,20], N-acetylcysteine [21,22] and glycine [23] alone (GluS), or combined with polydatin (GluReS), which acts as a precursor of resveratrol [24]. In order to provide a complete picture of the redox status before and after dietary supplementation, the antioxidant efficacy was evaluated by measuring ten thiol species and three endogenous vitamins. The anti-inflammatory effects were investigated by measuring the inflammatory marker neopterin [25].

Resveratrol (RV) is a well-known polyphenol which has gained wide attention for its anti-aging activity in animal models [26] and its observed ability to counteract the onset of many age-related comorbidities in humans [27], mainly through the activation of a family of deacetylating enzymes, the sirtuins [28,29]; RV however, has poor bioavailability. Polydatin (PD), which differs from RV only for a glycosidic residue, is endowed with a higher solubility that permits uptake by glucose receptors into the cells [30] where RV is then released via enzymatic hydrolysis. In addition, PD is more resistant to oxidation than RV [31] and also presents a higher free radical scavenging activity compared to RV [32]. Of interest, red wine contains greater amounts of PD than RV [33,34]. Similarly to RV, PD has been shown to activate sirtuins in a number of animal models [35–38]. In turn sirtuins are able to trigger phase II enzymes (anti-oxidant) [39] and enhance the glutathione system [40] activating the main enzymes involved in glutathione metabolism: GSH-Peroxidase (GPX) [41] and GSH-S-Transferase (GST) [42]. GPX and GST are potent mediators of the anti-oxidant and detoxifying function of GSH. Therefore, we hypothesized that the addition of PD to GSH precursors should boost the anti-inflammatory action and potentiate the anti-oxidant system, possibly leading to a synergistic effect. The antioxidant efficacy of the two dietary supplements GluS and GluReS was evaluated by the assessment of thiols redox status in erythrocytes, adopted as cellular model, and in plasma.

2. Materials and Methods

2.1. Study Design

To evaluate the anti-inflammatory and antioxidant effects of a supplement of glutathione precursors, with and without PD, we conducted a randomized clinical trial.

The healthy participants underwent a medical evaluation to determine his/her eligibility. The exclusion criteria were: (1) pregnancy; (2) regular use of drugs which inhibit the metabolism of homocysteine (such as methotrexate, anti-epileptic drugs); (3) serious disease; (4) chemotherapy. Each participant signed an informed consent.

Thirty adult men and women aged 45–75 years were randomly assigned to one of two treatment groups by adaptive minimization, balancing for age, gender and smoking status. The treatment groups received either GluReS (glutathione precursors + PD) or GluS (glutathione precursors only). Participants were blinded to the treatment (bottles were labeled A or B and capsules of GluReS and GluS were undistinguishable). Each treatment group was composed of 15 participants: a first group of seven males and eight females, mean age: 59 years ± 12 SD, who received GluReS (glutathione precursors + PD) and a second group of eight males and seven females, mean age: 60 years ± 10 SD who received GluS (glutathione precursors only). Five and four smokers were present in GluReS and GluS group, respectively. Anthropometric data of the two groups are reported in Table 1.

Table 1. Anthropometric data of the two groups.

	GluReS	GluS
Weight (kg)	68.65 (14.61)	75.36 (13.24)
Height (m)	1.69 (0.11)	1.70 (0.09)
BMI (kg/m^2)	23.80 (2.45)	25.87 (3.62)
Waistline (cm)	90.54 (8.05)	98.83 (11.95)
Heart rate	68.46 (9.39)	65.08 (8.55)
Sistolic blood pressure	122.92 (10.77)	125.58 (12.72)
Diastolic blood pressure	86.77 (4.73)	83.08 (8.71)

GluReS = Glutathione and Resveratrol Synthesis group; GluS = Glutathione Synthesis group. Values represent mean (±SD). BMI = body mass index.

Both dietary supplements were supplied as capsules, 2 capsules/day to be taken on a full stomach (each capsule after a meal).

The formulation of the two dietary supplements is shown below in Table 2.

Table 2. The formulation of the two dietary supplements.

	GluReS	GluS
	mg	mg
glutamine, α-ketoglutarate	217	217
N-acetylcysteine	210	210
glycine	105	105
alanine	126	126
sodium selenite	7	7
polydatin	35	—
total	700	665

Both dietary supplements were produced by Solimè Srl, (Milan, Italy) a company certified according to UNI EN ISO 9001:2015, which guarantees that the productions are carried out in a HACCP self-control system and in compliance with Good Manufacturing Practices (GMP) according to ISO 22716 and the indications of the regulations in force.

The participants did not interrupt consumption of usual drugs prescribed by their doctors and were asked to report adverse events or concomitant medications occurring during the study period. The participants were also asked to notify any changes in lifestyle or dietary habits.

The study period was eight weeks and each participant underwent a blood and urine sampling at the beginning (t0) and at the end (t1) of the study. Biological samples were obtained in the morning after an overnight fast. Approximately 5 mL of blood were drawn from the antecubital vein. The blood samples were collected in ethylenediaminetetraacetic acid dipotassium salt (K_2EDTA) vacutainer tubes (Becton Dickinson Company, Oxford UK), and blood was separated by centrifuge (5702R, Eppendorf, Germany) at 3000× *g* for 5 min at 4 °C. Samples of plasma and erythrocytes were then immediately stored in multiple aliquots at −80 °C until the analyses. Aliquots of the urine were stored at −20 °C until the analyses were performed.

This study was conducted in accordance with the Good Clinical Practice guidelines and the Declaration of Helsinki. Study approval was received from the IRCCS Ospedale San Raffaele (Milan, Italy) ethical committee.

2.2. High Performance Liquid Chromatography (HPLC) Determination of the Principal Circulating Thiols in Plasma and Erythrocytes

Total and reduced thiols were measured in plasma and erythrocytes following a validated method previously published, using two different procedures for preparation of samples [43]. In brief, two aliquots of each sample were prepared, one for the total and one for the reduced thiols. The latter was treated with Tris-(2-carboxyethyl)-phosphine hydrochloride (TCEP) as reducing agent, then both aliquots underwent a precipitation step with 10% trichloroacetic acid (TCA), followed by centrifugation at $14,000\times g$ for 10 min at 4 °C. Clear supernatant (100 μL) was incubated 90 min at room temperature with 4-fluoro-7-sulfamoylbenzofurazan (ABD-F, derivatizing agent) before chromatographic analysis. Thiols' separation was performed at room temperature through high pressure isocratic liquid chromatography on a Discovery analytical column (250 × 4.6 mm, Supelco, Sigma-Aldrich, Bellefonte, PA 16823-0048, USA) eluted with 0.1 M acetate buffer, pH 4.0–methanol, 81:19 (v/v) with a flow rate of 1.0 mL/min. Fluorescence intensity was measured by a Jasco fluorescent spectrophotometer with excitation and emission wavelengths of 390 and 510 nm, respectively. Sample concentration was obtained using a standard calibration curve. The units of measurement are μM. The concentration of oxidized forms was calculated as the difference between total and reduced thiols forms. Oxidized glutathione forms encompass the dimers (GSSG), the mixed disulfide forms (GSSR) and the protein mixed disulfide forms (ProSSG) [44].

2.3. HPLC Determination of Vitamin C (ascorbic acid), Vitamin A (retinol), Vitamin E (α-tocopherol)

HPLC determination of the three vitamins were performed using commercially available kits with European certification from Chromsystems Instruments and Chemicals GmbH (Grafelfing, Germany): for vitamin A and E order n. 34000, for vitamin C order n. 65065. Reproducibility: intra-assay for vitamin A coefficient of variation (CV) is 2.5%, for vitamin E CV is 2.4%, for vitamin C CV is 3.8%; inter-assay for vitamin A CV is 5.0%, for vitamin E CV is 4.9%, for vitamin C CV is 4.8%. Recovery for vitamin A is 106%, for vitamin E is 101%, for vitamin C is 97–103%. Linearity for vitamin A is 0.02–5.0 mg/L, for vitamin E is 0.25–45.0 mg/L, for vitamin C 0.4–100 mg/L. Limit of quantification (LLOQ) for vitamin A is 0.02 mg/L, for vitamin E is 0.25 mg/L, for vitamin C is 0.40 mg/L. Vitamin E concentrations (μM) were normalized to total cholesterol (mM) for a more accurate evaluation [45].

2.4. HPLC Determination of Urinary Neopterin

Urinary neopterin concentrations were measured by an isocratic HPLC method and were normalized to urine creatinine concentrations. Urine samples, stored at −20 °C were thawed and centrifuged at 13,000 rpm for 5 min at 4 °C; the supernatant was then adequately diluted with chromatographic mobile phase (15 mM of K_2HPO_4, pH 3.0). Neopterin and creatinine levels were measured using a Varian instrument (pump 240, autosampler ProStar 410) coupled to a fluorimetric detector (JASCO FP-1520, λ_{ex} = 355 nm and at λ_{em} = 450 nm) for neopterin detection and to an ultraviolet-visible detector (Shimadzu SPD 10-AV, λ = 240 nm) for creatinine determinations. Analytic separations were performed at 50 °C on a 5 μm Discovery C18 analytical column (250 × 4.6 mm I.D., Supelco, Sigma-Aldrich, Bellefonte, PA 16823-0048, USA) at flow rate of 0.9 mL/min. The calibration curves were linear over the range of 0.125^{-1} μmol/L and of 1.25^{-10} mmol/L for neopterin and creatinine levels, respectively. Inter-assay and intra-assay coefficients of variation were <5%.

2.5. Statistics

Data are expressed as mean ± SD and were analyzed using parametric statistics following mathematical confirmation of a normal distribution using Shapiro-Wilks W test. Experimental data were compared by ANOVA variance analysis followed by Tukey's multiple comparison test to further check the among group significance. Statistical analyses were performed using the software Prism 8 (GraphPad Prism 8.3, Software, Inc., San Diego, CA, USA). A p-value of <0.05 was considered statistically significant. Change Δ% estimation [(post value-pre value)/pre value) * 100] was used to compare the changes that occurred in the two study groups after the 8 weeks of diet supplementation and it is also reported in the text. The prospective calculation of the sample size was determined choosing the value of reduced glutathione [46]. At a power of 80% the number of participants of 13 was calculated, which is below the number of participants recruited for this study.

3. Results

All participants completed the eight weeks (t1) of dietary supplementation and no adverse effects were reported. Furthermore, no significant differences between the two groups examined at baseline were found.

The level of endogenous reduced thiols in erythrocytes at t0 and t1 are shown in Figure 1. A significant increase of reduced glutathione (GSH) was induced by both dietary supplements ($p < 0.001$ for GluRes and $p < 0.01$ for GluS) (Figure 1A); however, the increase was greater in GluReS compared to the GluS group (+40% and +32%, respectively). Cysteine (Cys) instead (Figure 1B) decreased significantly ($p < 0.001$) in both groups, as expected because it was efficiently used to produce GSH in the erythrocytes. The decrease was superior in group GluReS versus group GluS (−22% and −19%, respectively). Finally, a catabolite of GSH, cysteinylglycine (CysGly) was also measured (Figure 1C); a significant ($p < 0.05$ vs. $p < 0.001$) increase in both groups was found, and also in this case the increase was higher in group GluReS compared to group GluS (+32% versus +25%).

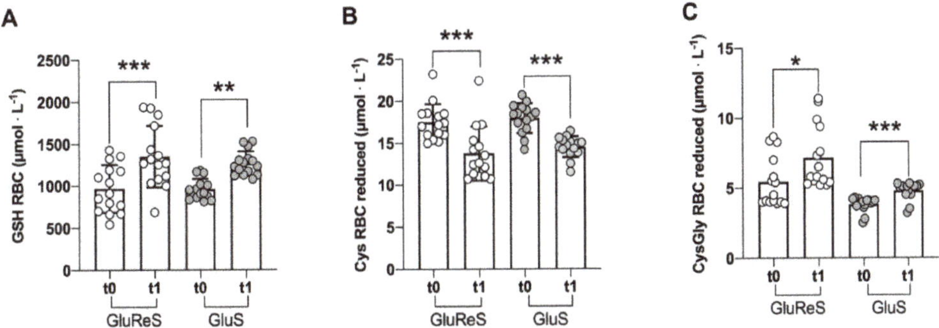

Figure 1. Scatter dot plot of reduced thiol species measured in erythrocytes before (t0) and after 8 weeks (t1) of dietary supplementation with GluReS (Glutathione and Resveratrol Synthesis) or GluS (Glutathione Synthesis). (**A**) reduced glutathione (GSH), (**B**) reduced cysteine (Cys), (**C**) reduced cysteinylglycine (CysGly); RBC = red blood cells. Data are expressed as the mean ± SD. Changes over time (t1 vs. t0) were significant at: * $p < 0.05$; ** $p < 0.01$; *** $p < 0.001$.

The results of the oxidized thiols in erythrocytes are depicted in Figure 2. A highly significant ($p < 0.0001$) decrease from t0 to t1 was observed in: oxidized glutathione (−56% vs. −79%, Figure 2A); oxidized Cys (−34% vs. −24%, Figure 2B) and oxidized CysGly (−44% vs. −47%, Figure 2C) for group GluReS compared to group GluS, respectively. The oxidized glutathione (Figure 2A) represents a total pool including not only the dimers (GSSG), but also the mixed disulfide (GSSR) and the protein mixed disulfide forms (ProSSG).

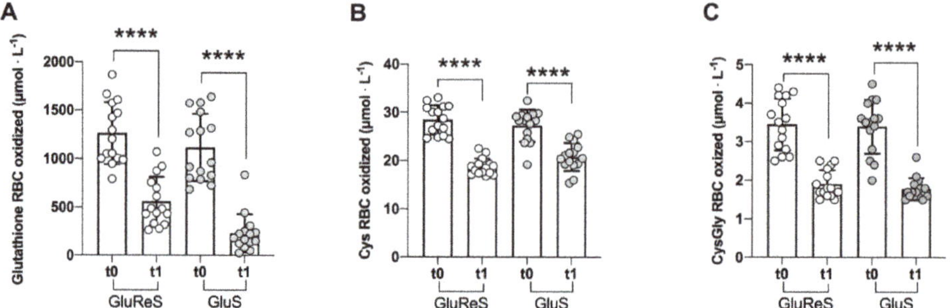

Figure 2. Scatter dot plot of oxidized thiol species measured in erythrocytes before (t0) and after eight weeks (t1) of dietary supplementation with GluReS or GluS. (**A**) oxidized glutathione, (**B**) cysteine (Cys), (**C**) cysteinylglycine (CysGly); RBC = red blood cells. Data are expressed as the mean ± SD. Change over time (t1 vs. t0) was significant at: **** $p < 0.0001$.

The reduced and oxidized thiols were evaluated also in plasma, except for glutathione due to its almost undetectable levels in plasma [9]. Figure 3 summarizes the data obtained at t0 and t1 for both treatment groups. Reduced Cys increased highly significantly ($p < 0.0001$) in GluReS group and significantly ($p < 0.01$) in the GluS group (+42%, vs. +16%, Figure 3A). The same occurred for the reduced CysGly catabolite which increased significantly ($p < 0.01$) in both groups (+45% vs. +24% Figure 3B). Conversely, oxidized Cys (Figure 3C) and CysGly (Figure 3D) declined significantly ($p < 0.001$) in both groups, with similar degrees for Cys (−28% GluReS and −27% GluS), whereas for oxidized CysGly the decrease was greater in GluS compared to GluReS (−37% vs. −30%, respectively).

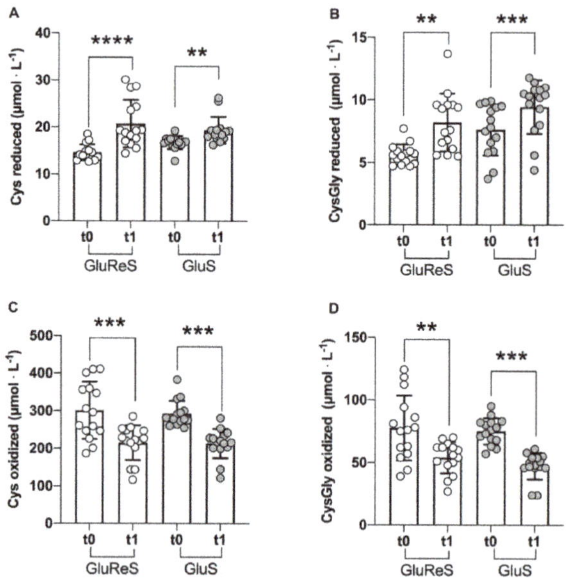

Figure 3. Scatter dot plot of reduced and oxidized thiol species measured in plasma before (t0) and after 8 weeks (t1) of dietary supplementation with GluReS or GluS. (**A**) reduced cysteine (Cys), (**B**) reduced cysteinylglycine (CysGly), (**C**) oxidized cysteine (Cys), (**D**) oxidized cysteinylglycine (CysGly). Data are expressed as the mean ± SD. Changes over time (t1 vs. t0) were significant at: ** $p < 0.01$; *** $p < 0.001$; **** $p < 0.0001$.

Table 3 summarizes the data displayed in Figures 1–3.

Table 3. Before and after concentrations (µM) of reduced and oxidized forms of glutathione, cysteine and cysteinylglycine in erythrocytes and plasma for the two groups.

Erythrocytes	GluReS			GluS		
	t0	t1	%Δ	t0	t1	%Δ
Glutathione						
reduced	967.59 ± 283.23	1349.87 ± 367.62	+40	958.42 ± 123.74	1265.09 ± 144.95	+32
oxidized	1262.95 ± 321.77	560.47 ± 251.47	−56	1115.35 ± 346.66	231.67 ± 196.89	−79
Cysteine						
reduced	17.51 ± 2.17	13.74 ± 3.24	−22	17.94 ± 1.77	14.49 ± 1.26	−19
oxidized	28.46 ± 3.04	18.72 ± 1.72	−34	27.24 ± 3.30	20.77 ± 2.89	−24
Cysteinylglycine						
reduced	5.43 ± 1.81	7.14 ± 2.26	+32	3.81 ± 0.51	4.77 ± 0.63	+25
oxidized	3.45 ± 0.67	1.92 ± 0.37	−44	3.40 ± 0.70	1.79 ± 0.28	−47
Plasma	GluReS			GluS		
	t0	t1	%Δ	t0	t1	%Δ
Cysteine						
reduced	14.59 ± 1.67	20.73 ± 5.05	+42	16.71 ± 1.36	19.35 ± 2.80	+16
oxidized	300.81 ± 75.58	215.48 ± 46.55	−28	292.28 ± 34.31	213.65 ± 39.17	−27
Cysteinylglycine						
reduced	5.63 ± 0.83	8.19 ± 2.30	+45	7.62 ± 2.06	9.45 ± 2.13	+24
oxidized	77.55 ± 25.95	53.95 ± 12.50	−30	74.99 ± 10.48	47.28 ± 10.72	−37

The difference between the t1 levels of reduced glutathione in erythrocytes observed in the two groups, i.e., 1349.87 ± 367.62 in the GluReS group versus 1265.09 ± 144.95 in the GluS group ($p < 0.013$) further supports the more efficient anti-oxidant activity of GluReS compared to GluS.

Since the redox status also has regulatory interactions with the vitamins' metabolic pathways, we evaluated whether changes occurred in the levels of endogenous vitamins C, A, and E. Noteworthy, endogenous vitamins C, A and E increased significantly in both groups (range $p < 0.01$–0.0001), with the GluReS group showing a much higher increase compared to the GluS group (Table 4). The Δ% of vitamin E refers to the ratio vitamin E:cholesterol.

Table 4. Summary of the change from t0 to t1 (Δ%) and significance (p) of the three vitamins for the two groups.

	vitamin C	p	vitamin A	p	vitamin E	p
	Δ%		Δ%		Δ%	
GluReS	+37	<0.01	+33	<0.01	+58	<0.0001
GluS	+11	<0.001	+14	<0.001	+39	<0.0008

Finally, we measured neopterin, an established inflammatory marker [47], in the urine of the participants (Figure 4). Levels of neopterin remained substantially the same in GluS group, whereas it diminished significantly ($p < 0.01$) in the GluReS (−30%).

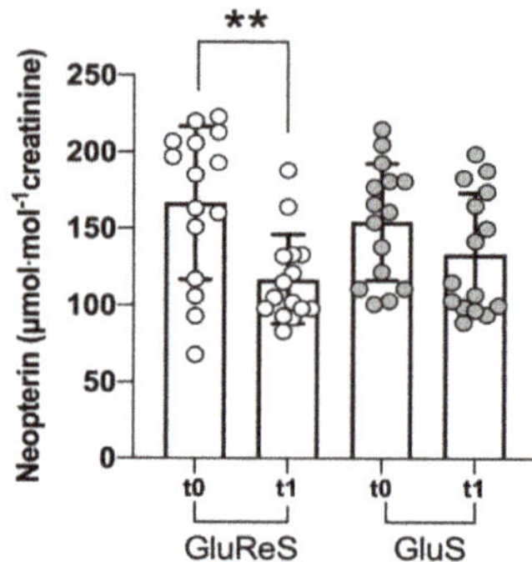

Figure 4. Scatter dot plot of neopterin levels measured in urine before (t0) and after 8 weeks (t1) of dietary supplementation with GluReS or GluS. Data are expressed as the mean ± SD. Changes over time (t1 vs. t0) was significant at ** $p < 0.01$.

4. Discussion

This study describes a before and after study of dietary intervention with two supplements; the first contains precursors for the endogenous synthesis of glutathione (GluS), whereas the second (GluReS) contains the same precursors in addition to polydatin, a precursor of resveratrol. Ten thiol species, three vitamins and the inflammatory marker neopterin were measured before and after eight weeks of supplementation. Both supplements appeared to improve redox status, as measured by thiol species, with the GluReS performing better overall. Additionally, endogenous vitamin concentrations of vitamins C, E and A were increased with both supplemental protocols, whereas GluReS only induced a significant reduction of neopterin.

Previous studies demonstrated that the metabolic pathways that lead to the generation of glutathione can be boosted by dietary supplementation with its precursors [18]. To the best of our knowledge however the potential synergy between the increase of sirtuin activity induced by resveratrol, as well as its precursor polydatin, and reduced glutathione (GSH) have not yet been reported. Figure 5 illustrates the pathways leading to GSH synthesis and degradation as well as the points of contact with the activity of polydatin.

The data reported here demonstrate that by providing the selected precursors both dietary supplements (GluReS and GluS) were successful in inducing in erythrocytes a significant increase of endogenous reduced glutathione (GSH) and a concomitant decrease of oxidized glutathione. It should be pointed out that our method measures all oxidized glutathione forms (dimers (GSSG), mixed disulfide forms (GSSR) and protein mixed disulfide forms (ProSSG)), but cannot distinguish among them, which is beyond the scope of our study. All changes in the thiol species occurring after eight weeks of dietary supplementation were highly significant in both arms, ranging from $p < 0.05$ to $p < 10^{-4}$, indicating a metabolic acceleration and a boost of intra-cellular (erythrocytes) and extra-cellular (plasma) antioxidant activity.

In both study groups the decrease in erythrocytes of both reduced and oxidized cysteine may indicate that the reduced form is used for the biosynthesis of GSH, decreasing the pool of molecules for oxidation. On the other hand, the plasma levels of oxidized cysteine, which represents a strong pro-oxidant molecule, were significantly decreased, whereas those of reduced cysteine were significantly increased. This increase in plasma can in turn facilitate the transport into the erythrocytes through γ-glutamyl binding. This binding and transport are probably due to the presence of glutamate which is synthesized and made available through the activity of the aminotransferases (ALT and AST) on the substrates α-ketoglutarate and glutamine, both provided by the dietary supplements. In both groups also the catabolite CysGly was affected with concomitant significant reduction of the oxidized form and increase of the reduced form occurring both in erythrocytes and plasma due to the activity of GGT.

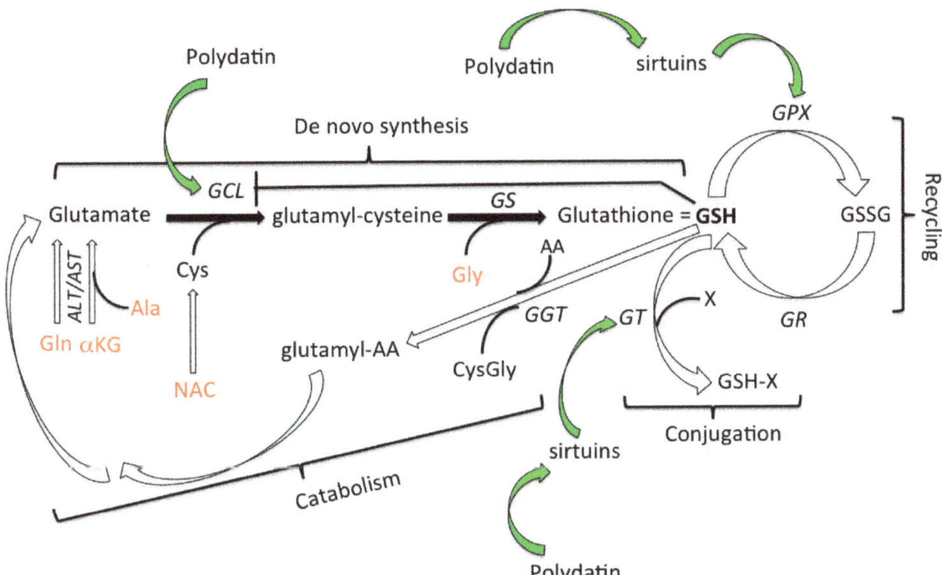

Figure 5. Schematic summary of glutathione synthesis, catabolism, regulation and putative connections with polydatin. Reduced glutathione (GSH) can be produced by continuous recycling mediated by the enzymes glutathione peroxidase (GPX) and glutathione reductase (GR). GPX uses GSH as substrate and oxidizes it to its dimeric oxidized form (GSSG), reducing peroxides and acid peroxides to water and alcohol. GR brings back GSSG to its reduced form GSH which will again be oxidized by GPX, thus leading to the antioxidant status necessary for cellular homeostasis. Instead de novo synthesis of GSH is mediated by glutamate-cysteine ligase (GCL) and glutathione synthetase (GS) when the building blocks (Glutamate, Cys = Cysteine, Gly = Glycine) are available in sufficient quantities. A negative feed-back mechanism exists since GSH inhibits GCL, avoiding unnecessary biosynthesis of GSH. In order to perform its detoxification function GSH is consumed by its conjugation to drugs or other xenobiotics mediated by glutathione S-transferase (GT), whereas its catabolism is initiated by γ-glutamyl transpeptidase (GGT) which yields the CysteinylGlycine (CysGly) dipeptide. Polydatin may impact GSH metabolism either augmenting the quantity or activation of GCL through the Nrf2 pathway or by triggering sirtuins which in turn modulate the activity of GPX and GT. The precursors present in the GluS and GluReS dietary supplements are depicted in red. GSH-X = reduced glutathione bound to a xenobiotic substance; X = xenobiotic substance; AA = aminoacids; NAC = N-acetylcysteine; Ala = alanine; Gln = glutamine; αKG = α-ketoglutarate.

Although both study arms behaved in the same manner towards a positive redox outcome from t0 to t1, GluReS was substantially superior to GluS group, in nine out of the 10 thiol species evaluated. The higher performance of GluReS supplementation can be ascribed to the presence of polydatin that provides additional reducing equivalents owing to its antioxidant activity [48] carried out by the hydroxyl group OH in position C4. In addition, polydatin may also act indirectly by activating sirtuins [35–38], which in turn stimulate phase II detoxifying enzymes and enhance the activity of GST and GPX [41–49]. Consumption of reduced glutathione (GSH) molecules through GST activity induces reactivation of GSH de novo synthesis, suggesting that dietary supplement GluReS synergizes with GST through the precursors-mediated biosynthesis of new GSH molecules. While GST is responsible for the detoxifying activity of GSH, the antioxidant action of GSH is carried out by GPX. In our study the reduced and oxidized thiol species demonstrated that GluReS was superior to GluS, suggesting an increased enzymatic activity of GPX induced by polydatin as compared to GR. GluS resulted superior to GluReS only on the reduction of oxidized glutathione at t1, as if the action of GR was delayed.

This result could be explained by the fact that GR maintains the same level of activity in the two groups, but in GluReS it is the activity of GPX that appears accelerated due to the presence of polydatin. Finally, polydatin could also activate GLC activity through the NF-E2-related factor 2 (Nrf2) pathway [50] which coordinates the antioxidant and phase II detoxification enzymes to adapt to different stress conditions. Both resveratrol and polydatin have been shown to activate the Nrf2 pathway in different murine models [51,52].

A second important result of our study was the significant increase of endogenous vitamins C, E, and A in both groups at t1. These results can be explained by the fact that the newly synthetized reduced glutathione (GSH) is able to regenerate ascorbic acid (vitamin C) from its precursor dehydroascorbic acid (DHA) through the action of the enzyme dehydroascorbate reductase. Otherwise DHA would irreversibly pass to 2,3-diketogulonic acid, causing a loss of vitamin C. The reduction of DHA by GSH is a very active pathway and leads to almost total recovery of vitamin C. Vitamin C in turn is able to transform the α-tocopherol radical into α-tocopherol (vitamin E), concomitantly neutralizing the free radicals of carotenoids which could have a pro-oxidative effect if not removed by vitamin C. Vitamin E in turn protects from oxidation β-carotene and vitamin A localized intracellularly and in the circulating lipoproteins, thus leading to an increase of vitamin A. GluReS was definitely superior to GluS in upregulating the endogenous levels of vitamins C, E, A, most likely due to the boost of GSH with its effect on the vitamins' metabolism. Noteworthy, in age-related macular degeneration, vitamins C, E and β-carotene have been shown to be effective in retarding the degenerative process [53] and within the aging population (British participants over 65 years) plasma levels of vitamin C and E have been found to reduce the risk for all-cause mortality, and respiratory mortality, respectively [54].

The significant increase of endogenous vitamins (C, E, A), especially in the polydatin plus glutathione precursors group, is an original and important finding since exogenously provided vitamins may be given in excess to well-nourished people [55] and could possibly be harmful [56]. On the same line, the fact that GSH itself when it has reached the necessary levels establishes a negative feed-back on its de novo synthesis through inhibition of the GCL enzyme [57] implies that providing precursors for endogenous synthesis of GSH could represent a safe, and efficient, strategy.

At last, a significant drop of the inflammatory marker neopterin measured in urine was observed only in the GluReS group. This result can be explained by the anti-inflammatory effect of PD through inhibition of NF-kB activity documented in vitro [58] and in various animal models of inflammation-induced damage of various cells or organs [59–62] including the kidney [63].

A limitation to this study is represented by the lack of a placebo-controlled arm. Indeed uncontrolled before and after studies have been demonstrated to often be confounded, which leads to an overestimation of the intervention's effectiveness [64], thus these results, although innovative and significant, require cautious interpretation and further validation.

The age range of our participants was 45–75 years, an important time to implement strategies aiming towards a healthy aging with disease-free years to live. Dietary supplements containing both anti-inflammatory factors such as PD and precursors effective in maintaining GSH levels appear to be a valuable approach. The maintenance of a balanced redox status and of an adequate level of vitamins along with the control of subclinical inflammation can be leveraged for the prevention of chronic degenerative diseases associated with aging. Depletion of cellular GSH induces ferroptosis [65], an iron-dependent non-apoptotic form of cell death [66] and low levels of GSH are found in HIV-infected individuals [67] and in a number of age-related diseases [57], such as cataract [68], Alzheimer's [69], diabetes and cancer [70]. Finally, a recent paper by Lian and colleagues reported that GSH de novo synthesis, but not GSH recycling, is required for activation and proliferation of murine T lymphocytes [71], underlying the importance of providing GSH precursors to maintain an immune-competent status, and providing an additional potential application of GluS and/or GluReS supplements in the prevention of infectious diseases.

5. Conclusions

This study provides preliminary but strong evidence for the use of GSH precursors combined with polydatin to maintain the redox status into an ideal range while at the same time activating sirtuins which, in addition to their synergistic effects on the redox status, exert a multitude of beneficial effects on cellular metabolism and immune functions. These results should encourage further testing of GluReS in randomized placebo-controlled studies on the prevention of age-associated diseases and the optimization of immunocompetency.

Author Contributions: Writing, P.B.; experimental performance, C.D.; data curation, A.V.; visualization, S.M.-S.; resources, M.M.; supervision, A.B.; conceptualization, R.A. All authors have read and agreed to the published version of the manuscript.

Funding: This research received no external funding.

Conflicts of Interest: A.B. and R.A. are founders and scientific officers of SoLongevity srl which provided GluS and GluReS. P.B. is a life science research consultant who also acts as scientific officer of SoLongevity srl.

References

1. Pomatto, L.C.; Davies, K.J. Adaptive homeostasis and the free radical theory of ageing. *Free. Radic. Biol. Med.* **2018**, *124*, 420–430. [CrossRef] [PubMed]
2. Franceschi, C.; Capri, M.; Monti, D.; Giunta, S.; Olivieri, F.; Sevini, F.; Panourgia, M.P.; Invidia, L.; Celani, L.; Scurti, M.; et al. Inflammaging and anti-inflammaging: A systemic perspective on aging and longevity emerged from studies in humans. *Mech. Ageing Dev.* **2007**, *128*, 92–105. [CrossRef]
3. Finkel, T.; Holbrook, N.J. Oxidants, oxidative stress and the biology of ageing. *Nat. Cell Biol.* **2000**, *408*, 239–247. [CrossRef] [PubMed]
4. Lindahl, T. Instability and decay of the primary structure of DNA. *Nat. Cell Biol.* **1993**, *362*, 709–715. [CrossRef] [PubMed]
5. Rikans, L.E.; Hornbrook, K. Lipid peroxidation, antioxidant protection and aging. *Biochim. Biophys. Acta (BBA) Mol. Basis Dis.* **1997**, *1362*, 116–127. [CrossRef]
6. Stadtman, E.R. Protein oxidation and aging. *Science* **1992**, *257*, 1220–1224. [CrossRef] [PubMed]
7. Biswas, S.K. Does the Interdependence between Oxidative Stress and Inflammation Explain the Antioxidant Paradox? *Oxidative Med. Cell. Longev.* **2016**, *2016*, 5698931. [CrossRef] [PubMed]
8. Hawkins, C.L.; Davies, M.J. Detection, identification, and quantification of oxidative protein modifications. *J. Biol. Chem.* **2019**, *294*, 19683–19708. [CrossRef] [PubMed]
9. Shan, X.; Aw, T.Y.; Jones, D.P. Glutathione-dependent projection against oxidative injury. *Pharmacol. Ther.* **1990**, *47*, 61–71. [CrossRef]
10. DePonte, M. The Incomplete Glutathione Puzzle: Just Guessing at Numbers and Figures? *Antioxid. Redox Signal.* **2017**, *27*, 1130–1161. [CrossRef]
11. Lu, S.C. Glutathione synthesis. *Biochim. Biophys. Acta (BBA) Gen. Subj.* **2013**, *1830*, 3143–3153. [CrossRef] [PubMed]
12. Toledano, M.B.; Huang, M.-E. The Unfinished Puzzle of Glutathione Physiological Functions, an Old Molecule That Still Retains Many Enigmas. *Antioxid. Redox Signal.* **2017**, *27*, 1127–1129. [CrossRef] [PubMed]
13. Inal, M.E.; Sunal, E.; Kanbak, G. Age-related changes in the glutathione redox system. *Cell Biochem. Funct.* **2002**, *20*, 61–66. [CrossRef] [PubMed]
14. Liu, H.; Wang, H.; Shenvi, S.; Hagen, T.M.; Liu, R.-M. Glutathione Metabolism during Aging and in Alzheimer Disease. *Ann. N. Y. Acad. Sci.* **2004**, *1019*, 346–349. [CrossRef]
15. Rebrin, I.; Sohal, R.S. Pro-oxidant shift in glutathione redox state during aging. *Adv. Drug Deliv. Rev.* **2008**, *60*, 1545–1552. [CrossRef] [PubMed]
16. Wu, G.; Fang, Y.-Z.; Yang, S.; Lupton, J.R.; Turner, N.D. Glutathione Metabolism and Its Implications for Health. *J. Nutr.* **2004**, *134*, 489–492. [CrossRef]
17. Sinha, R.; Sinha, I.; Calcagnotto, A.; Trushin, N.; Haley, J.S.; Schell, T.D.; Richie, J.P. Oral supplementation with liposomal glutathione elevates body stores of glutathione and markers of immune function. *Eur. J. Clin. Nutr.* **2017**, *72*, 105–111. [CrossRef]
18. Gould, R.L.; Pazdro, R. Impact of Supplementary Amino Acids, Micronutrients, and Overall Diet on Glutathione Homeostasis. *Nutrition* **2019**, *11*, 1056. [CrossRef]

19. Whillier, S.; Garcia, B.; Chapman, B.E.; Kuchel, P.W.; Raftos, J.E. Glutamine and α-ketoglutarate as glutamate sources for glutathione synthesis in human erythrocytes. *FEBS J.* **2011**, *278*, 3152–3163. [CrossRef]

20. Ellinger, J.J.; Lewis, I.A.; Markley, J.L. Role of aminotransferases in glutamate metabolism of human erythrocytes. *J. Biomol. NMR* **2011**, *49*, 221–229. [CrossRef]

21. Atkuri, K.R.; Mantovani, J.J.; Herzenberg, L.A. N-Acetylcysteine—A safe antidote for cysteine/glutathione deficiency. *Curr. Opin. Pharmacol.* **2007**, *7*, 355–359. [CrossRef] [PubMed]

22. Rushworth, G.F.; Megson, I.L. Existing and potential therapeutic uses for N-acetylcysteine: The need for conversion to intracellular glutathione for antioxidant benefits. *Pharmacol. Ther.* **2014**, *141*, 150–159. [CrossRef] [PubMed]

23. Sekhar, R.V.; McKay, S.V.; Patel, S.G.; Guthikonda, A.P.; Reddy, V.T.; Balasubramanyam, A.; Jahoor, F. Glutathione Synthesis Is Diminished in Patients with Uncontrolled Diabetes and Restored by Dietary Supplementation with Cysteine and Glycine. *Diabetes Care* **2010**, *34*, 162–167. [CrossRef]

24. Du, Q.-H.; Peng, C.; Zhang, H. Polydatin: A review of pharmacology and pharmacokinetics. *Pharm. Biol.* **2013**, *51*, 1347–1354. [CrossRef] [PubMed]

25. Beretta, A.; Accinni, R.; Dellanoce, C.; Tonini, A.; Cardot, J.-M.; Bussière, A. Efficacy of a Standardized Extract ofPrunus mumein Liver Protection and Redox Homeostasis: A Randomized, Double-Blind, Placebo-Controlled Study. *Phytotherapy Res.* **2016**, *30*, 949–955. [CrossRef]

26. Bhullar, K.S.; Hubbard, B.P. Lifespan and healthspan extension by resveratrol. *Biochim. Biophys. Acta (BBA)-Mol. Basis Dis.* **2015**, *1852*, 1209–1218. [CrossRef] [PubMed]

27. Berman, A.Y.; Motechin, R.A.; Wiesenfeld, M.Y.; Holz, M.K. The therapeutic potential of resveratrol: A review of clinical trials. *NPJ Precis. Oncol.* **2017**, *1*, 1–9. [CrossRef]

28. Howitz, K.T.; Bitterman, K.J.; Cohen, H.Y.; Lamming, D.W.; Lavu, S.; Wood, J.G.; Zipkin, R.E.; Chung, P.; Kisielewski, A.; Zhang, L.-L.; et al. Small molecule activators of sirtuins extend Saccharomyces cerevisiae lifespan. *Nat. Cell Biol.* **2003**, *425*, 191–196. [CrossRef]

29. Grabowska, W.; Sikora, E.; Bielak-Zmijewska, A. Sirtuins, a promising target in slowing down the ageing process. *Biogerontology* **2017**, *18*, 447–476. [CrossRef]

30. Henry-Vitrac, C.; Desmoulière, A.; Girard, D.; Mérillon, J.-M.; Krisa, S. Transport, deglycosylation, and metabolism of trans-piceid by small intestinal epithelial cells. *Eur. J. Nutr.* **2006**, *45*, 376–382. [CrossRef]

31. Regev-Shoshani, G.; Shoseyov, O.; Bilkis, I.; Kerem, Z. Glycosylation of resveratrol protects it from enzymic oxidation. *Biochem. J.* **2003**, *374*, 157–163. [CrossRef]

32. Su, D.; Cheng, Y.; Liu, M.; Liu, D.; Cui, H.; Zhang, B.-L.; Zhou, S.; Yang, T.; Mei, Q. Comparision of Piceid and Resveratrol in Antioxidation and Antiproliferation Activities In Vitro. *PLoS ONE* **2013**, *8*, e54505. [CrossRef] [PubMed]

33. Moreno-Labanda, J.F.; Mallavia, R.; Pérez-Fons, L.; Lizama, V.; Saura, A.D.; Micol, V. Determination of Piceid and Resveratrol in Spanish Wines Deriving fromMonastrell(*Vitis vinifera* L.) Grape Variety. *J. Agric. Food Chem.* **2004**, *52*, 5396–5403. [CrossRef]

34. Weiskirchen, S.; Weiskirchen, R. Resveratrol: How Much Wine Do You Have to Drink to Stay Healthy? *Adv. Nutr.* **2016**, *7*, 706–718. [CrossRef] [PubMed]

35. Zeng, Z.; Chen, Z.; Xu, S.; Zhang, Q.; Wang, X.; Gao, Y.; Zhao, K.-S. Polydatin Protecting Kidneys against Hemorrhagic Shock-Induced Mitochondrial Dysfunction via SIRT1 Activation and p53 Deacetylation. *Oxidative Med. Cell. Longev.* **2016**, *2016*, 1737185. [CrossRef] [PubMed]

36. Zeng, Z.; Yang, Y.; Dai, X.; Xu, S.; Li, T.; Zhang, Q.; Zhao, K.-S.; Chen, Z. Polydatin ameliorates injury to the small intestine induced by hemorrhagic shock via SIRT3 activation-mediated mitochondrial protection. *Expert Opin. Ther. Targets* **2016**, *20*, 645–652. [CrossRef]

37. Zhang, M.; Wang, S.; Cheng, Z.; Xiong, Z.; Lv, J.; Yang, Z.; Li, T.; Jiang, S.; Gu, J.; Sun, D.; et al. Polydatin ameliorates diabetic cardiomyopathy via Sirt3 activation. *Biochem. Biophys. Res. Commun.* **2017**, *493*, 1280–1287. [CrossRef]

38. Cao, K.; Lei, X.; Liu, H.; Zhao, H.; Guo, J.; Chen, Y.; Xu, Y.; Cheng, Y.; Liu, C.; Cui, J.; et al. Polydatin alleviated radiation-induced lung injury through activation of Sirt3 and inhibition of epithelial-mesenchymal transition. *J. Cell. Mol. Med.* **2017**, *21*, 3264–3276. [CrossRef]

39. Singh, C.K.; Chhabra, G.; Ndiaye, M.A.; Garcia-Peterson, L.M.; Mack, N.J.; Ahmad, N. The Role of Sirtuins in Antioxidant and Redox Signaling. *Antioxid. Redox Signal.* **2018**, *28*, 643–661. [CrossRef]

40. Someya, S.; Yu, W.; Hallows, W.C.; Xu, J.; Vann, J.M.; Leeuwenburgh, C.; Tanokura, M.; Denu, J.M.; Prolla, T.A. Sirt3 Mediates Reduction of Oxidative Damage and Prevention of Age-Related Hearing Loss under Caloric Restriction. *Cell* **2010**, *143*, 802–812. [CrossRef]

41. Ren, Y.; Du, C.; Shi, Y.; Wei, J.; Wu, H.; Cui, H. The Sirt1 activator, SRT1720, attenuates renal fibrosis by inhibiting CTGF and oxidative stress. *Int. J. Mol. Med.* **2017**, *39*, 1317–1324. [CrossRef] [PubMed]

42. Tao, N.-N.; Zhou, H.-Z.; Tang, H.; Cai, X.-F.; Zhang, W.-L.; Ren, J.-H.; Zhou, L.; Chen, X.; Chen, K.; Li, W.-Y.; et al. Sirtuin 3 enhanced drug sensitivity of human hepatoma cells through glutathione S-transferase pi 1/JNK signaling pathway. *Oncotarget* **2016**, *7*, 50117–50130. [CrossRef] [PubMed]

43. Dellanoce, C.; Cozzi, L.; Zuddas, S.; Pratali, L.; Accinni, R. Determination of different forms of aminothiols in red blood cells without washing erythrocytes. *Biomed. Chromatogr.* **2013**, *28*, 327–331. [CrossRef] [PubMed]

44. Svardal, A.M.; Mansoor, M.A.; Ueland, P.M. Determination of reduced, oxidized, and protein-bound glutathione in human plasma with precolumn derivatization with monobromobimane and liquid chromatography. *Anal. Biochem.* **1990**, *184*, 338–346. [CrossRef]

45. Ford, L.; Farr, J.; Morris, P.; Berg, J. The value of measuring serum cholesterol-adjusted vitamin E in routine practice. *Ann. Clin. Biochem. Int. J. Lab. Med.* **2006**, *43*, 130–134. [CrossRef] [PubMed]

46. Faul, F.; Erdfelder, E.; Lang, A.-G.; Buchner, A. G*Power 3: A flexible statistical power analysis program for the social, behavioral, and biomedical sciences. *Behav. Res. Methods* **2007**, *39*, 175–191. [CrossRef] [PubMed]

47. Murr, C.; Widner, B.; Wirleitner, B.; Fuchs, D. Neopterin as a Marker for Immune System Activation. *Curr. Drug Metab.* **2002**, *3*, 175–187. [CrossRef] [PubMed]

48. Jayalakshmi, P.; Devika, P.T. Assessment of in vitro antioxidant activity study of polydatin. *J. Pharmacogn. Phytochem.* **2019**, *8*, 55–58.

49. Webster, B.R.; Lu, Z.; Sack, M.N.; Scott, I. The role of sirtuins in modulating redox stressors. *Free. Radic. Biol. Med.* **2012**, *52*, 281–290. [CrossRef]

50. Huang, Y.; Li, W.; Su, Z.-Y.; Kong, A.-N.T. The complexity of the Nrf2 pathway: Beyond the antioxidant response. *J. Nutr. Biochem.* **2015**, *26*, 1401–1413. [CrossRef]

51. Kim, E.N.; Lim, J.H.; Kim, M.Y.; Ban, T.H.; Jang, I.A.; Yoon, H.E.; Park, C.W.; Chang, Y.S.; Choi, B.S. Resveratrol, an Nrf2 activator, ameliorates aging-related progressive renal injury. *Aging* **2018**, *10*, 83–99. [CrossRef] [PubMed]

52. Tang, S.; Tang, Q.; Jin, J.; Zheng, G.; Xu, J.; Huang, W.; Li, X.; Shang, P.; Liu, H.-X. Polydatin inhibits the IL-1β-induced inflammatory response in human osteoarthritic chondrocytes by activating the Nrf2 signaling pathway and ameliorates murine osteoarthritis. *Food Funct.* **2018**, *9*, 1701–1712. [CrossRef] [PubMed]

53. Age-Related Eye Disease Study Research Group A Randomized, Placebo-Controlled, Clinical Trial of High-Dose Supplementation With Vitamins C and E, Beta Carotene, and Zinc for Age-Related Macular Degeneration and Vision Loss. *Arch. Ophthalmol.* **2001**, *119*, 1417–1436. [CrossRef] [PubMed]

54. Bates, C.J.; Hamer, M.; Mishra, G.D. Redox-modulatory vitamins and minerals that prospectively predict mortality in older British people: The National Diet and Nutrition Survey of people aged 65 years and over. *Br. J. Nutr.* **2010**, *105*, 123–132. [CrossRef] [PubMed]

55. Ford, K.; Jorgenson, D.; Landry, E.J.; Whiting, S.J. Vitamin and mineral supplement use in medically complex, community-living, older adults. *Appl. Physiol. Nutr. Metab.* **2019**, *44*, 450–453. [CrossRef] [PubMed]

56. Guallar, E.; Stranges, S.; Mulrow, C.; Appel, L.J.; Miller, E.R. Enough Is Enough: Stop Wasting Money on Vitamin and Mineral Supplements. *Ann. Intern. Med.* **2013**, *159*, 850–851. [CrossRef]

57. Lu, S.C. Regulation of glutathione synthesis. *Mol. Asp. Med.* **2009**, *30*, 42–59. [CrossRef]

58. Lou, T.; Jiang, W.; Xu, D.; Chen, T.; Fu, Y. Inhibitory Effects of Polydatin on Lipopolysaccharide-Stimulated RAW 264.7 Cells. *Inflammation* **2015**, *38*, 1213–1220. [CrossRef]

59. Jiang, Q.; Yi, M.; Guo, Q.; Wang, C.; Wang, H.; Meng, S.; Liu, C.; Fu, Y.; Ji, H.; Chen, T. Protective effects of polydatin on lipopolysaccharide-induced acute lung injury through TLR4-MyD88-NF-κB pathway. *Int. Immunopharmacol.* **2015**, *29*, 370–376. [CrossRef]

60. Jiang, K.-F.; Zhao, G.; Deng, G.-Z.; Wu, H.-C.; Yin, N.-N.; Chen, X.-Y.; Qiu, C.-W.; Peng, X.-L. Polydatin ameliorates Staphylococcus aureus-induced mastitis in mice via inhibiting TLR2-mediated activation of the p38 MAPK/NF-κB pathway. *Acta Pharmacol. Sin.* **2016**, *38*, 211–222. [CrossRef]

61. Zhao, G.; Jiang, K.; Wu, H.; Qiu, C.; Deng, G.; Peng, X. Polydatin reducesStaphylococcus aureuslipoteichoic acid-induced injury by attenuating reactive oxygen species generation and TLR2-NFκB signalling. *J. Cell. Mol. Med.* **2017**, *21*, 2796–2808. [CrossRef] [PubMed]

62. Ye, J.; Piao, H.; Jiang, J.; Jin, G.; Zheng, M.; Yang, J.; Jin, X.; Sun, T.; Choi, Y.H.; Li, L.; et al. Polydatin inhibits mast cell-mediated allergic inflammation by targeting PI3K/Akt, MAPK, NF-κB and Nrf2/HO-1 pathways. *Sci. Rep.* **2017**, *7*, 11895. [CrossRef] [PubMed]

63. Chen, L.; Lan, Z. Polydatin attenuates potassium oxonate-induced hyperuricemia and kidney inflammation by inhibiting NF-κB/NLRP3 inflammasome activation via the AMPK/SIRT1 pathway. *Food Funct.* **2017**, *8*, 1785–1792. [CrossRef] [PubMed]

64. Grimshaw, J.; Campbell, M.; Eccles, M.; Steen, N. Experimental and quasi-experimental designs for evaluating guideline implementation strategies. *Fam. Pr.* **2000**, *17*, S11–S16. [CrossRef] [PubMed]

65. Yu, X.; Long, Y.C. Crosstalk between cystine and glutathione is critical for the regulation of amino acid signaling pathways and ferroptosis. *Sci. Rep.* **2016**, *6*, 30033. [CrossRef]

66. Dixon, S.J.; Lemberg, K.M.; Lamprecht, M.R.; Skouta, R.; Zaitsev, E.M.; Gleason, C.E.; Patel, D.N.; Bauer, A.J.; Cantley, A.M.; Yang, W.S.; et al. Ferroptosis: An Iron-Dependent Form of Nonapoptotic Cell Death. *Cell* **2012**, *149*, 1060–1072. [CrossRef]

67. Borges-Santos, M.D.; Moreto, F.; Pereira, P.C.M.; Ming-Yu, Y.; Burini, R.C. Plasma glutathione of HIV+ patients responded positively and differently to dietary supplementation with cysteine or glutamine. *Nutrition* **2012**, *28*, 753–756. [CrossRef]

68. Fan, X.; Monnier, V.M.; Whitson, J. Lens glutathione homeostasis: Discrepancies and gaps in knowledge standing in the way of novel therapeutic approaches. *Exp. Eye Res.* **2017**, *156*, 103–111. [CrossRef]

69. Mandal, P.K.; Shukla, D.; Tripathi, M.; Ersland, L. Cognitive Improvement with Glutathione Supplement in Alzheimer's Disease: A Way Forward. *J. Alzheimer's Dis.* **2019**, *68*, 531–535. [CrossRef]

70. Teskey, G.; Abrahem, R.; Cao, R.; Gyurjian, K.; Islamoglu, H.; Lucero, M.; Martinez, A.; Paredes, E.; Salaiz, O.; Robinson, B.; et al. Glutathione as a Marker for Human Disease. *Adv. Clin. Chem.* **2018**, *87*, 141–159. [CrossRef]

71. Lian, G.; Gnanaprakasam, J.R.; Wang, T.; Wu, R.; Chen, X.; Liu, L.; Shen, Y.; Yang, M.; Yang, J.J.; Chen, Y.; et al. Glutathione de novo synthesis but not recycling process coordinates with glutamine catabolism to control redox homeostasis and directs murine T cell differentiation. *eLife* **2018**, *7*. [CrossRef] [PubMed]

Publisher's Note: MDPI stays neutral with regard to jurisdictional claims in published maps and institutional affiliations.

Article

In Vivo Anti-Inflammation Potential of *Aster koraiensis* Extract for Dry Eye Syndrome by the Protection of Ocular Surface

Sung-Chul Hong [1], Jung-Heun Ha [2,3], Jennifer K. Lee [4], Sang Hoon Jung [5,6] and Jin-Chul Kim [1,*]

[1] Natural Informatics Research Center, Korea Institute of Science and Technology (KIST), Gangneung 25451, Korea; schong@kist.re.kr

[2] Research Center for Industrialization of Natural Neutralization, Dankook University, Cheonan 31116, Korea; ha@dankook.ac.kr

[3] Department of Food Science and Nutrition, Dankook University, Cheonan 31116, Korea

[4] Food Science & Human Nutrition Department, University of Florida, Gainesville, FL 32611, USA; leejennifer@ufl.edu

[5] Natural Product Research Center, Korea Institute of Science and Technology (KIST), Gangneung 25451, Korea; shjung@kist.re.kr

[6] Division of Bio-Medical Science and Technology, KIST School, Korea University of Science and Technology (UST), Daejeon 34113, Korea

[*] Correspondence: jckim@kist.re.kr; Tel.: +82-33-650-3515

Received: 23 September 2020; Accepted: 17 October 2020; Published: 23 October 2020

Abstract: Dry eye syndrome (DES) is a corneal disease often characterized by an irritating, itching feeling in the eyes and light sensitivity. Inflammation and endoplasmic reticulum (ER) stress may play a crucial role in the pathogenesis of DES, although the underlying mechanism remains elusive. *Aster koraiensis* has been used traditionally as an edible herb in Korea. It has been reported to have wound-healing and inhibitory effects against insulin resistance and inflammation. Here, we examined the inhibitory effects of inflammation and ER stress by *A. koraiensis* extract (AKE) in animal model and human retinal pigmented epithelial (ARPE-19) cells. Oral administration of AKE mitigated DE symptoms, including reduced corneal epithelial thickness, increased the gap between lacrimal gland tissues in experimental animals and decreased tear production. It also inhibited inflammatory responses in the corneal epithelium and lacrimal gland. Consequently, the activation of NF-κB was attenuated by the suppression of cyclooxygenase-1 (COX-1) and cyclooxygenase-2 (COX-2). Moreover, AKE treatment ameliorated TNF-α-inducible ocular inflammation and thapsigargin (Tg)-inducible ER stress in animal model and human retinal pigmented epithelial (ARPE-19) cells. These results prove that AKE prevents detrimental functional and histological remodeling on the ocular surface and in the lacrimal gland through inhibition of inflammation and ER stress, suggesting its potential as functional food material for improvement of DES.

Keywords: functional food; dry eye; *Aster koraiensis*; animal model and human retinal pigmented epithelial (ARPE-19) cells

---

1. Introduction

Dry eye (DE) has been identified as a common disease of the eye, which is induced by the failure of tear production and tear retention in the ocular surface [1]. Dry eye syndrome (DES), also known as keratoconjunctivitis sicca, is primarily caused by genetic factors, autoimmunity, and external environmental insults. A previous cohort study estimated that approximately 14.5% of the US population suffers from DE [2]. DE is mainly caused by the increased osmolality of the tear film and immune response on the ocular surfaces [3]. When tears lose their integrity or excessively evaporate, the tear film becomes thinner and unstable [4]. The weakened tear film results in damage with increased inflammatory responses on ocular surfaces. The tear film is the primary defensive line of the eye against foreign pathogens and abiotic factors, and it also maintains ocular homeostasis [5,6].

Furthermore, inflammation induced by an imbalance of ocular homeostasis triggers the recruitment of pro-inflammatory immune cells. Consequently, the lacrimal unit remodels its structures and the tear film properties [7–9]. Growing evidence suggests that DE-related ocular surface inflammation is mediated by immune cells [10]. Infiltrating inflammatory cells from peripheral circulatory blood were determined to secrete inflammatory cytokines such as interleukin (IL)-1, 6, 8, and 10 and tumor necrosis factor (TNF)-α on ocular surfaces. These cytokines degrade the corneal epithelial barrier and further induce apoptosis in the conjunctiva and lacrimal gland. Thus, an inflammatory response impedes the integrity of the ocular surfaces and accelerates the progression of DE [11–13].

The endoplasmic reticulum (ER) is a cellular organelle located in the cytoplasm. The ER has been known to metabolize proteins and lipids (biosynthesis, packaging, and secretion), and it stores calcium ions, which play a role as a cellular signal regulator. ER stress occurs by protein misfolding, including glycosylation or disulfide bond formation, protein overexpression, or mutations in protein s [14]. Severe intracellular Ca^{2+} dysregulations can promote cell death through apoptosis [15]. ER stress has received growing attention in many pathophysiological disorders, such as cardiovascular diseases, neurodegenerative diseases, inflammatory bowel disease, and rheumatoid arthritis [16]. Recently, it was suggested that ER stress is also related to DES [17].

Conventional therapeutic agents such as artificial tears, anti-inflammatory drugs, and corticosteroids have been determined to help DE patients in mitigating uncomfortable ocular symptoms and improving their clinical and pathological conditions for the short term [18–20]. However, this modality provides only temporary symptomatic relief, and up to two-thirds of DES sufferers complain of persisting symptoms despite such treatments [21]. Due to the limitations of medical treatment, a new therapeutic approach for DES is needed. An alternative method that can be presented is improvement through ingestion of functional food, which is a non-medical treatment method. Functional food derived from natural products was reported to have low side effects and to be safe for long-term intake [22–24]. In particular, it has preventive effects such as prevention of recurrence by improving the physical constitution during long-term ingestion [25]. In this study, *Aster koraiensis* extract (AKE) was selected as a candidate material of functional food to improve DES symptoms. *A. koraiensis*, also known as Korean starwort, is an herbaceous perennial plant of the Asteraceae family. Because *A. koraiensis* is widely distributed in most regions of Korea [26], it has been used as traditional herbal medicine and food item in the country [27,28]. *A. koraiensis* has been reported to inhibit diabetic inflammation [29,30] and enhance wound healing on the skin and during bronchitis [31]. Moreover, chlorogenic acid and 3,5-di-O-caffeoylquinic acid, both isolated from *A. koraiensis*, have been used to relieve symptoms of diabetic-related diseases [32,33].

In this study, we produced an extract to examine the improving effect of *A. koraiensis*, a medical edible herb crop in Korea, on DE. The functionality of *A. koraiensis* extract was evaluated in an animal model in which scopolamine-induced DE. To investigate the detailed mechanism of action, we confirmed the possibility of improving DE in *A. koraiensis* via animal tissues and human retinal pigmented epithelial (ARPE-19) cells. In this way, we try to prove the value that *A. koraiensis* can be used as a functional food material.

2. Materials and Methods

2.1. Plant Material and Preparation of A. koraiensis Extracts

A. koraiensis was collected from Jeongseon-gun, Gangwon-do, Republic of Korea, in September 2016. The plant's aerial parts, including the flowers and leaves, were thoroughly washed with water to remove impurities and were further dried in the shade for 1 month. The medicinal plant mixture was extracted from the dried plants with ethanol (EtOH) by maceration at room temperature for

3 days. *A. koraiensis* ethanol extracts (AKE) were combined and concentrated by evaporation in a rotary evaporator at 60–70 °C. The EtOH extracts were then freeze-dried.

2.2. Antioxidative Activities of AKE

The antioxidative capacity of AKE related to either electron or radical scavenging was scrutinized with three different analytical methods and backgrounds; (1) A 1,1-diphenyl-2-picrylhydrazyl (DPPH), (2) the ferric reducing antioxidant power (FRAP), and the Trolox equivalent antioxidant capacity (TEAC). DPPH assay was performed to assess the stable DPPH radical generating capacity by AKE. FRAP assay was tested to understand reducing activity from ferric to ferrous iron by AKE. TEAC assay was intended to assess radical scavenging cation ABTS+ (2,2'-azinobis (3-ethylbenzothiazoline-6-sulfonic acid) by radical quenching or electron donation. Each evaluation was performed through the following experimental methods.

A 1,1-diphenyl-2-picrylhydrazyl (DPPH) assay was conducted as described by Serpen et al. [34] and Thaipong [35], with slight modifications. Briefly, 0.1 mL of 200 M DPPH reagent (Sigma-Aldrich Co., St. Louis, MO, USA) was added to 0.1 mL of each sample in 96-well plates. After incubation in the dark for 30 min, the absorbance was measured at 520 nm using a microplate reader (Spectramax M2E; Thermo Fisher Scientific, Waltham, MA, USA). Ascorbic acid (Sigma-Aldrich) as a DPPH-scavenging compound was used as a standard. Assay results were expressed in mg ascorbic acid/g.

The ferric reducing antioxidant power (FRAP) was assessed via the method developed by Benzie and Strain [36]; however, it had slight modifications. Briefly, acetic acid buffer (pH 3.6, 23 mM) was made by dissolving sodium acetate (Sigma-Aldrich) in acetic acid (Sigma-Aldrich). A 10 mM solution of 2,4,6-tripyridyl-s-triazine (TPTZ) was made by mixing 40 mM HCl (Sigma-Aldrich) and TPTZ (Sigma-Aldrich). A total of 25 µL of the AKE extract was then added to the FRAP reagent (acetic acid, 10 mM; TPTZ, 20 mM; $FeCl_3 \cdot 6H_2O$ = 10:1:1) and incubated in the dark at 37 °C for 15 min. The absorbance was measured at 593 nm using a microplate reader. $FeSO_4$ (Sigma-Aldrich) was employed as a standard in this assay. Assay results were expressed in nM $FeSO_4$/mL.

The Trolox equivalent antioxidant capacity (TEAC) was measured as described by Oki et al. [37] and Re et al. [38], with minor modifications. Briefly, 200 µL of 2,2'-azino-bis (3-ethylbenzothiazoline -6-sulfonic acid; ABTS) reagent (Sigma-Aldrich) was added to 10 µL of each sample in 96-well plates. After a 60-min incubation in the dark, the absorbance was measured at 405 nm using a microplate reader. The reaction rate was calibrated using Trolox equivalent (TE). Assay results were expressed in g TE/mL.

2.3. Total Polyphenol and Flavonoid Contents of AKE

Total polyphenol content was measured using the method described by Alves et al. [39], with minor modifications. Briefly, 10 µL of Folin–Denis reagent (Sigma-Aldrich) was added to 160 µL of each sample in 96-well plates. After 8 min, 30 µL of sodium carbonate (Showa Chemical Industry, Tokyo, Japan) was added to the mixture. After incubation in the dark for 2 h, the absorbance was measured at 765 nm using the Spectramax M2E microplate reader. Gallic acid (Sigma-Aldrich) was also used as the standard curve to extrapolate gallic acid equivalent (GAE). Results were expressed in mg GAE/mL.

Total flavonoid content was analyzed using slight modifications of the methods previously described by Pourmorad et al. [40] and Marinova et al. [41]. In brief, 400 µL of each sample was mixed with 1200 µL of EtOH (Junsei Chemical Co., Tokyo, Japan) and 240 µL of distilled water. Subsequently, 80 µL of 10% aluminum nitrate (Sigma-Aldrich) was added, followed by 80 µL of 1.0 M potassium acetate (Sigma-Aldrich). The mixture was allowed to stand in the dark for 40 min, and the absorbance was measured at 415 nm using the microplate reader. Quercetin (Sigma-Aldrich) was used to construct the standard curve against quercetin equivalents (QE), and the assay results were expressed in mg QE/mL.

2.4. Animal Experiment and Induction of DE Model

All procedures received approval from the Institutional Animal Care and Use Committee (IACUC) of Korea Institute of Science and Technology (KIST): KIST No. 2020-002, Gangneung Institute, and they were performed according to the Association for Research in Vision and Ophthalmology (ARVO) statement for the Use of Animals in Ophthalmic and Vision Research. Mice were accommodated in the animal room with air conditioning, temperatures of 22 ± 2 °C, humidity of 50 ± 10%, and 12 h light/12 h dark circadian cycles. Food and water were supplied ad libitum. Mice were acclimatized for 1 week and were later assigned into five groups composed of seven male 6-week-old BALB/c mice. Experimental DE in mice was achieved by twice-daily intraperitoneal (i.p.) injection of 200 μL (2.5 mg/mL) of phosphate buffered saline (PBS)-diluted scopolamine (Sigma-Aldrich). Groups of mice were orally administered with AKE once per day at concentrations of 0 (as vehicle control), 10, 50, or 100 mg/kg in 200 μL EtOH. In the control group, only PBS buffer was injected without scopolamine. In the case of AKE, 0 mg/kg in 200 μL EtOH was administered when orally administered as in the DE group. After 2 weeks, tear production was quantified by a standard Schirmer's test strip placed in the lower one-third of the temporal eyelid before the eye was closed for 1 min. After the strip was removed, the length of the wet point was measured in millimeters in order to determine the Schirmer's test value. For measurements of the tear breakup time (TBUT), 5 μL of sodium fluorescein was instilled into the eye and photographed. Corneal surface staining was performed to assess the extent of corneal surface changes. The corneal surface was observed and scored after the administration of one drop of 3% fluorescein (Sigma-Aldrich, St. Louis, MO, USA) into the inferior lateral conjunctival sac. The staining of the cornea was evaluated in a blinded manner. Mice are euthanized through cervical dislocation. When removing the cornea, click the mouse's eyelid and use forceps to cut out the eyeball's optic nerve that pops out and removes the cornea. After removing the crystalline lens inside the cornea, store in a freezer at −80 °C. After pulling the mouse's lower jaw with forceps for removal of the lacrimal gland, cut off the epidermis of the pulled part so that the lower part of the mouse's face can be seen. A lacrimal gland located near the lower jaw of the mouse can be secured. The secured lacrimal gland tissue is stored frozen at −80 °C.

2.5. Histology

The corneal epidermal tissue and lacrimal gland tissues were collected, fixed in 10% formaldehyde, and processed for paraffin embedding and sectioning. Sections were stained with hematoxylin and eosin (H & E) and examined a microscope (TE-2000U, Nikon, Tokyo, Japan) at X40. Central corneal epithelial thickness was evaluated in 5 sections for each cornea.

2.6. Western Blot Analysis

Total protein lysate was extracted from the corneal epidermal tissue and lacrimal glands of each mouse group and from APRE-19 cells after washing with ice-cold PBS (Invitrogen, Carlsbad, CA, USA) three times. Radioimmune precipitation assay (RIPA) buffer (Pierce, Rockford, IL, USA) was applied to the samples with protease inhibitors (Cell Signaling Technology, Danvers, MA, USA) and phosphatase inhibitor (PMSF, Thermo Fisher Scientific, Waltham, MA, USA) at 4 °C. After lysis, the supernatant was collected after centrifugation at 12,000× g for 15 min at 4 °C. Protein lysates were loaded and separated using SDS-PAGE gels (Bio-RAD, Hercules, CA, USA) and transferred to polyvinylidene difluoride (PVDF) membranes (Bio-RAD) using a wet-transfer method at 4 °C. Membranes were blocked with 5% non-fat milk in tris-buffered saline and Tween-20 (TBS-T) for 1 h to prevent nonspecific binding. Blots were incubated with primary and secondary antibodies at room temperature, developed with chemiluminescence reagent (Pierce), and detected and analyzed using LAS-4000 (General Electric Image Quant LAS 4000 Biomolecular Imager; GE Healthcare, Chicago, IL, USA) in grayscale. Experimental protein band intensity on blots was normalized to the intensity of glyceraldehyde 3-phosphate dehydrogenase (GAPDH), which did not vary significantly by treatment. Primary and secondary antibodies are summarized in Table 1.

Table 1. Antibodies for western blot analysis.

	Antibody	Dilution Factor	Corporation
	phospho-p65	1:1000	Cell signaling
	p65	1:1000	Cell signaling
	phospho-ERK	1:1000	Cell signaling
	ERK	1:1000	Santa Cruz
	phospho-JNK	1:1000	Cell signaling
	JNK	1:1000	Santa Cruz
	phospho-p38	1:1000	Cell signaling
Primary antibody	p38	1:1000	Santa Cruz
	phospho-AMPK	1:1000	Cell signaling
	AMPK	1:1000	Santa Cruz
	phospho-IκB	1:1000	Cell signaling
	IκB	1:1000	Santa Cruz
	COX-1	1:1000	Cell signaling
	COX-2	1:1000	Cell signaling
	GAPDH	1:2000	Cell signaling
Secondary antibody	Goat anti-mouse-HRP	1:2000	Santa Cruz
	Goat anti-rabbit-HRP	1:5000	Thermo scientific

ERK, extracellular signal regulated kinase; JNK, c-Jun N-terminal kinase; AMPK, 5′ adenosine monophosphate -activated protein kinase; IκB, inhibitor of nuclear factor kappa B; COX-1, cyclooxygenase-1; COX-2, cyclooxygenase-2; GAPDH, glyceraldehyde 3-phosphate dehydrogenase; HRP, horseradish peroxidase.

2.7. Real-Time PCR (RT-qPCR)

Total RNA was isolated from corneal epidermal tissue, lacrimal glands, and ARPE-19 cells with the RNase mini kit (Qiagen, Valencia, CA, USA). Genomic DNA was removed by digestion with DNase I (Qiagen). Reverse transcription was performed using 1 μg of mRNA per sample with the RevertAid First Strand cDNA Synthesis Kit (Thermo Fisher Scientific, Waltham, MA, USA). Gene expression was assessed with a SYBR Green PCR mixture with gene-specific oligonucleotide primers using an AB 7500 real-time PCR machine (Thermo Fisher Scientific, Waltham, MA, USA). Primer sequences and parameters are described in Table 2. The expression of the target genes was normalized to the expression of β-Actin and glyceraldehyde 3-phosphate dehydrogenase (GAPDH), which was not significantly altered by treatments.

Table 2. RT-qPCR primer sequences (5′ to 3′).

Transcript	Forward Primer	Reverse Primer	Annealing Temp. (°C)
IL-1β	TCATTGTGGCTGTGGAGAAG	GGTGTGCCGTCTTTCATTAC	53.8
TNF-α	CTCAGATCATCTTCTCAA	CAGAGCAATGACTCCAAA	55.1
TGF-β	GAAAGCCCTGTATTCCGTCTCCTT	CAACAATTCCTGGCGTTACCTTGG	53.8
IFN-γ	AGCGGCTGACTGAACTCAGATTGTA	GTCACAGTTTTCAGCTGTATAGGG	61.2
IL-6	TTCCCTACTTCACAAGTC	GGTTTGCCGAGTAGATCT	52.9
IL-23	CAAGCAGAACTGGCTGTTGTC	GCACCAGCGGGACATATGAA	60.2
MMP-9	CACAACCGACGACGACGAGTTGTG	CTGTGGTGAGGCCGAATAG	65.0
β-Actin	TTGTTACCAACTGGGACGACATGG	GATCTTGATCTTCATGGTGCTAGG	59.2
GAPDH	ATGGTGAAGGTCGGTGTG	ACCAGTGGATGCAGGGAT-	58.0

IL-1β, interleukin 1 beta; TNF-α, Tumor necrosis factor alpha; TGF-β, Transforming growth factor beta; IFN-γ, Interferon gamma; IL-6, interleukin 6; IL-23, interleukin 23; MMP-9, Matrix metallopeptidase 9.

2.8. ARPE-19 Cell Culture

Human retinal epithelial ARPE-19 cells (American Type Culture Collection, ATCC) were cultured in DMEM/F-12 media (Gibco, Carlsbad, CA, USA) containing 10% FBS (HyClone Laboratories, Logan, UT, USA) and 1% penicillin/streptomycin (HyClone Laboratories). To induce an inflammatory and endoplasmic response in ARPE-19 cells, 2×10^5 cells per well were cultured in 6 well plates. After culturing for 24 h, 200 µL media solution with 10 µg/mL of TNF-α or 5 µmol/L thapsigargin (Tg) was applied to the cells in the presence or absence of AKE (0, 0.1, 1, and 10 µg/mL) for 12 h. In the Control group case, the experiment was conducted with doubles without any inflammation-inducing factors or AKE.

2.9. Intracellular Calcium Release

Intracellular calcium $[Ca^{2+}]_i$ level was measured using a Fluo-4 NW calcium assay kit (F36206; Thermo Fisher Scientific, Waltham, MA, USA), according to the manufacturer's protocol. Briefly, 1×10^5 ARPE-19 cells per well were cultured in 96 well plates for 24 h, and ER stress was induced by the treatment with Tg and AKE (0, 0.1, 1, and 10 µg/mL) for 12 h. Cells were then incubated with a cell-permeable calcium indicator (Flow 4 A) for 1 hr before treatment of Tg (final concentration, 5 µmol/L). The $[Ca^{2+}]_i$ levels were accessed by measuring the fluorescent intensity using a microplate spectrophotometer (Bio-Tek Power Wave XS, Winooski, VT, USA). Image software (Bio-Tek software, Gen5) was applied for the subsequent quantitative analysis.

2.10. VEGF-α Secretion

1×10^5 ARPE-19 cells per well were cultured in 96 well plates for 24 h. In order to measure the vascular endothelial growth factor (VEGF)-α secretion, Tg with AKE (0, 0.1, 1, or 10 µg/mL) were dissolved in the medium and treated with 100 µL each, incubated for 24 h, and then doubled and collected. VEGF-α protein secretion in the media was quantified using a commercial ELISA kit (BMS277-2; Invitrogen) as per the manufacturer's instructions using a microplate spectrophotometer (Bio-Tek Power Wave XS, Winooski, VT, USA).

2.11. Statistical Analysis

Results are presented as means ± standard deviation (SD). Data were analyzed statistically using one-way ANOVA with Tukey's post hoc analysis. For $[Ca^{2+}]_i$ determination, two-way ANOVA with Tukey's post hoc test was used. Data were analyzed using a two-factor ANOVA test. If analytical results showed significant time X treatment interactions ($p < 0.05$), Tukey's post hoc multiple comparisons test was applied. $p < 0.05$ was defined as statistically significant and was further indicated by a filled asterisk or number sign. All statistical analyses were performed using the GraphPad Prism 8 (San Diego, CA, USA).

3. Results

3.1. Antioxidative Effects and Polyphenol and Flavonoid Contents of AKE

Plants that are rich in secondary metabolites, including phenolics and flavonoids, have been identified to possess antioxidant properties afforded by their chemical structures and redox potentials [42]. As antioxidant activity is multifactorial and associated with several mechanisms [43], we performed three complementary tests to measure the antioxidant activity of AKE's DPPH radical scavenging activity, ferric reducing antioxidant power (FRAP), and Trolox equivalent antioxidant capacity (TEAC). The AKE was prepared at different concentrations: 0.1, 0.5, 1, 5, and 10 mg/mL. As shown in Table 3, the DPPH radical scavenging activity of AKE at concentrations of 10, 5, and 1 mg/mL was determined to be significantly higher than at concentrations of 0.5 and 0.1 mg/mL (10 ≒ 5 ≒ 1 mg/mL > 0.5 ≒ 0.1 mg/mL). FRAP and TEAC were increased as the concentration of AKE increased in a dose-dependent manner. The total polyphenol content increased dose-dependently as the concentration of AKE increased (Table 4). The total flavonoid content increased as the concentration increased in the following order: 10 mg/mL > 5 mg/mL > 1 ≒ 0.5 ≒ 0.1 mg/mL). Thus, the amount of polyphenol and flavonoid is correlated with the antioxidant capacities. Moreover, the antioxidant activity is associated with a high concentration of extracts.

Table 3. DPPH radical scavenging activity, FRAP, and TEAC of *A. koraiensis* ethanol extracts (AKE).

	0.1 mg/mL	0.5 mg/mL	1 mg/mL	5 mg/mL	10 mg/mL
DPPH radical scavenging activity (mg ascorbic acid/g)	30.8 ± 2.55 [b]	80.0 ± 3.05 [b]	112.7 ± 3.02 [a]	117.8 ± 0.58 [a]	118.4 ± 0.55 [a]
FRAP (nM FeSO$_4$/mL)	0.2 ± 0.01 [e]	1.0 ± 0.02 [d]	2.0 ± 0.10 [c]	7.8 ± 0.41 [b]	9.7 ± 0.07 [a]
TEAC (g TE/mL)	0.2 ± 0.01 [e]	0.4 ± 0.02 [d]	0.9 ± 0.10 [c]	2.7 ± 0.77 [b]	2.9 ± 0.01 [a]

AKE, *A. koraiensis* ethanol extracts; DPPH, 2,2-diphenyl-1-picrylhydrazyl; FRAP, ferric reducing antioxidant power; TEAC, Trolox equivalent antioxidant capacity; TE, Trolox equivalent. Values are means ± SD, n = 4. Data were analyzed by one-way ANOVA analysis followed by Tukey's posthoc test. Means labeled without a common letter differ, $p < 0.001$ (DPPH radical scavenging activity) and $p < 0.01$ (FRAP and TEAC).

Table 4. Total polyphenol and flavonoid contents of *A. koraiensis* ethanol extracts (AKE).

	0.1 mg/mL	0.5 mg/mL	1 mg/mL	5 mg/mL	10 mg/mL
Total polyphenol content (mg GAE/mL)	0.2 ± 0.00 [e]	0.6 ± 0.01 [d]	0.9 ± 0.01 [c]	1.2 ± 0.01 [b]	2.6 ± 0.03 [a]
Total flavonoids content (mg QE/mL)	2.2 ± 0.58 [c]	5.8 ± 1.37 [c]	15.5 ± 0.61 [c]	32.7 ± 1.51 [b]	336.9 ± 11.06 [a]

AKE, *A. koraiensis* ethanol extracts; GAE, 2 gallic acid equivalent; QE, quercetin equivalents; Values are means ± SD, n = 4. Data were analyzed by one-way ANOVA analysis followed by Tukey's post hoc test. Means labeled without a common letter differ, $p < 0.01$ (polyphenol and flavonoid).

3.2. Effects of Aster koraiensis Ethanol Extracts on Eye Damage and Tear Production

In the scopolamine-induced mouse model of DE, scopolamine was observed to trigger the breakup of tear film, decrease tear production, irritate the lacrimal gland, and shrink the corneal epithelial cells [5]. To identify the effect of the AKE on DE, we administered AKE (0, 10, 50, or 100 mg/kg) orally once per day for 14 days to groups of mice with experimental DE. Eyes were stained with fluorescein in order to observe eye injury quantitatively. Under blue light, a green dot indicated the level of cornea damage (Figure 1A, black arrows). Quantitative analysis showed that AKE moderately ameliorated corneal damage in the DE mice in a dose-dependent manner (Figure 1B). The typical symptoms of DE include the quick dissipation of the tear film and a reduced amount of tear production [44]. To examine the mechanism by which AKE treatment inhibits corneal damage, tear breakup time (TBUT) and tear volume were observed. In DE mice, TBUT was reduced significantly, and AKE treatment at 100 mg/kg presented a reverse TBUT (Figure 1C). Additionally, the average volume of tear secretion was quantified using Schirmer's test. AKE treatment moderately enhanced tear production in a dose-dependent manner (Figure 1D). Taken together, these data show that AKE inhibited scopolamine-inducible corneal damage with increases in tear production and tear film stability.

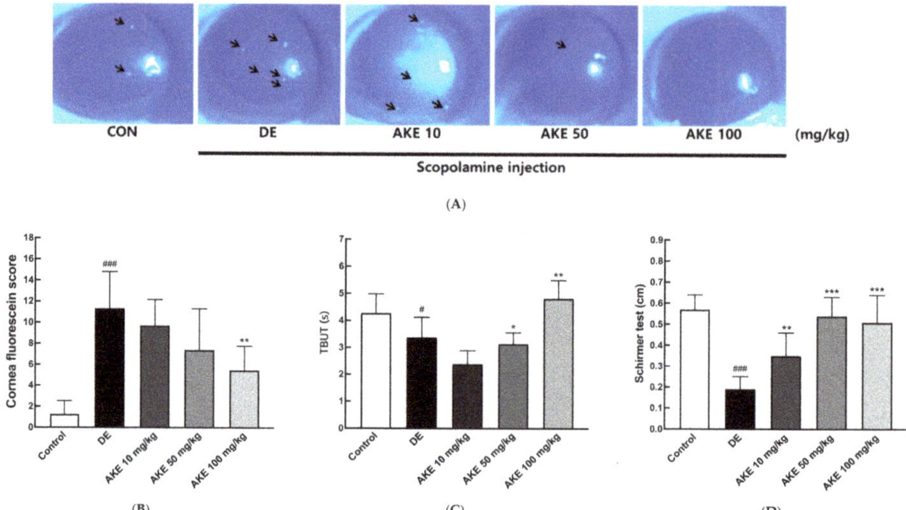

Figure 1. Effects of *A. koraiensis* ethanol extracts (AKE) on eye damage and tear production. Dry eye (DE) was reportedly induced by scopolamine injection, and AKE was administered at 0 (control; CON which was not induced by scopolamine injection, DE), 10 (AKE 10), 50 (AKE 50), or 100 mg/kg (AKE 100). (**A**) Representative images of corneal fluorescein staining. Black arrow indicates ocular damage spots of corneal fluorescein staining. (**B**) Quantitative analysis of images in A. (**C**) Tear breakup time (TBUT) was measured in seconds and analyzed. (**D**) Tear volume was measured using Schirmer's test. In the graph, each bar represents mean ± SD of n = 7 mice per group. [#] $p < 0.05$, [###] $p < 0.001$ versus CON; * $p < 0.05$, ** $p < 0.01$, *** $p < 0.001$ versus DE. Data were analyzed statistically using one-way ANOVA followed by Tukey's post hoc test.

3.3. Histological Alterations of the Corneal Epithelial and Lacrimal Glands Following AKE Treatment in the DE Mouse Model

When DE symptoms develop, the corneal epithelial tissue thins to a delicate slim stratum [45]. In order to evaluate the effect of AKE on corneal epithelium from DE mice, we measured the thickness of corneal epithelial cells by histology. In Figure 2A, the dark pink area represents the corneal epithelial layer. This was significantly reduced in the DE group compared with the control (CON) group. AKE treatment inhibited the thinning of the corneal epithelial layer at doses of 50 and 100 mg/kg (Figure 2A,B).

Damage to the lacrimal glands is another general symptom of DE [46,47]. Lacrimal gland histopathology during AKE treatment of DE mice revealed a decrease in the number of infiltrating immune cells in the tissue, compared with the DE mice without AKE treatment (Figure 3C). Additionally, gaps in the glands were narrowed by the administration of AKE (Figure 3C). These results suggest that AKE is beneficial to corneal epithelial cells and lacrimal glands weakened during DE. Because these immune infiltrates are found to be common during DE, we next elucidated immune regulation by AKE treatment in experimental DE.

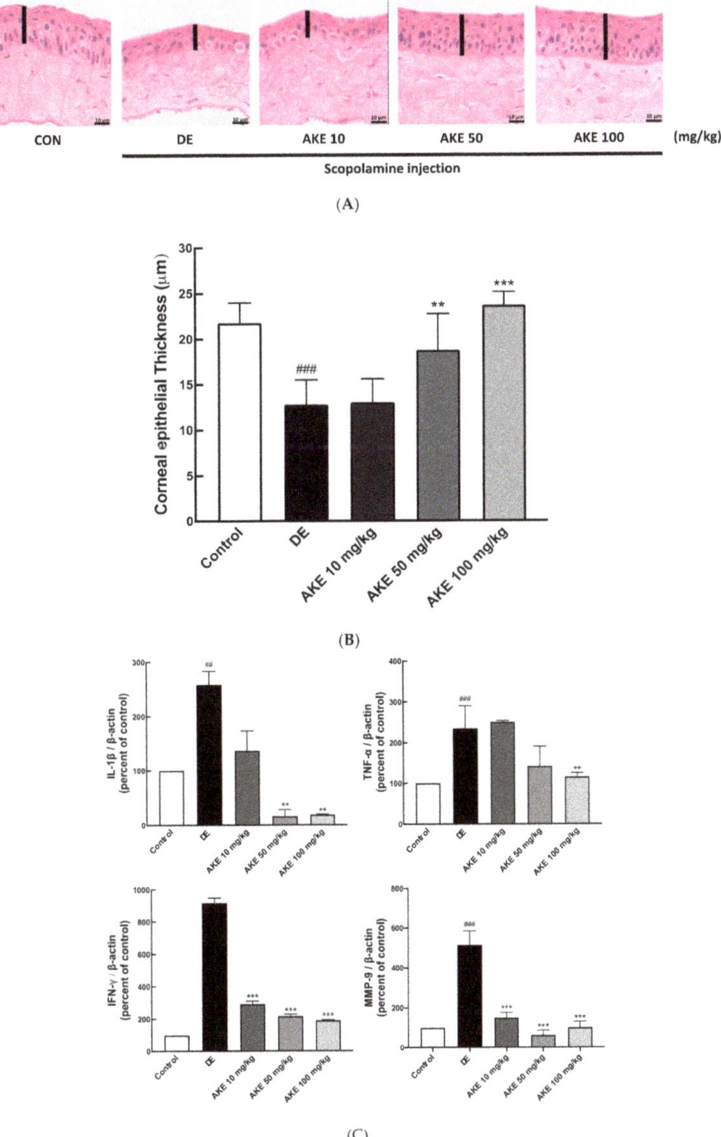

Figure 2. Effects of *A. koraiensis* ethanol extracts (AKE) on the corneal epithelial thickness and inflammatory cytokine expression. Dry eye (DE) was reportedly induced by scopolamine injection, and AKE was administered at 0 (control; CON which was not induced by scopolamine injection, DE), 10 (AKE 10), 50 (AKE 50), or 100 mg/kg (AKE 100). (**A**) Representative H and E staining images of histological sections of corneas. (**B**) Quantitative analysis of A. (**C**) Expression of inflammatory cytokines was accessed by real-time PCR (RT-qPCR) for each target gene (IL-1β, TNF-α, IFN-γ, and MMP-9). β-actin was used as an internal control. Each bar represents the mean ± SD of *n* = 7 mice per group. [##] *p* < 0.01, [###] *p* < 0.001 versus CON; ** *p* < 0.01, *** *p* < 0.001 versus DE. Data were analyzed statistically using one-way ANOVA followed by Tukey's post hoc test. IL-1β, interleukin 1 beta; TNF-α, Tumor necrosis factor alpha; IFN-γ, Interferon gamma; MMP-9, Matrix metallopeptidase 9.

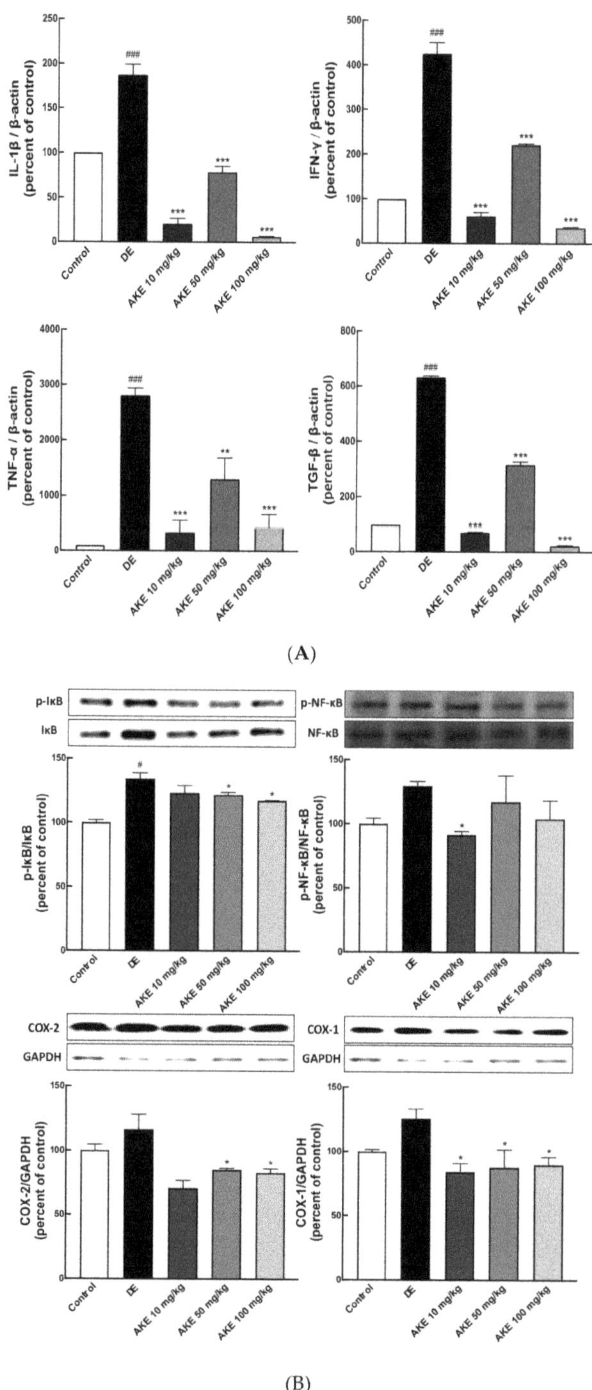

(**A**)

(B)

Figure 3. *Cont.*

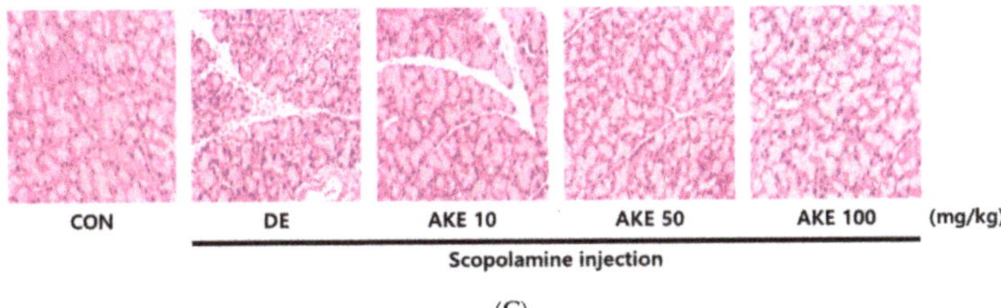

Figure 3. Effects of *A. koraiensis* ethanol extracts (AKE) on the inflammatory response and gap between tissues in the lacrimal gland. Dry eye (DE) was reportedly induced by scopolamine injection, and AKE was administered at 0 (control; CON which was not induced by scopolamine injection, DE), 10 (AKE 10), 50 (AKE 50), or 100 mg/kg (AKE 100). (**A**) Expression of inflammatory cytokines was accessed using RT-qPCR for each target gene (IL-1β, TNF-α, IFN-γ, and Transforming growth factor beta (TGF-β)). β-Actin was used as an internal control. (**B**) Inflammatory proteins were analyzed by western blot (WB). (Upper panels) Representative figures of WB for phosphorylated inhibitor of nuclear factor kappa B (IkB) and total IkB, phosphorylated nuclear factor kappa B (NF-κB), and total NF kB were targeted. Quantitative analysis revealed the ratio of phosphorylated protein/total protein. (Lower panels) The representative figures and their analyses of WB for cyclooxygenase (COX)-1 and COX-2. GAPDH was used for the loading control. (**C**) Representative H & E-stained histological sections of lacrimal glands. The white area in the image shows the gaps between tissues in the lacrimal gland. Each bar represents the mean ± SD of n = 7 mice per group. [#] $p < 0.05$, [###] $p < 0.001$ versus CON; * $p < 0.05$, *** $p < 0.001$ versus DE. Data were analyzed statistically using one-way ANOVA followed by Tukey's post hoc test.

3.4. AKE Suppressed Immune Responses in the Corneal Epithelium and Lacrimal Glands of DE Model Mice

Inflammatory responses are closely related to the integrity of the ocular surfaces and structure of ocular units [7,13]. To investigate the role of an immune response during the pathology of DE, the relative mRNA expression levels of pro-inflammatory cytokines (interleukin 1 beta (IL-1β), Tumor necrosis factor alpha (TNF-α), Interferon gamma (IFN-γ), and Matrix metallopeptidase 9 (MMP-9)) were assessed in the corneal tissue and lacrimal gland from DE mice. All pro-inflammatory cytokines were found to be upregulated in the DE mice compared with the control mouse group. In the corneal tissues, AKE treatment inhibited the expression of each of these genes. In particular, IFN-γ and MMP-9 were remarkably attenuated, even at a low dose (10 mg/kg) of AKE (Figure 2C, Figure 3A). In lacrimal glands, AKE suppressed the expression of inflammatory cytokines IL-1β, TNF-α, IFN-γ, and TGF-β (Figure 3A). Especially, when 50 mg/kg of AKE was administered, cytokine expression did not decrease in proportion to the concentration in lacrimal gland tissues. However, AKE treatment experiment groups were found to reduce compared to the DE group that induced DE. Taken together, these data suggest that AKE treatment protected the eyes of the DE mice via effective anti-inflammatory activity in the corneal epithelium and lacrimal glands.

3.5. AKE Inhibited Inflammatory Response in Lacrimal Gland

To examine the molecular signaling pathway to be regulated by pro-inflammatory cytokines, the IκB/nuclear factor kappa B (NF-κB) axis, one of the main inflammations signaling pathways, was studied using western blot analysis of lacrimal gland tissue. Inactive NF-κB was observed to locate in the cytosol, bound with its inhibitor IκB. Phosphorylation of IκB by IκB kinases (IKKs) results in the degradation of IκB. Subsequently, NF-κB translocates to the nucleus, where it regulates gene expression concerning inflammation [48]. Optimal induction of the NF-κB target genes also requires the phosphorylation of NF-κB proteins, such as p65 [49]. AKE treatment has suppressed phosphorylation of IκB-α, implying a decrease in immune reaction. NF-κB has also showed a similar trend to that of IκB-α, but no significant difference between the control group and the DE group (Figure 3B). The target genes of the IκB/NF-κB axis increased in DE mice and were inhibited by AKE treatment (Figure 3B). Therefore, the immune response in lacrimal glands was markedly attenuated by AKE consumption.

3.6. AKE Inhibited Inflammatory Reaction and ER Stress on ARPE-19 Cells

Ocular inflammation has been known to occur not only on the corneal surface and lacrimal gland area in DE, but it also causes retinal damage. Two well-studied retinal damage-induced diseases are uveitis and diabetic retinopathy (DR) [50,51]. Thus, we observed the effect of AKE on a retinal damage model. The ARPE-19 cell line is a widely used cell line for eye study, while TNF-α is a pro-inflammatory cytokine that induces inflammation in ARPE-19 cells [52]. Treatment with AKE prevented TNF-α-inducible inflammation in ARPE-19 cells with attenuation of TNF-α, IL-1β, IL-6, IL-8, and MMP-9 mRNA expression (Figure 4A). In addition, post-translational modification by phosphorylation of mitogen-activated protein kinase (MAPK) was examined, which is a key pathway signal transduction in inflammation. Phosphorylation of p-38 and p-ERK proteins was markedly attenuated in AKE-treated ARPE-19 cells (Figure 4B, Figure S1). By contrast, retinopathy is closely related to increased ocular ER stress activated in the retina and retinal endothelial cells under diabetic and hypoxic conditions [53,54].

Given the potent anti-inflammatory properties of AKE, we asked whether AKE would be effective in preventing other ocular stresses, such as oxidative stress or ER stress. To address this question, we induced ER stress in ARPE-19 cells by treatment with Tg (5 μmol/L) in the presence or absence of AKE. VEGF-α has been identified as a proangiogenic factor involved in the pathophysiology of some ocular diseases with neovascularization [55]. The level of VEGF-α is a general marker of eye disease. At a 10 μg/mL dose, AKE significantly decreased Tg-inducible VEGF-α secretion into the media (Figure 4C). Loss of cellular homeostasis and disruption of Ca^{2+} signaling can lead to the activation of ER stress responses in both the reticular network and cytoplasmic compartments [56]. Tg treatment has remarkably increased $[Ca^{2+}]_i$ in ARPE-19 cells, implying that ER stress has also increased. AKE treatment decreased $[Ca^{2+}]_i$ release from ARPE-19 cells compared with the Tg group, suggesting that AKE may prevent Tg-induced ocular ER stress in ARPE-19 cells (Figure 4D). Taken together, AKE can attenuate inflammatory responses and Tg-inducible ER stress in human retinal epithelial cells.

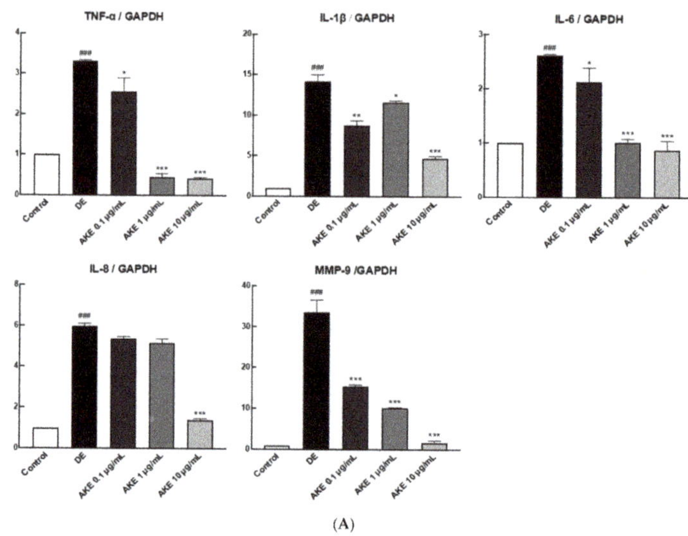

(A)

Figure 4. *Cont.*

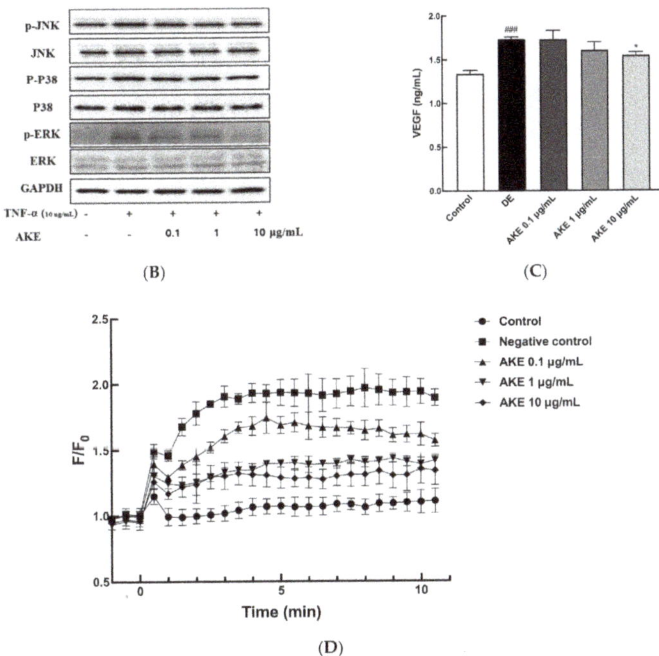

Figure 4. Effects of *A. koraiensis* ethanol extracts (AKE) on ER stress and the inflammatory response in animal model and human retinal pigmented epithelial (ARPE-19) cells. TNF-α-inducible pro-inflammatory gene expression in AKE-treated ARPE-19 cells. To induce an inflammatory reaction, phosphate buffered saline (PBS) or 10 µg/mL of TNF- α was treated, and AKE was treated in a dose-dependent manner (0, 0.1, 1, and 10 mg/mL). (**A**) Inflammatory cytokines were measured by qRT-PCR in order to validate the expression of each target gene (TNF-a, IL-1b, IL-6, IL-8, and MMP-9). glyceraldehyde 3-phosphate dehydrogenase (GAPDH) was used as an internal control. (**B**) The intracellular inflammatory signaling (MAPK) was analyzed using western blot (WB) with specific antibodies such as phospho-c-Jun N-terminal kinase (p-JNK), JNK, p-P38, P38, phospho-extracellular signal regulated kinase (p-ERK), and ERK. GAPDH was used for loading control. (C and D) endoplasmic reticulum (ER) stress was induced in ARPE-19 cells by thapsigargin (Tg), and AKE was treated with 0 (control; CON), 0.1 µg/mL (AKE 0.1), 1 µg/mL treatment (AKE 1), and 10 µg/mL treatment (AKE 10). (**C**) Tg-induced VEGF-α expression was measured using ELISA. (**D**) Tg-induced calcium ion efflux measurement using fluorescence. Data from three independent experiments have been presented as bar graphs showing mean ± SD. ### $p < 0.001$ versus CON; * $p < 0.05$, ** $p < 0.01$, *** $p < 0.001$ versus DE. Data were analyzed statistically using one-way ANOVA, followed by Tukey's test. For $[Ca^{2+}]_i$ determination, two-way ANOVA with repeated measures was used, followed by Tukey's hoc test. A, TC, D) Negative control to DE.

4. Discussion

Dry eye syndrome (DES) has been defined as a chronic eye disease associated with aging, hormonal changes, inflammation, and autoimmune diseases [57]. The eyes of a patient with DE contain upregulation of inflammatory cytokines, such as IL-8, IL-6, IL-1β, TGF-β, and TNF-α, and infiltration of immune cells [58]. Therefore, the inhibition of an inflammatory reaction is one of the clinical strategies in treating DE [59]. Multiple inflammation-suppressive therapeutic agents, including cyclosporine A (immunosuppressive agent), corticosteroids (steroid), and tetracycline (anti-inflammatory drug), have been widely used in clinical practice to treat patients with DES. However, these agents only partly attenuate clinical symptoms; not all agents relieve the complete burden of DE symptoms. Thus, natural sources have been considered that produce fewer adverse side effects and promote easy usage of agents [60,61], and would be an alternative material for the treatment of DE symptoms.

A. koraiensis has been traditionally used as a treatment for diabetes and inflammatory-related diseases in Korea. A previous study has reported that *A. koraiensis* plays a vital role in retinal angiogenesis and oxidative stress-inducible retinopathy [33]. Inflammation-induced by ocular surface stress can lead to an immune response in order to generate additional damage and amplify inflammation [62]. In this study, we investigated whether AKE reduces ocular inflammation and ER stress. Our results demonstrated that oral administration of AKE to DE mice reversed the symptoms as it increased TBUT and improved tear production from lacrimal glands. Moreover, AKE treatment modestly augmented histological remodeling in the corneal epithelium and lacrimal glands from DE mice to a level similar to that of the control group. Atrophy of the lacrimal gland and loss of acinar cells leads to an osmotic imbalance in the tear film and propels the infiltration of inflammatory cells and expression of inflammatory cytokines, such as IL-1β, TNF-α, IFN-γ, and MMP-9, on the ocular surface and lacrimal glands [63–65]. The activation of NF-κB signaling was reported to stimulate inflammatory chemokines and cytokines, including IL-1β, IL-6, TNF-α, C-X-C motif chemokine 5 (CXCL5), C-X-C motif chemokine 8 (CXCL8), and monocyte chemoattractant protein 1 (MCP1) [66,67]. Additionally, IL-1β also activates COX-1 and COX-2, which are responsible for the synthesis of prostaglandins and contribute to the regulation of the inflammatory response [68].

Treatment with AKE (50 mg/kg) has restored tear production and acinar cells in the lacrimal gland of mice with experimental DE. Moreover, AKE significantly inhibited ocular inflammation in these mice through the downregulation of inflammatory mediators such as IL-1β, TNF-α, IFN-γ, and TGF-β at the transcriptional level in corneal tissue and lacrimal glands. It also suppressed the phosphorylation of i nhibitory κBa (IκBa) at a translational level. However, other inflammatory proteins, NF-κB, COX-1, and COX-2, did not alter dramatically, although AKE was effective in mitigating scopolamine-inducible DE in vivo. These findings suggest that AKE partially suppresses ocular inflammation and may affect other ocular stresses, such as oxidative and ER stress. Among the functional fractions of AKE, chlorogenic acids (CGAs) and 3,5-O-dicaffeoylquinic acids were identified to possess anti-inflammatory effects [69,70]. The anti-inflammatory response of AKE may rely on these functional chemicals. However, additional studies with specific functional components from the fractionated AKE should be performed to elucidate the active components of AKE.

Chronic ocular inflammation in DE also causes retinal damage. In a TNF-α-induced retinal inflammation (RI) model, AKE attenuated IL-1β, TNF-α, IL-6, IL-8, and MMP-9 in a dose-dependent manner. Along with NF-κB, the MAPK pathway plays a crucial role in inflammation, infiltration of innate immune cells, antigen presentation, and upregulation of inflammatory molecules [71,72]. Activation of the MAPK pathway was markedly attenuated in AKE-treated ARPE-19 cells. Together, these data suggest that AKE may reduce local inflammation in ocular tissues by suppressing the activation of inflammatory cytokines and the MAPK pathway. RI has been determined as a prominent signature in the pathogenesis of age-related macular degeneration, DR, and uveitis [73]. The anti-inflammatory effects of AKE need further investigation to reduce the burden of RI as well as DE.

ER stress is another primary intracellular event with chronic local inflammation [74]. In DR and retinal degeneration, increased ocular neovascularization and apoptosis were observed along with elevation of ER stress [75,76]. In the present study, AKE treatment was also determined to affect the inhibition of ER stress-related cellular pathologies. Additionally, we demonstrated that AKE has sufficient potency to inhibit the expression of VEGF-α and increase [Ca^{2+}]$_i$ levels in a model stress-triggered system. Previously, it was reported that VEGF-α and corneal lymphangiogenesis are increased in the DE mouse model [77]. Type 2 diabetes mellitus often triggers patients to have ophthalmic complications, including corneal abnormalities, glaucoma, iris neovascularization, cataracts, and neuropathies [78]. DR is the most common medical complication of these ophthalmic complications as it damages the blood vessels in the retina [79]. Indeed, DR is associated with increased intraocular levels of VEGF-α [78]. However, Shin et al. [80] reported that general nutrients and antioxidant bioactive materials such as polyphenolic substances in *A. koraiensis* possessed radical scavenging activity and reducing power. The current report promotes the study of the antioxidant ability by AKE as one of the underlying mechanisms to inhibit DE. As summarized in Tables 3 and 4, a high concentration of AKE was associated with greater antioxidative activities. A study investigating the protective effects of muscadine grape polyphenols (MGPs) demonstrated that the major polyphenols (quercetin, ellagic acid, myricetin, and kaempferol) in MGPs effectively attenuated ocular inflammation and ER stress [81]. In the present study, we demonstrated that AKE mitigated symptoms of DE, including corneal epithelium thinning, lacrimal gland tissue gap formation, and tear production decline. The underlying mechanism is that AKE treatment inhibited inflammatory responses to attenuate the activation of IκB. Moreover, AKE treatment ameliorated retinal inflammation and ER stress.

5. Conclusions

In conclusion, the oral administration of AKE could be an effective treatment for DES or inflammation caused by ocular disorders. Especially, substances derived from natural products are in the limelight as raw materials for functional foods in that they can be easily processed as raw materials for food, and their safety is generally guaranteed. Therefore, if it is used as a functional material that anyone can easily accept and as a material for health foods for improving eye health, it can be judged that it is beneficial for reducing inflammation and alleviating DE. However, before it can be used as an actual food ingredient, evaluating its health effects in a human test, such as repetition and genetic toxicity, is required. If we do more of these studies, we can expect that they can be used as a convenient natural resource to improve DES.

Supplementary Materials: The following are available online at http://www.mdpi.com/2072-6643/12/11/3245/s1, Figure S1: The Inflammatory signaling (MAPK) were analyzed by western blot (WB) for supplementary data Figure 4B.

Author Contributions: Conceptualization, S.H.J. and J.-C.K.; Methodology, S.-C.H. and J.-C.K.; Software S.-C.H.; Validation, S.-C.H. and J.-C.K.; Formal Analysis, S.-C.H.; Investigation, S.-C.H. and J.-C.K.; Resources S.H.J. and J.-C.K.; Data Curation, S.-C.H. and J.-H.H.; Writing—Original Draft Preparation, S.-C.H., J.-H.H. and J.-C.K.; Writing—Review and Editing, S.-C.H., J.-H.H., J.K.L. and J.-C.K.; Visualization, S.-C.H.; Supervision J.-C.K.; Project Administration, J.-C.K. and S.H.J.; Funding Acquisition J.-C.K. and J.S.H. All authors have read and agreed to the published version of the manuscript.

Funding: This research was financially supported by Korea Institute of Science and Technology (KIST) internal project (2Z05310) and the Ministry of Ocean and Fisheries in Korea (20190015).

Acknowledgments: This study is supported by Eye Health Food & Drugs Exploration of Natural materials (2Z05310) from KIST, Korea, and in part by "Development of Advanced Process for the Production of Eye Health Materials Using *Tetraselmis chuii*", the Ministry of Ocean and Fisheries in Korea.

Conflicts of Interest: The authors confirm that article content has no conflict of interest.

References

1. Phadatare, S.P.; Momin, M.; Nighojkar, P.; Askarkar, S.; Singh, K.K. A comprehensive review on dry eye disease: Diagnosis, medical management, recent developments, and future challenges. *Adv. Pharm.* **2015**, *2015*, 704946. [CrossRef]

2. Paulsen, A.J.; Cruickshanks, K.J.; Fischer, M.E.; Huang, G.-H.; Klein, B.E.; Klein, R.; Dalton, D.S. Dry eye in the beaver dam offspring study: Prevalence, risk factors, and health-related quality of life. *Am. J. Ophthalmol.* **2014**, *157*, 799–806. [CrossRef] [PubMed]

3. Stahl, U.; Willcox, M.; Stapleton, F. Osmolality and tear film dynamics. *Clin. Exp. Optom.* **2012**, *95*, 3–11. [CrossRef] [PubMed]

4. Zhang, X.; Qu, Y.; He, X.; Ou, S.; Bu, J.; Jia, C.; Wang, J.; Wu, H.; Liu, Z.; Li, W. Dry eye management: Targeting the ocular surface microenvironment. *Int. J. Mol. Sci.* **2017**, *18*, 1398. [CrossRef]

5. Lee, L.; Garrett, Q.; Flanagan, J.; Chakrabarti, S.; Papas, E. Genetic factors and molecular mechanisms in dry eye disease. *Ocul. Surf.* **2018**, *16*, 206–217. [CrossRef]

6. Liu, K.C.; Huynh, K.; Grubbs, J.; Davis, R.M. Autoimmunity in the pathogenesis and treatment of keratoconjunctivitis sicca. *Curr. Allergy Asthma Rep.* **2014**, *14*, 403. [CrossRef]

7. Baudouin, C.; Irkeç, M.; Messmer, E.M.; Benítez-del-Castillo, J.M.; Bonini, S.; Figueiredo, F.C.; Geerling, G.; Labetoulle, M.; Lemp, M.; Rolando, M. Clinical impact of inflammation in dry eye disease: Proceedings of the ODISSEY group meeting. *Acta Ophthalmol.* **2018**, *96*, 111–119. [CrossRef]

8. Chen, Y.; Zhang, X.; Yang, L.; Li, M.; Li, B.; Wang, W.; Sheng, M. Decreased PPAR-γ expression in the conjunctiva and increased expression of TNF-α and IL-1β in the conjunctiva and tear fluid of dry eye mice. *Mol. Med. Rep.* **2014**, *9*, 2015–2023. [CrossRef]

9. VanDerMeid, K.R.; Su, S.P.; Ward, K.W.; Zhang, J.-Z. Correlation of tear inflammatory cytokines and matrix metalloproteinases with four dry eye diagnostic tests. *Investig. Ophthalmol. Vis. Sci.* **2012**, *53*, 1512–1518. [CrossRef]

10. Kunert, K.S.; Tisdale, A.S.; Stern, M.E.; Smith, J.; Gipson, I.K. Analysis of topical cyclosporine treatment of patients with dry eye syndrome: Effect on conjunctival lymphocytes. *Arch. Ophthalmol.* **2000**, *118*, 1489–1496. [CrossRef]

11. Rolando, M.; Stern, M.E.; Calonge, M. Modern perspectives on dry eye disease. *Eur. Ophthalmic Rev.* **2017**, *11*, 2–6.

12. Stern, M.E.; Schaumburg, C.S.; Pflugfelder, S.C. Dry eye as a mucosal autoimmune disease. *Int. Rev. Immunol.* **2013**, *32*, 19–41. [CrossRef]

13. Wei, Y.; Asbell, P.A. The core mechanism of dry eye disease (DED) is inflammation. *Eye Contact Lens* **2014**, *40*, 248. [CrossRef]

14. Xu, C.; Bailly-Maitre, B.; Reed, J.C. Endoplasmic reticulum stress: Cell life and death decisions. *J. Clin. Investig.* **2005**, *115*, 2656–2664. [CrossRef]

15. Bahar, E.; Kim, H.; Yoon, H. ER stress-mediated signaling: Action potential and Ca^{2+} as key players. *Int. J. Mol. Sci.* **2016**, *17*, 1558. [CrossRef] [PubMed]

16. Hotamisligil, G.S. Endoplasmic reticulum stress and the inflammatory basis of metabolic disease. *Cell* **2010**, *140*, 900–917. [CrossRef]

17. Seo, Y.; Ji, Y.; Lee, S.; Shim, J.; Noh, H.; Yeo, A.; Park, C.; Park, M.; Chang, E.; Lee, H. Activation of HIF-1 α (hypoxia inducible factor-1 α) prevents dry eye-induced acinar cell death in the lacrimal gland. *Cell Death Dis.* **2014**, *5*, e1309. [CrossRef]

18. Barabino, S.; Chen, Y.; Chauhan, S.; Dana, R. Ocular surface immunity: Homeostatic mechanisms and their disruption in dry eye disease. *Prog. Retin. Eye Res.* **2012**, *31*, 271–285. [CrossRef]

19. Lin, H.; Yiu, S.C. Dry eye disease: A review of diagnostic approaches and treatments. *Saudi J. Ophthalmol.* **2014**, *28*, 173–181. [CrossRef] [PubMed]

20. Messmer, E.M. The pathophysiology, diagnosis, and treatment of dry eye disease. *Dtsch. Ärztebl. Int.* **2015**, *112*, 71. [CrossRef]

21. Downie, L.E.; Keller, P.R. A pragmatic approach to the management of dry eye disease: Evidence into practice. *Optom. Vis. Sci.* **2015**, *92*, 957–966. [CrossRef]

22. Sun, N.-N.; Wu, T.-Y.; Chau, C.-F. Natural dietary and herbal products in anti-obesity treatment. *Molecules* **2016**, *21*, 1351. [CrossRef]

23. Granato, D.; Barba, F.J.; Bursać Kovačević, D.; Lorenzo, J.M.; Cruz, A.G.; Putnik, P. Functional foods: Product development, technological trends, efficacy testing, and safety. *Annu. Rev. Food Sci. Technol.* **2020**, *11*, 93–118. [CrossRef]

24. Aguilar-Toalá, J.; Hernández-Mendoza, A.; González-Córdova, A.; Vallejo-Cordoba, B.; Liceaga, A. Potential role of natural bioactive peptides for development of cosmeceutical skin products. *Peptides* **2019**, *122*, 170170.

25. Hooshmand, S.; Arjmandi, B.H. Dried plum, an emerging functional food that may effectively improve bone health. *Ageing Res. Rev.* **2009**, *8*, 122–127. [CrossRef] [PubMed]

26. Park, S.-H.; Sim, Y.-B.; Kim, S.-M.; Kang, Y.-J.; Lee, J.-K.; Lim, S.-S.; Kim, J.-K.; Suh, H.-W. Antinociceptive profiles and mechanisms of orally administered *Aster Koraiensis* extract in the mouse. *J. Med. Plants Res.* **2011**, *5*, 6267–6272.

27. Ahn, D. *Illustrated Book of Korean Medicinal Herbs*; Kyohaksa: Seoul, Korea, 1998; Volume 497, pp. 23–115.

28. Ko, J.; Lee, K. Effect of plant growth regulators on growth and flowering of potted *Lychnis cognata*, *Aster koraiensis* and *Campanula takesimana*. *RDA J. Agric. Sci. Korea Repub.* **1996**, *38*, 627–632.

29. Lee, J.; Lee, Y.M.; Lee, B.W.; Kim, J.-H.; Kim, J.S. Chemical constituents from the aerial parts of *Aster koraiensis* with protein glycation and aldose reductase inhibitory activities. *J. Nat. Prod.* **2012**, *75*, 267–270. [CrossRef]

30. Sohn, E.; Kim, J.; Kim, C.-S.; Kim, Y.S.; Jang, D.S.; Kim, J.S. Extract of the aerial parts of *Aster koraiensis* reduced development of diabetic nephropathy via anti-apoptosis of podocytes in streptozotocin-induced diabetic rats. *Biochem. Biophys. Res. Commun.* **2010**, *391*, 733–738. [CrossRef]

31. Hyun, S.-W.; Kim, J.; Jo, K.; Kim, J.S.; Kim, C.-S. *Aster koraiensis* extract improves impaired skin wound healing during hyperglycemia. *Integr. Med. Res.* **2018**, *7*, 351–357. [CrossRef] [PubMed]

32. Kim, J.; Jo, K.; Lee, I.-S.; Kim, C.-S.; Kim, J.S. The extract of *Aster Koraiensis* prevents retinal Pericyte apoptosis in diabetic rats and its active compound, chlorogenic acid inhibits AGE formation and AGE/RAGE interaction. *Nutrients* **2016**, *8*, 585. [CrossRef] [PubMed]

33. Kim, J.; Lee, Y.M.; Jung, W.; Park, S.-B.; Kim, C.-S.; Kim, J.S. *Aster koraiensis* extract and chlorogenic acid inhibit retinal angiogenesis in a mouse model of oxygen-induced retinopathy. *Evid. Based Complementary Altern. Med.* **2018**, *2018*, 4916497. [CrossRef] [PubMed]

34. Serpen, A.; Gökmen, V.; Fogliano, V. Total antioxidant capacities of raw and cooked meats. *Meat Sci.* **2012**, *90*, 60–65. [CrossRef]

35. Thaipong, K.; Boonprakob, U.; Crosby, K.; Cisneros-Zevallos, L.; Byrne, D.H. Comparison of ABTS, DPPH, FRAP, and ORAC assays for estimating antioxidant activity from guava fruit extracts. *J. Food Compos. Anal.* **2006**, *19*, 669–675. [CrossRef]

36. Benzie, I.F.; Strain, J. [2] Ferric reducing/antioxidant power assay: Direct measure of total antioxidant activity of biological fluids and modified version for simultaneous measurement of total antioxidant power and ascorbic acid concentration. In *Methods in Enzymology*; Academic Press: Cambridge, MA, USA, 1999; Volume 299, pp. 15–27.

37. Oki, T.; Nagai, S.; Yoshinaga, M.; Nishiba, Y.; Suda, I. Contribution of β-carotene to radical scavenging capacity varies among orange-fleshed sweet potato cultivars. *Food Sci. Technol. Res.* **2006**, *12*, 156–160. [CrossRef]

38. Re, R.; Pellegrini, N.; Proteggente, A.; Pannala, A.; Yang, M.; Rice-Evans, C. Antioxidant activity applying an improved ABTS radical cation decolorization assay. *Free Radic. Biol. Med.* **1999**, *26*, 1231–1237. [CrossRef]

39. Alves, R.C.; Costa, A.S.; Jerez, M.; Casal, S.; Sineiro, J.; Nunez, M.J.; Oliveira, B. Antiradical activity, phenolics profile, and hydroxymethylfurfural in espresso coffee: Influence of technological factors. *J. Agric. Food Chem.* **2010**, *58*, 12221–12229. [CrossRef] [PubMed]

40. Pourmorad, F.; Hosseinimehr, S.; Shahabimajd, N. Antioxidant activity, phenol and flavonoid contents of some selected Iranian medicinal plants. *Afr. J. Biotechnol.* **2006**, *5*, 1142–1145.

41. Marinova, D.; Ribarova, F.; Atanassova, M. Total phenolics and total flavonoids in Bulgarian fruits and vegetables. *J. Univ. Chem. Technol. Metallurgy* **2005**, *40*, 255–260.

42. Kasote, D.M.; Katyare, S.S.; Hegde, M.V.; Bae, H. Significance of antioxidant potential of plants and its relevance to therapeutic applications. *Int. J. Biol. Sci.* **2015**, *11*, 982. [CrossRef]

43. El Jemli, M.; Kamal, R.; Marmouzi, I.; Zerrouki, A.; Cherrah, Y.; Alaoui, K. Radical-scavenging activity and ferric reducing ability of *Juniperus thurifera* (L.), *J. oxycedrus* (L.), *J. phoenicea* (L.) and *Tetraclinis articulata* (L.). *Adv. Pharmacol. Sci.* **2016**, *2016*, 6392656.

44. Javadi, M.-A.; Feizi, S. Dry eye syndrome. *J. Ophthalmic Vis. Res.* **2011**, *6*, 192.

45. Dienes, L.; Kiss, H.J.; Perényi, K.; Nagy, Z.Z.; Acosta, M.C.; Gallar, J.; Kovács, I. Corneal sensitivity and dry eye symptoms in patients with keratoconus. *PLoS ONE* **2015**, *10*, e0141621. [CrossRef] [PubMed]

46. Conrady, C.D.; Joos, Z.P.; Patel, B.C. The lacrimal gland and its role in dry eye. *J. Ophthalmol.* **2016**, *2016*, 7542929. [CrossRef]

47. Tiwari, S.; Ali, M.J.; Vemuganti, G.K. Human lacrimal gland regeneration: Perspectives and review of literature. *Saudi J. Ophthalmol.* **2014**, *28*, 12–18. [CrossRef] [PubMed]

48. Khwaja, A. Akt is more than just a Bad kinase. *Nature* **1999**, *401*, 33–34. [CrossRef] [PubMed]

49. Viatour, P.; Merville, M.P.; Bours, V.; Chariot, A. Phosphorylation of NF-kappaB and IkappaB proteins: Implications in cancer and inflammation. *Trends Biochem. Sci.* **2005**, *30*, 43–52. [CrossRef] [PubMed]

50. Sánchez-Chávez, G.; Hernández-Ramírez, E.; Osorio-Paz, I.; Hernández-Espinosa, C.; Salceda, R. Potential role of endoplasmic reticulum stress in pathogenesis of diabetic retinopathy. *Neurochem. Res.* **2016**, *41*, 1098–1106. [CrossRef]

51. Bose, T.; Diedrichs-Möhring, M.; Wildner, G. Dry eye disease and uveitis: A closer look at immune mechanisms in animal models of two ocular autoimmune diseases. *Autoimmun. Rev.* **2016**, *15*, 1181–1192. [CrossRef]

52. An, E.; Gordish-Dressman, H.; Hathout, Y. Effect of TNF-α on human ARPE-19-secreted proteins. *Mol. Vis.* **2008**, *14*, 2292.

53. Salminen, A.; Kauppinen, A.; Hyttinen, J.M.; Toropainen, E.; Kaarniranta, K. Endoplasmic reticulum stress in age-related macular degeneration: Trigger for neovascularization. *Mol. Med.* **2010**, *16*, 535–542. [CrossRef] [PubMed]

54. Li, J.; Wang, J.J.; Yu, Q.; Wang, M.; Zhang, S.X. Endoplasmic reticulum stress is implicated in retinal inflammation and diabetic retinopathy. *FEBS Lett.* **2009**, *583*, 1521–1527. [CrossRef]

55. Krzystolik, M.G.; Afshari, M.A.; Adamis, A.P.; Gaudreault, J.; Gragoudas, E.S.; Michaud, N.A.; Li, W.; Connolly, E.; O'Neill, C.A.; Miller, J.W. Prevention of experimental choroidal neovascularization with intravitreal anti–vascular endothelial growth factor antibody fragment. *Arch. Ophthalmol.* **2002**, *120*, 338–346. [CrossRef]

56. Krebs, J.; Agellon, L.B.; Michalak, M. Ca^{2+} homeostasis and endoplasmic reticulum (ER) stress: An integrated view of calcium signaling. *Biochem. Biophys. Res. Commun.* **2015**, *460*, 114–121. [CrossRef]

57. Pflugfelder, S.C.; de Paiva, C.S. The pathophysiology of dry eye disease: What we know and future directions for research. *Ophthalmology* **2017**, *124*, S4–S13. [CrossRef] [PubMed]

58. Clayton, J.A. Dry eye. *N. Engl. J. Med.* **2018**, *378*, 2212–2223. [CrossRef]

59. Kang, W.S.; Jung, E.; Kim, J. *Aucuba japonica* extract and aucubin prevent desiccating stress-induced corneal epithelial cell injury and improve tear secretion in a mouse model of dry eye disease. *Molecules* **2018**, *23*, 2599. [CrossRef] [PubMed]

60. Martin, E.; Oliver, K.M.; Pearce, E.I.; Tomlinson, A.; Simmons, P.; Hagan, S. Effect of tear supplements on signs, symptoms and inflammatory markers in dry eye. *Cytokine* **2018**, *105*, 37–44. [CrossRef]

61. Deveci, H.; Kobak, S. The efficacy of topical 0.05% cyclosporine A in patients with dry eye disease associated with Sjögren's syndrome. *Int. Ophthalmol.* **2014**, *34*, 1043–1048. [CrossRef]

62. McMonnies, C.W. Conjunctival tear layer temperature, evaporation, hyperosmolarity, inflammation, hyperemia, tissue damage, and symptoms: A review of an amplifying cascade. *Curr. Eye Res.* **2017**, *42*, 1574–1584. [CrossRef]

63. Guzmán, M.; Keitelman, I.; Sabbione, F.; Trevani, A.S.; Giordano, M.N.; Galletti, J.G. Mucosal tolerance disruption favors disease progression in an extraorbital lacrimal gland excision model of murine dry eye. *Exp. Eye Res.* **2016**, *151*, 19–22. [CrossRef] [PubMed]

64. De Paiva, C.S.; Corrales, R.M.; Villarreal, A.L.; Farley, W.J.; Li, D.-Q.; Stern, M.E.; Pflugfelder, S.C. Corticosteroid and doxycycline suppress MMP-9 and inflammatory cytokine expression, MAPK activation in the corneal epithelium in experimental dry eye. *Exp. Eye Res.* **2006**, *83*, 526–535. [CrossRef] [PubMed]

65. Aragona, P.; Aguennouz, M.H.; Rania, L.; Postorino, E.; Sommario, M.S.; Roszkowska, A.M.; De Pasquale, M.G.; Pisani, A.; Puzzolo, D. Matrix metalloproteinase 9 and transglutaminase 2 expression at the ocular surface in patients with different forms of dry eye disease. *Ophthalmology* **2015**, *122*, 62–71. [CrossRef] [PubMed]

66. Liu, T.; Zhang, L.; Joo, D.; Sun, S.-C. NF-κB signaling in inflammation. *Signal Transduct. Target. Ther.* **2017**, *2*, 1–9. [CrossRef]

67. Tornatore, L.; Thotakura, A.K.; Bennett, J.; Moretti, M.; Franzoso, G. The nuclear factor kappa B signaling pathway: Integrating metabolism with inflammation. *Trends Cell Biol.* **2012**, *22*, 557–566. [CrossRef]

68. Ricciotti, E.; FitzGerald, G.A. Prostaglandins and inflammation. *Arterioscler. Thromb. Vasc. Biol.* **2011**, *31*, 986–1000. [CrossRef]

69. Liang, N.; Kitts, D.D. Role of chlorogenic acids in controlling oxidative and inflammatory stress conditions. *Nutrients* **2016**, *8*, 16. [CrossRef]

70. Abdel Motaal, A.; Ezzat, S.M.; Tadros, M.G.; El-Askary, H.I. In vivo anti-inflammatory activity of caffeoylquinic acid derivatives from *Solidago virgaurea* in rats. *Pharm. Biol.* **2016**, *54*, 2864–2870. [CrossRef]

71. Newton, K.; Dixit, V.M. Signaling in innate immunity and inflammation. *Cold Spring Harb. Perspect. Biol.* **2012**, *4*, a006049. [CrossRef]

72. Kyriakis, J.M.; Avruch, J. Mammalian MAPK signal transduction pathways activated by stress and inflammation: A 10-year update. *Physiol. Rev.* **2012**, *2*, 807–869. [CrossRef]

73. Whitcup, S.M.; Nussenblatt, R.B.; Lightman, S.L.; Hollander, D.A. Inflammation in retinal disease. *Int. J. Inflamm.* **2013**, *2013*, 724648. [CrossRef] [PubMed]

74. Sprenkle, N.T.; Sims, S.G.; Sánchez, C.L.; Meares, G.P. Endoplasmic reticulum stress and inflammation in the central nervous system. *Mol. Neurodegener.* **2017**, *12*, 1–18. [CrossRef] [PubMed]
75. Kroeger, H.; Chiang, W.C.; Felden, J.; Nguyen, A.; Lin, J.H. ER stress and unfolded protein response in ocular health and disease. *FEBS J.* **2019**, *286*, 399–412. [CrossRef]
76. Zhang, S.X.; Sanders, E.; Fliesler, S.J.; Wang, J.J. Endoplasmic reticulum stress and the unfolded protein responses in retinal degeneration. *Exp. Eye Res.* **2014**, *125*, 30–40. [CrossRef] [PubMed]
77. Goyal, S.; Chauhan, S.K.; El Annan, J.; Nallasamy, N.; Zhang, Q.; Dana, R. Evidence of corneal lymphangiogenesis in dry eye disease: A potential link to adaptive immunity? *Arch. Ophthalmol.* **2010**, *128*, 819–824. [CrossRef]
78. Titchenell, P.M.; Antonetti, D.A. Using the past to inform the future: Anti-VEGF therapy as a road map to develop novel therapies for diabetic retinopathy. *Diabetes* **2013**, *62*, 1808–1815. [CrossRef] [PubMed]
79. Stolar, M. Glycemic control and complications in type 2 diabetes mellitus. *Am. J. Med.* **2010**, *123*, S3–S11. [CrossRef]
80. Shin, E.H.; Park, S.J. Component analysis and antioxidant activity of *Aster koraiensis* Nakai. *J. Korean Soc. Food Sci. Nutr.* **2014**, *43*, 74–79. [CrossRef]
81. Ha, J.-H.; Shil, P.K.; Zhu, P.; Gu, L.; Li, Q.; Chung, S. Ocular inflammation and endoplasmic reticulum stress are attenuated by supplementation with grape polyphenols in human retinal pigmented epithelium cells and in C57BL/6 mice. *J. Nutr.* **2014**, *144*, 799–806. [CrossRef]

Publisher's Note: MDPI stays neutral with regard to jurisdictional claims in published maps and institutional affiliations.

 *nutrients*

Article

Chronic Oleoylethanolamide Treatment Decreases Hepatic Triacylglycerol Level in Rat Liver by a PPARγ/SREBP-Mediated Suppression of Fatty Acid and Triacylglycerol Synthesis

Adele Romano [1,†], Marzia Friuli [1,†], Laura Del Coco [2,†], Serena Longo [2], Daniele Vergara [2], Piero Del Boccio [3,4], Silvia Valentinuzzi [3,4], Ilaria Cicalini [4,5], Francesco P. Fanizzi [2,*], Silvana Gaetani [1,‡] and Anna M. Giudetti [2,*,‡]

[1] Department of Physiology and Pharmacology "V. Erspamer", Sapienza University of Rome, P.le Aldo Moro 5, 00185 Rome, Italy; adele.romano@uniroma1.it (A.R.); marzia.friuli@uniroma1.it (M.F.); silvana.gaetani@uniroma1.it (S.G.)

[2] Department of Biological and Environmental Sciences and Technologies, University of Salento, Via Prov.le Lecce-Monteroni, 73100 Lecce, Italy; laura.delcoco@unisalento.it (L.D.C.); serena.longo@unisalento.it (S.L.); daniele.vergara@unisalento.it (D.V.)

[3] Department of Pharmacy, University "G. d'Annunzio" of Chieti-Pescara, 66100 Chieti, Italy; piero.delboccio@unich.it (P.D.B.); silvia.valentinuzzi@unich.it (S.V.)

[4] Center for Advanced Studies and Technology (CAST), University "G. d'Annunzio" of Chieti-Pescara, 66100 Chieti, Italy; ilaria.cicalini@unich.it

[5] Department of Medicine and Aging Science, University "G. d'Annunzio" of Chieti-Pescara, 66100 Chieti, Italy

[*] Correspondence: fp.fanizzi@unisalento.it (F.P.F.); anna.giudetti@unisalento.it (A.M.G.); Tel.: +39-0832-299-265 (F.P.F.); +39-0832-298-679 (A.M.G.); Fax: +39-83-298-626 (F.P.F.); +39-0832-298-626 (A.M.G.)

[†] These authors contributed equally to this work.

[‡] These authors contributed equally to this work.

Citation: Romano, A.; Friuli, M.; Del Coco, L.; Longo, S.; Vergara, D.; Del Boccio, P.; Valentinuzzi, S.; Cicalini, I.; Fanizzi, F.P.; Gaetani, S.; et al. Chronic Oleoylethanolamide Treatment Decreases Hepatic Triacylglycerol Level in Rat Liver by a PPARγ/SREBP-Mediated Suppression of Fatty Acid and Triacylglycerol Synthesis. *Nutrients* 2021, 13, 394. https://doi.org/10.3390/nu13020394

Academic Editor: Lindsay Brown

Received: 29 December 2020

Accepted: 23 January 2021

Published: 27 January 2021

Abstract: Oleoylethanolamide (OEA) is a naturally occurring bioactive lipid belonging to the family of N-acylethanolamides. A variety of beneficial effects have been attributed to OEA, although the greater interest is due to its potential role in the treatment of obesity, fatty liver, and eating-related disorders. To better clarify the mechanism of the antiadipogenic effect of OEA in the liver, using a lipidomic study performed by ^{1}H-NMR, LC-MS/MS and thin-layer chromatography analyses we evaluated the whole lipid composition of rat liver, following a two-week daily treatment of OEA (10 mg kg^{-1} i.p.). We found that OEA induced a significant reduction in hepatic triacylglycerol (TAG) content and significant changes in sphingolipid composition and ceramidase activity. We associated the antiadipogenic effect of OEA to decreased activity and expression of key enzymes involved in fatty acid and TAG syntheses, such as acetyl-CoA carboxylase, fatty acid synthase, diacylglycerol acyltransferase, and stearoyl-CoA desaturase 1. Moreover, we found that both SREBP-1 and PPARγ protein expression were significantly reduced in the liver of OEA-treated rats. Our findings add significant and important insights into the molecular mechanism of OEA on hepatic adipogenesis, and suggest a possible link between the OEA-induced changes in sphingolipid metabolism and suppression of hepatic TAG level.

Keywords: lipid metabolism; oleoylethanolamide; peroxisome proliferator-activated receptorγ; NMR spectroscopy; sphingolipids

1. Introduction

Oleoylethanolamide (OEA) is a naturally occurring bioactive lipid belonging to the family of N-acylethanolamides that has received great attention in the last two decades for its biological properties [1,2].

Diet-derived oleic acid promotes OEA formation in the small intestine of different species including rats and mice [3–5]; the membrane protein CD36, a multiligand class B scavenger receptor located on cell surface lipid rafts, plays a pivotal role in OEA biosynthesis by acting as a biosensor for food-derived oleic acid and facilitating OEA mobilization [6,7].

A variety of effects are attributed to exogenous administered OEA spanning in different domains, from neuroprotection [8–10] to memory [11], from inflammation to mood disorders [12–14], from the regulation of satiety to glucose homeostasis and lipid metabolism [8,9].

The majority of OEA's biological functions explains its potential interest as a pharmacological target for the treatment of obesity and eating-related disorders [1,15,16]. Therefore, as a drug, OEA reduces food intake and body weight gain [3,17,18] in both lean and obese rodents. These effects are primarily related to the activation of the peroxisome proliferator-activated receptor-α (PPAR-α) [3–7], for which OEA shows high affinity.

PPARα regulates several aspects of lipid metabolism [19], and in keeping with this, it was demonstrated that OEA, by recruiting PPARα, reduces serum cholesterol and triacylglycerol (TAG) levels and has beneficial effects on the high-fat diet-induced non-alcoholic fatty liver disease (NAFLD) in rats, by stimulating fatty acid β-oxidation, and by inhibiting lipogenesis [20].

OEA can influence sphingolipid metabolism in mice [21]. Indeed, OEA increases ceramide levels in cell cultures, via inhibition of ceramidase [22–24], an enzyme that catalyzes the degradation of ceramide to sphingosine and fatty acids. Ceramide and ceramide-derived sphingolipids are structural components of membranes and have been linked to insulin resistance, oxidative stress, inflammation [7–9] and then to hepatic steatosis.

It was reported that ceramide can influence TAG homeostasis, and hepatic steatosis throughout PPARγ [19], a member of a nuclear hormone superfamily.

Aberrant hepatic PPARγ expression can stimulate hepatic lipogenesis [25] and induce steatosis in mice hepatocytes [26,27], by up-regulating proteins involved in lipid uptake, and TAG storage such as CD36, monoacylglycerol O-acyltransferase 1, and stearoyl-CoA desaturase 1 (SCD1) [25,28].

PPARγ and CD36 mRNA expression are up-regulated in high-fat diet-induced liver steatosis in mice [29]. CD36 expression has been associated with insulin resistance in humans with type 2 diabetes [30,31] and increased hepatic *Cd36* gene expression was reported to increase fatty uptake, TAG accumulation [32,33] and fatty liver [32].

To clarify the mechanism of the antilipogenic effect of OEA in the liver, by using different approaches, such as ^{1}H-NMR, LC-MS/MS and thin-layer chromatography (TLC), we performed a lipidomic analysis of the whole hepatic lipid composition. We found that OEA induced a significant decrease in ceramidase activity and profoundly impacted hepatic lipid composition by significantly increasing sphingomyelin, 24:0ceramide, dihydroceramide, and sphingosine with a concomitant significant decrease in glucosylceramide level. Moreover, a significant decrease in TAG level was also measured. We found that the antilipogenic effect of OEA was associated with a decreased activity and expression of the key enzymes involved in fatty acid and TAG syntheses, such as acetyl-CoA carboxylase (ACC), fatty acid synthase (FAS), diacylglycerol acyltransferase (DGAT), and SCD1. Moreover, we found that PPARγ and SREBP-1 protein expression was significantly reduced in the liver of OEA-treated rats. A possible link between OEA-induced alterations in sphingolipid metabolism and PPARγ-mediated suppression of enzymes involved in TAG synthesis was proposed.

2. Materials and Methods

2.1. Animals, Diet and Chronic Treatments

Adult male Wistar-Han rats (250–300 g at the beginning of the study) were housed in single cages under controlled conditions of temperature and humidity (T = 22 $\pm$ 2 °C; 60% of relative humidity) and were kept on a 12 h light/dark cycle. Rats were fed for 11 weeks with a rodent diet (D12450B, Research Diets Inc., New Brunswick, NJ, USA) containing 3.82 kcal/g, which were distributed among carbohydrates, proteins, and fats according to the following percentages: 70%, 20%, and 10%. The rats were accustomed to handling and injections for 7 days before the beginning of the experiments. Rats, matched for body weight, were randomly divided into two different groups (12 rats per group) and housed individually in metal cages (30 × 30 × 30 cm). One group was treated with vehicle (VEH) and the other with OEA for 2 weeks. Both groups had free access to both food and water.

Both OEA and VEH solutions were freshly prepared on each test day and administered about 10 min before dark onset by following our previous protocols [34–36]. OEA (Sigma-Aldrich, St. Louis, MI, USA) was prepared in the laboratory [37] and administered by intraperitoneal injection (i.p.) at the dosage of 10 mg kg^{-1} in a vehicle of saline/polyethylene glycol/Tween 80 (90/5/5, *v/v*). Theses dosage, vehicle and route of administration were chosen on the basis of the extensive scientific literature published by our and other research groups during the last 20 years. The i.p. route of administration was the most reliable to obtain a lower dosage variability, as compared to oral administration of OEA mixed in the rodent diet, and the highest bioavailability of OEA with the less stressful manipulation of the animals. Animal body weight was monitored on daily basis. As expected from previous observations [1], OEA-treated rats showed a significant decrease of body weight gain as compared to the VEH rats (Figure S1).

At the end of the 2-week treatment period, animals were sacrificed, their livers immediately collected, washed in ice-cold phosphate-buffered saline, snap frozen in 2-metylbutane (−60 °C) and stored at −80 °C until analyzed. All experiments were carried out in accordance with the European directive 2010/63/UE governing animal welfare and with the Italian Ministry of Health guidelines for the care and use of laboratory animals.

2.2. ^1H NMR Spectroscopy

Hepatic total lipids were extracted using the Bligh and Dyer procedure. Lipid extracts (VEH and OEA samples) were analyzed by using 600 μL of deuterated chloroform (CDCl$_3$) and transferred to a 5 mm NMR tube, using tetramethylsilane (TMS, δ = 0.00) as an internal standard. 1D 1H and 2D ^{1}H J-resolved, ^{1}H–^{1}H COSY Correlation Spectroscopy, ^{1}H–^{13}C HSQC Heteronuclear, and ^{1}H–^{13}C HMBC, Multiple Bond Correlation, spectra were acquired at 300 K on a Bruker Avance III NMR spectrometer (Bruker, Biospin, Milan, Italy), operating at 600.13 MHz for ^{1}H observation, equipped with a TCI cryoprobe incorporating a z-axis gradient coil and automatic tuning-matching (ATM). The following parameters were used for ^{1}H NMR spectrum: 64 K data points, spectral width of 20.0276 Hz, 64 scans with a 2 s repetition delay, 90° power pulse (p1) 7.06 μsec and power level −8.05 dB. The acquisition and processing of spectra were performed using Topspin 3.5 software (Bruker Biospin, Milan, Italy). Resonances of fatty acids and metabolites were assigned according with data available in the literature [38–40].

2.3. Tissue Collection and Lipid Analysis

For sphingolipid and neutral lipid analyses by TLC, 0.1 mg of tissue homogenate proteins was used. The developing system was composed of toluene:methanol (70:30, *v:v*) for sphingolipid analysis and hexane:ethyl ether:acetic acid (70:30:1, *v:v*) for neutral lipids. Ceramide and sphingomyelin were identified by comparison with specific standards (C18 ceramide and sphingomyelin (d18:1/12:0) from Avanti Polar). After separation, plates were sprayed uniformly with 8% cupric sulfate in 8% aqueous phosphoric acid, allowed to dry 10 min at room temperature, and then placed into a 145 °C oven for 10 min as reported in [41]. Band intensities were quantified using Image LabTM Version 6.0.1 2017 (Bio-Rad Laboratories, Inc., Segrate (MI)—Italy) software.

Hepatic TAG level was also determined using an enzymatic colorimetric kit (RANDOX Laboratories), following manufactory instructions.

2.4. LC-MS/MS Analysis

Sphingolipid analysis was performed with LC-MS/MS following the previously described method [42]. Briefly, lipid extracts from hepatic homogenates were dried down and reconstituted in 1000 μL di CHCl$_3$:CH$_3$OH (2:1, *v:v*), vortexed and centrifuged for

15 min at +4 °C. The supernatant was subsequently diluted 1:10 with a solution formed by 50% solvent A (H_2O with Formic acid 0.1%) and 50% solvent B (CH3OH:iPrOH:ACN (4:4:1, *v:v*) with Formic acid 0.1%. A volume of 10 μL of Internal Standard solution (Avanti Polar), containing d17So1P 0.1 μg/mL, C17Cer 0.01 μg/mL, C17GlcCer 1.0 μg/mL, was added to 90 μL of sample diluted as described above. After vortexing and centrifuging (5 min at +4 °C) 90 μL of the sample was transferred in vials for subsequent LC-MS/MS analysis.

The LC-MS/MS system consists of an HPLC Alliance HT 2795 Separations Module coupled to Quattro UltimaPt ESI tandem quadrupole mass spectrometer (Waters Corporation, Milford, MA, USA), operating in positive ion mode. The chromatographic separation of analytes was performed using Ascentis Express Fused-Core C18 2.7 μm, 7.5 cm × 2.1 mm column. In a total run time of 25 min, the elution of ceramides was achieved through a gradient of mobile phases, starting from 50% to 100% of methanol:isopropanol:acetonitrile (4:1:1, *v:v*) with Formic acid 0.1% (solvent B), water with Formic acid 0.1% was used as solvent A. The flow rate was 0.20 mL/min. The capillary voltage was 3.8 kV, source temperature was 120 °C, desolvation temperature was 400 °C, and the collision cell gas pressure was 3.62×10^{-3} mbar argon. Multiple reaction monitoring (MRM) functions for detection of sphingolipids are reported in Table S1.

2.5. Liver Microsome Isolation and DGAT Activity Measurements

At the end of the experimental period, livers from VEH and OEA rats were removed and suspended in a medium containing 250 mM sucrose, 1.0 mM Tris-HCl (pH 7.4), 0.5 mM EGTA. In the same medium, the liver was gently homogenized with a Potter-Elvehjem homogenizer. Microsomes were isolated by differential centrifugation as in [43] with a final ultra-centrifugation step at 40,000× *g* for 1 h. The pellet of this centrifugation, corresponding to the microsomal fraction, was suspended in the same sucrose buffer.

DGAT activity was measured on the microsomal fraction as described in [44] using endogenous diacylglycerols as substrates in the presence of [1-^{14}C]palmitoyl-CoA (240 Bq/mol). The incubation was terminated after 1 min by the addition of 2 mL of methanol/chloroform (2/1, *v/v*). After lipid extraction, TAG were isolated by TLC on Silica G as reported in [44]. The silica, containing the TAG fraction, was scraped from the plate for radioactivity measurements.

2.6. Assay of Hepatic Enzymatic Activities

For sphingomyelin synthase assay, the liver was homogenized in a buffer containing 50 mM Tris-HCl, 1 mM EDTA, 5% and sucrose and centrifuged at 2700× *g* for 10 min. The supernatant was then used for the activity assay. The mix reaction contained 50 mM Tris-HCl (pH 7.4), 25 mM KCl, C6-NBDceramide (0.1 mg/mL), and PC (0.01 mg/mL). The reaction was started with 100 μg proteins and the mixture was incubated at 37 °C for 2 h. Lipids were extracted in chloroform/methanol (2/1, *v/v*), dried under nitrogen, and separated by TLC. For the fluorescence measurement, plates were scanned with ChemiDoc™ MP System with Image Lab™ Software.

For neutral and acid sphingomyelinase assay, the liver was homogenized in a buffer containing 50 mM Na acetate, 5 mM $MgCl_2$, 1 mM ETA and 0.5% triton X-100, pH4.5. The enzymatic assay was conducted with a Colorimetric Sphingomyelinase Assay Kit furnished by MERK (Italy).

Ceramidase activity was measured as reported in [45], by incubating 25 μg of protein from liver homogenate with 100 μM N-lauroyl ceramide (Nu-Chek Prep) as substrate in assay buffer (125 mM sucrose, 0.01 mM EDTA, 125 mM Na acetate, and 3 mM DTT, pH 4.5) for 1h min at 37 °C. Reactions were stopped by the addition of a mixture of chloroform/methanol (2:1, *v:v*). The organic phases were collected, dried under N2, and the N-lauroyl amount was measured by LC-MS/MS.

ACC activity was determined as the incorporation of radiolabelled acetyl-CoA into fatty acids in a coupled assay with FAS reaction as described [46]. The reaction was carried out at 37 °C for 8 min. To determine FAS activity, malonyl-CoA was included in the ACC assay mixture, while adenosine triphosphate (ATP), butyryl-CoA, and FAS were omitted. The assay was allowed to proceed for 10 min.

2.7. Western Blot Analyses

Proteins were extracted from the whole liver homogenate using RIPA lysis buffer (15 mM Tris-HCl, 165 mM-NaCl, 0.5% Na-deoxycholate, 1% Triton X-100 and 0.1% SDS), with a protease inhibitor cocktail (1:1000; Sigma-Aldrich, St. Louis, MI, USA) and 1 mM-PMSF (phenylmethanesuOEAnyl fluoride solution). Total protein levels of the lysate were determined using the Bradford method (Bio-Rad Laboratories). After boiling for 5 min, proteins were loaded and separated by SDS-polyacrylamide gel. The samples were then transferred to a nitrocellulose membrane (Bio-Rad Laboratories) and blocked at room temperature for 1 h using 5% (w/v) non-fat milk in TBS-Tris buffer (Tris-buffered saline (TBS) plus 0.5% (v/v) Tween-20, TTBS). The membranes were incubated overnight at 4 °C with primary antibodies against ACC (Cell Signaling #3676, Rabbit 1:1000), FAS (Cell Signaling #3180, Rabbit 1:1000), DGAT1 (Novus Biologicals #NB110-41487, Rabbit 1:1000), DGAT2 (Novus Biologicals #NBP1-71701, Mouse 1:1000), PPARγ (Santa Cruz #sc-7273, Mouse 1:500), sterol regulatory element binding protein-1 (SREBP-1) (Santa Cruz #sc-365513, Mouse 1:500), CD36 (Santa Cruz #sc-7309, Mouse 1:1000), and SCD1 (Santa Cruz #sc-58420, Mouse 1:1000). β-actin (Cell Signaling #8457, Rabbit 1:1000), was used to determine loading fairness. Western blotting analyses were performed using Amersham ECL Advance Western Blotting Detection Kit (GE Healthcare, Little ChaOEAnt, UK) and detection was made using a VersaDoc Image System (Bio-Rad Laboratories, Hercules, CA, USA). β-actin was used to determine loading fairness. After washing with TTBS, the blots were incubated with peroxidase-conjugated monoclonal anti-rabbit secondary antibodies (Sigma-Aldrich, St. Louis, MI, USA) at 1:10.000 dilutions at room temperature for 1–2 h. The blots were then washed thoroughly in TTBS. Western blotting analyses were performed using Amersham ECL Advance Western Blotting Detection Kit (GE Healthcare, Little ChaOEAnt, UK). Densitometric analysis of the immunoblots was performed using Image Lab™ Version 6.0.1 2017 (Bio-Rad Laboratories, Inc., Segrate (MI)—Italy) software.

2.8. Hepatic Cell Tretaments

The human hepatocellular carcinoma cell line HLF was maintained in Dulbecco's minimum essential medium eagle (DMEM) low glucose with 10% fetal bovine serum (FBS), 100 U/mL penicillin, 100 µg/mL streptomycin, and 2 mM glutamine. Cells were cultured at 37 °C with 5% partial pressure of CO_2 in a humidified atmosphere. Cells were treated for 2 h with OEA or Carmofur (Cayman Chemical) at the concentration of 10 µM and 5 µM, respectively. OEA and Carmofur were dissolved in dimethyl sulfoxid (DMSO) at a 10 mM concentration. At the end of the incubation time total proteins were extracted using RIPA lysis buffer, with a protease inhibitor cocktail and 1 mM-PMSF. Total protein lysate levels were determined using the Bradford method. After boiling for 5 min, proteins were loaded and separated by SDS-polyacrylamide gel and probed with PPARγ antibody.

2.9. Multivariate Statistical Analyses

The ^{1}H-NMR spectra of lipid extracts (ZG Bruker pulse program experiments) were used for multivariate statistical analysis. Each spectrum was segmented in fixed rectangular buckets of 0.04 ppm width and successively integrated, by using Amix 3.9.13 (Bruker Biospin, Milano, Italy) software. The spectral regions between 7.45–7.00, 2.00–1.50 ppm, due to the residual peaks of solvents (chloroform and residual water), were discarded. The total sum normalization and the Pareto scaling procedure (performed by dividing the mean-centered data by the square root of the standard deviation) were then applied to the whole data to minimize small differences due to sample concentration and/or experimental conditions among samples [47]. Unsupervised Principal Component Analysis (PCA) and Orthogonal Partial Least Squares Discriminant Analysis (OPLS-DA) were applied to examine the intrinsic variation in the data using Metaboanalyst software [48]. The validation of statistical models was performed and further evaluated by using the internal 10-fold cross-validation and with the permutation test (100 permutations). Two parameters, R^2 (the total variations in the data) and Q^2 (the predictive ability of the models) were analyzed to describe the goodness of the statistical models [49]. The box and whisker plots, obtained for the discriminant metabolites found by multivariate analyses, summarize the normalized bucket values (box limits indicate the range of the central 50% of the data, with a central line marking the median value; the notch indicates the 95% confidence interval around the median of each group).

Metabolic results are reported as means $\pm$ standard error of the mean (SEM). The comparison between the two groups was made using Student's t-test. Differences between groups were considered statistically significant when $p < 0.05$.

3. Results

3.1. ^{1}NMR-Hepatic Lipidomic Analysis and Identification of Lipid Classes

An overall study of the lipid composition of hepatic homogenate from VEH and OEA rats was performed by ^{1}NMR. 2D COSY, HSQC, HMBC and *J-resolved* spectra were randomly performed on samples and used to accurately assign the lipid classes present in samples (saturated, unsaturated fatty acids, phospholipids). The identified NMR signals of lipid extracts and related assignments are reported in Table 1 and Figure S2.

Table 1. Different lipid classes, with the corresponding resonance assignments, were identified by the NMR analysis. DAGP = diacylglycerophospholipids; FA = fatty acid chain; MUFA = monounsaturated fatty acids; SL = sphingolipids; PC = phosphatidylcholine; PG = phosphatidylglycerol; PUFA = polyunsaturated fatty acids; TAG = triacylglycerol; UFA = unsaturated fatty acids.

Resonance	^{1}H NMR Signal	Chemical Shift (ppm)	Lipid Class
1	$-C18H_3$	0.69	Cholesterol
2	$-CH_3$	0.89	Total FA
3	$-C19H_3$	1.02	Cholesterol
4	$-(CH_2)_n-$	1.25	Total FA
5	$=CHCH_2CH_2(CH_2)-$	1.30	Total FA
6	$-CO-CH_2CH_2-$	1.62	Total FA
7	$-CH_2HC=$	2.02	UFA
8	$-CH_2HC=$	2.07	UFA
9	$-CO-CH_2-$	2.37–2.23	Total FA
10	$=CHCH_2CH=$	2.74	*n*-6 PUFA
11	$=CHCH_2CH=$	2.88	*n*-3 PUFA
12	$-N^+(CH_3)_3$	3.35–3.15	PC/PE/SL
13	CH_2CHCH_2	3.74	PG
14	PC and DAGP signals	4.5–3.5	PC
15	$>C_2H$ in glycerol backbone	5.21	DAGP
16	$>C_2H$ in glycerol backbone	5.26	TAG
17	$-HC=CH-$ in fatty acyl chain PUFA and MUFA	5.42–5.29	UFA
18	$OH-CH-CH=CH-$	5.74	SL

The ^{1}H NMR spectrum (Figure S2) can be divided into three broad regions: 3.0–0.65 ppm for fatty acids and sterol methyl and methylene resonances (A); 5.00–3.00 ppm for phospholipids head groups and glycerol backbone proton resonances (B); 6.00–5.00 ppm for vinyl protons resonances for fatty acids and sphingolipids (C) [38,50]. Two signals of cholesterol (CH_3 groups) were identified in the spectral region between 1.05 and 0.5 ppm. In particular, distinctive singlets of the methyl groups of cholesterol are identified at 1.02 and 0.69 ppm. A characteristic contribution from the choline head groups was found in phosphatidylcholine (PC), phosphatidylethanolamine (PE) and sphingolipids (SL), with sharp singlet signals at approximately 3.35–3.15 ppm. In particular, the signals at 3.35–3.32 and 3.17 ppm of methyl groups ($-N^+(CH_3)_3$) were ascribable to PC/PE and SM, respectively. Moreover, the presence of PC and PE were confirmed by resonances at 3.60 and 4.07 ppm, respectively, while the multiplets at 5.74 ppm were diagnostic of the characteristic sphingenine moiety vinyl protons (HO-CH-**CH=CH**-) and, therefore, suggested the presence of sphingolipids, including sphingomyelin, as also reported in the literature for specific systems [40,51,52]. All diacylglycerophospholipids (DAGP) are represented by the backbone glycerol sn-2 proton multiplet at ~5.21 ppm, while TAG show characteristic signals at 5.27, 4.29–4.27 and 4.15 ppm, partially overlapped with other moieties peaks. Among DAGP, phosphatidylglycerol (PG) signals appeared at 3.74 ppm [38,52]. Finally, the signal at 2.38 ppm appears for the presence of *n*-3 polyunsaturated fatty acids together with other NMR signals at 0.87, 1.30, 1.59 ppm [50].

3.2. Multivariate Analysis of VEH and OEA Liver Samples

A multivariate statistical approach was applied to the NMR data, without removing any of the observations as outliers, to reveal the general trend or data grouping of the samples. Unsupervised Principal Component Analysis (PCA) was performed to investigate the differences between samples, after the pre-processing treatment of the NMR spectra. In the PCA score plot of the corresponding model (Figure 1a), three principal components (t[1]/t[2]/t[3]) explained more than 80% of the total variance ($R^2X = 0.81$, $Q^2 = 0.45$).

OEA and VEH samples resulted well separated in the PCA score scatterplot, in particular when the t[2] and t[3] principal components were observed. The separation of OEA and VEH samples as two specific classes was further confirmed by the supervised PLS-DA analysis (Figure S3), which gave a satisfactory model ($R^2 = 0.95$, $Q^2 = 0.81$).

The OPLS-DA model ($R^2X = 0.79$, $R^2Y = 0.98$, $Q^2 = 0.90$, Figure 1b), obtained from ^{1}H NMR lipid extracts, showed a clear-cut separation of samples, which were clearly distinct along the predictive t[1] axis. The analysis of NMR signals responsible for the class separation was reported as box and whisker plots, which summarize the normalized values obtained for the buckets containing discriminant metabolite signals (Figure 1c). Relative increased values of phospholipids (such as PC/PE/SL), polyunsaturated fatty acids (PUFA), among which linoleic acid and cholesterol were found in OEA vs. VEH samples. Vice versa, a significantly decreased amount of TAG, and PG in OEA vs. VEH rats was observed.

3.3. Hepatic Lipid Analysis by TLC and LC-MS/MS

By NMR spectroscopy we identified specific characteristic sphingenine moieties, suggesting the possible presence of sphingomyelin [40,51,52]. On the basis of the NMR result, we investigated on the different sphingolipid species by TLC and LC-MS/MS analysis.

TLC analysis (Figure 2a) revealed a significant increase in the amount of both ceramide (Figure 2b) and sphingomyelin (Figure 2c) in OEA versus VEH rats.

To verify whether the changes in the hepatic level of ceramides and sphingomyelin was linked to changes in the activity of main enzymes involved in sphingolipid metabolism, we measured the activity of sphingomyelinase (neutral and acid forms), sphingomyelin synthase (Figure 2d) and ceramidase (Figure 2e). We found a significant decrease in the ceramidase activity in OEA vs. VHE animals. No significant changes were instead measured in either neutral and acid sphingomyelinase or sphingomyelin synthase activities.

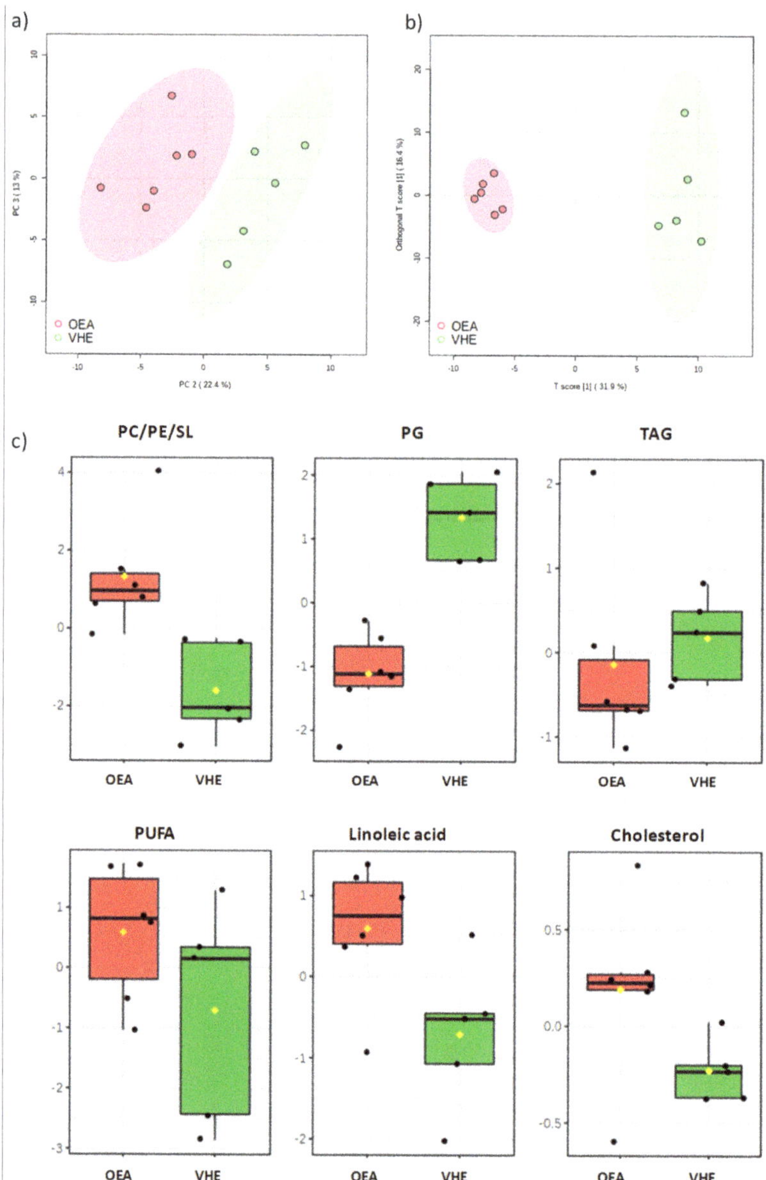

Figure 1. (**a**) Principal component analysis (PCA) score plot for OEA and VEH samples. (**b**) t[1]/to[1] orthogonal partial least squares discriminant analysis (OPLS-DA) score scatter plot for OEA and VEH samples, and (**c**) discriminant metabolites between the two groups are reported in the corresponding box and whisker plots, which summarize the normalized bucket values. The mean value for each group is indicated with a yellow diamond, while the notch indicates the 95% confidence interval around the median of each group, with dots placed past the line edges to indicate outliers. PC = phosphatidylcholine; PE = phosphatidylethanolamine; SL = sphingolipids; PG = phosphatidylglycerol; TAG = triacilglycerols; PUFA = polyunsaturated fatty acids.

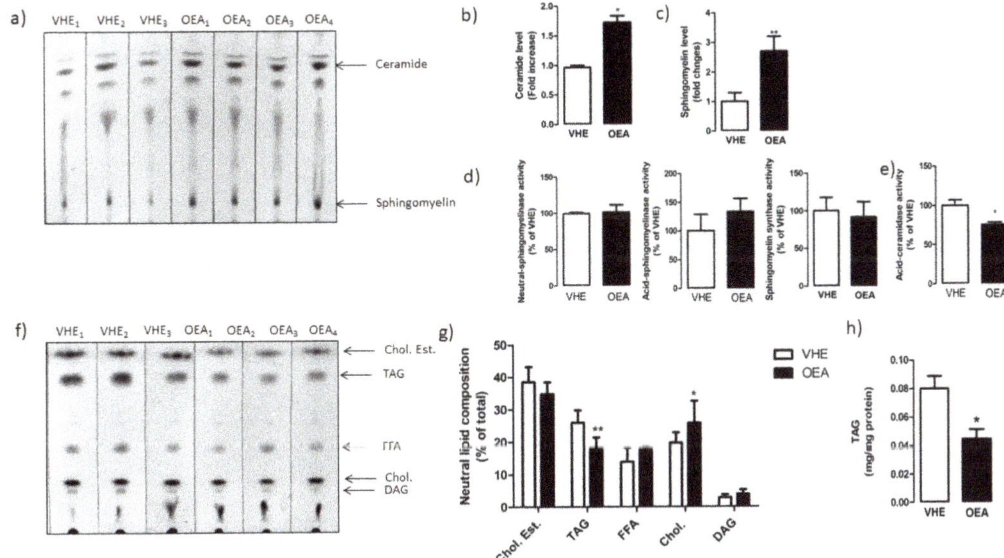

Figure 2. Hepatic sphingolipid and neutral lipid analysis. Total lipids from VEH and OEA groups were extracted and separated by TLC. (**a**) Representative TLC separation of sphingolipids. (**b,c**) Quantification of ceramide and sphingomyelin by densitometry analysis. Ceramide and sphingomyelin level is reported as fold change of OEA versus VHE. (**d**) Neutral and acid sphingomyelinase and sphingomyelin synthase activities reported as percentages of values measured in VEH. (**e**) Ceramidase activity assayed in liver homogenates. (**f**) Representative TLC separation of neutral lipids. (**g**) Neutral lipids are represented as percentage of the total. (**h**) Triacylglycerol (TAG) amount was spectrophotometrically quantified with a specific enzymatic assay. In the figure, the mean $\pm$ SEM of values obtained from three different analyses is reported. ** $p < 0.005$; * $p < 0.05$. Abbreviations: Chol. Est. = cholesterol ester; TAG = triacylglycerols; FFA = free fatty acids; DAG = diacylglycerols.

Moreover, by TLC analysis we also evaluated the level of neutral lipids (Figure 2f) in the liver of OEA vs. VHE rats. According to the NMR analysis, a reduced level of TAG and increased level of cholesterol was found in the liver of OEA-treated rats with respect to VHE. No significant changes were instead measured in the level of cholesterol esters, free fatty acids, and diacylglycerols (Figure 2g). By using an enzymatic colorimetric assay, we also measured the amount of total hepatic TAG in VEH and OEA groups of rats. We found a reduced (about 40%) hepatic amount of TAG in OEA versus VEH animals (0.074 ± 0.01 mg/mg protein in VHE versus 0.044 ± 0.01 mg/mg protein in OEA; $p < 0.05$) (Figure 2h). These results confirmed data obtained by NMR analysis.

To gain insight into the molecular mechanism of the OEA effect on sphingolipid pathway (Figure 3a), using LC-MS/MS we measured the hepatic concentrations of long (C16-C20) and very long ($\geq$C24) chain ceramides as well as of their precursors, dihydroceramides. We observed a significant accumulation of both dihydroceramide (d18:0/16:0) (C16dHCer) and C24:0ceramide in OEA as compared to VHE (Figure 3b). Sphingosine, the bioactive amino-alcohol backbone of sphingolipids, was up-regulated in the liver of OEA rats as well. Moreover, the amount of C16glucosilceramide showed a decrease in OEA with respect to VHE.

3.4. Activity and Expression of Triacylglycerol and Fatty Acid Synthesis Enzymes

In the liver, TAG synthesis is catalyzed by two main DGAT isoforms, DGAT1 and DGAT2. Although both DGAT enzymes synthetize TAG from fatty-acyl-CoA and diacylglycerols, they have different roles and regulatory mechanisms [43].

We investigated whether OEA chronically administered to rats might impact the activity and the expression of hepatic DGAT enzymes.

Western blot analyses (Figure 4a) revealed a significant decrease of both hepatic DGAT1 and DGAT2 expressions in OEA- with respect to VEH-treated rats (Figure 4b). Moreover, DGAT activity, measured in the liver microsomal fraction, was significantly decreased in OEA with respect to VEH rats (Figure 4f).

SCD1 synthesizes oleic acid from stearate, a conversion that facilitates the biosynthesis of TAG and other neutral lipids. SCD1 expression is highly correlated with liver steatosis [53]. Western blot analysis (Figure 4a) revealed that the hepatic protein level of SCD1 was significantly lower in OEA versus VEH rats (Figure 4c).

DGAT can catalyze TAG synthesis by using de novo synthetized fatty acids or fatty acids taken-up from the bloodstream. The de novo synthesis of fatty acids is catalyzed by ACC, which uses the ATP-dependent carboxylation of acetyl-CoA to produce malonyl-CoA and FAS, which catalyzes the synthesis of palmitoyl-CoA, using malonyl-CoA as substrate.

ACC and FAS protein expressions and activities were analyzed in the liver of VEH and OEA rats. The result obtained by western blotting (Figure 4a) demonstrated that both ACC and FAS protein levels were statistically reduced in OEA versus VEH (Figure 4d). This data was consistent with a significantly reduced activity of both ACC and FAS measured in OEA versus VEH rats (Figure 4f).

Moreover, we also measured, by western blot analysis (Figure 4a) the hepatic level of CD36, a protein that facilitates the transport of long-chain fatty acids in several cell types [54]. We found that the hepatic protein level of CD36 decreased about 2-fold in OEA versus VEH rats (Figure 4e).

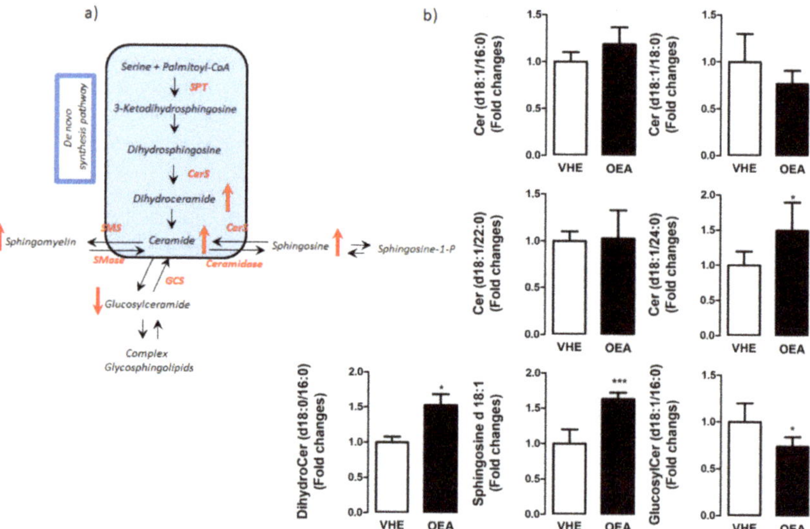

Figure 3. Hepatic sphingolipid analysis by LC-MS/MS. (**a**) Schematic representation of the sphingolipid pathway. De novo sphingolipid synthesis starts with the condensation of serine and palmitoyl-CoA by the rate-limiting enzyme, serine palmitoyltransferase (SPT) to 3-keto-dihydrosphingosine. Following further modification to dihydrosphingosine, ceramide synthases (CerS) convert dihydrosphingosine to dihydroceramide which is then desaturated to generate ceramides. Ceramides can be glycosylated by glucosylceramide synthase (GCS) to glucosylceramides which can serve as an intermediary for other glycosphingolipids, used as a substrate for sphingomyelin synthesis by sphingomyelin synthase (SMS) or deacylated by ceramidases to form sphingosine and subsequently, through the action of sphingosine kinases generate sphingosine-1- phosphate. (**b**) Distribution of sphingolipid species in the liver. In the figure, the mean ± SEM of values obtained from five different samples is reported; *** $p < 0.001$; * $p < 0.05$. Red arrows indicated metabolites that are up and down-regulated in OEA vs. VHE rats.

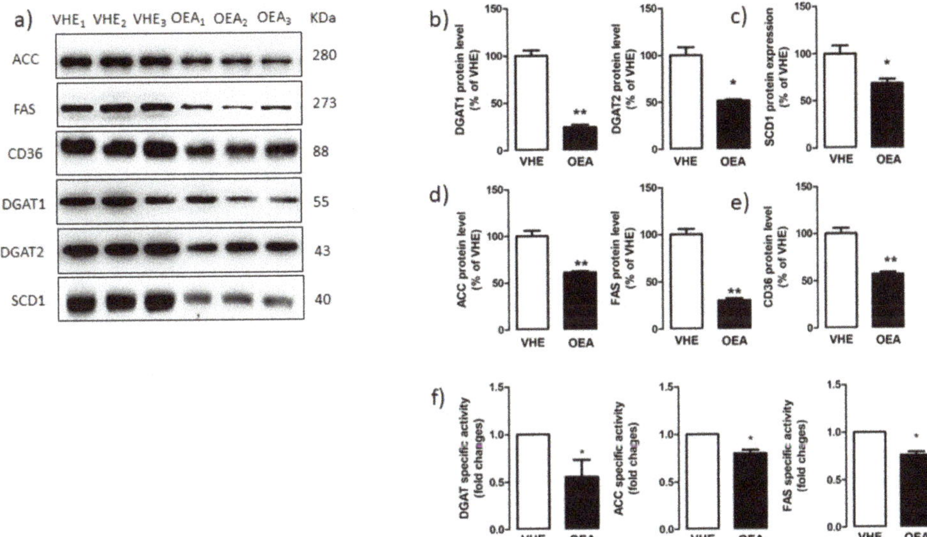

Figure 4. Hepatic fatty acid and triacylglycerol synthesis enzymes. (**a**) Proteins extracted from hepatic VEH and OEA samples were subjected to SDS-page electrophoresis. Membranes were analyzed by immunoblotting using specific antibodies for acetyl-CoA carboxylase (ACC), and fatty acid synthase (FAS), CD36, diacylglycerol acyltransferase 1 (DGAT1), DGAT2, and stearoyl-CoA desaturase 1 (SCD1). Each blot was normalized to the proper specific β-actin. (**b–e**) Blot signals were quantified by densitometric analysis and reported as % of the vehicle (VHE). (**f**) DGAT, ACC and FAS specific activities were assayed as reported in the Material and Methods section and reported as fold changes with respect to values measured in VEH rats. Values are the mean ± SEM of five different experiments. * $p < 0.05$; ** $p < 0.005$.

3.5. Hepatic PPARγ and SREBP-1 Protein Expression

PPARγ regulates several proteins associated with fatty acid synthesis and TAG storage [19]. Moreover, SREBP-1 represents the master transcription factor regulating lipogenic enzyme expression in the liver [55]. We assessed whether the down-regulation of the lipogenic proteins upon OEA treatment could depend on PPARγ and/or SREBP-1c.

By western blot analysis (Figure 5a) we measured the PPARγ and SREBP-1 hepatic protein content in both VEH and OEA groups of rats. We found a significantly decreased protein level of both PPARγ (Figure 5b) and SREBP-1 (Figure 5c) in the liver of OEA rats with respect to VEH rats. The suppressive effect of OEA on PPAγ expression was also demonstrated in vitro by incubating HLF hepatic cells for 2 h with 10 µM OEA (Figure 5d,e). Moreover, the ceramidase inhibitor Carmofur, which increases OEA cell level by inhibiting fatty acid hydrolases [56], induced both alone and in association with OEA, a strong decrease in PPARγ expression (Figure 5d,e).

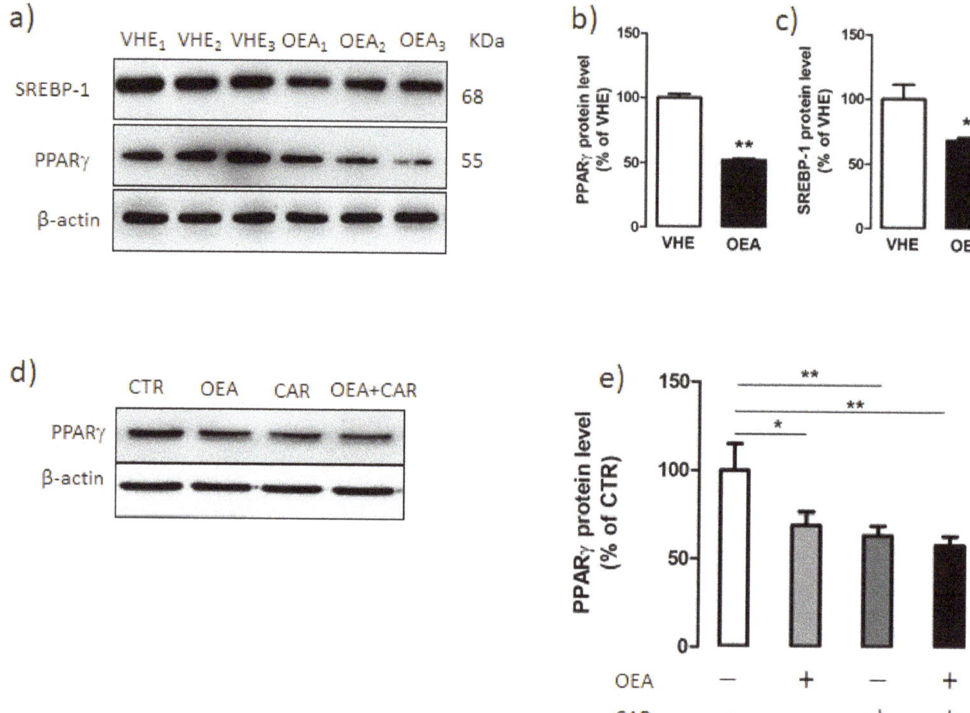

Figure 5. Hepatic PPARγ and SREBP-1 protein expression. (**a**) Hepatic proteins from VEH- and OEA-treated rats were subjected to SDS-page electrophoresis. Membranes were then subjected to immunoblotting using specific antibodies for sterol regulatory binding protein-1 (SREBP-1) and peroxisome proliferator-activated receptor γ (PPARγ). (**b,c**) Signals were quantified by densitometric analysis and expressed as percent of the vehicle (VHE). (**d,e**) PPARγ expression in HLF cells treated for 2 h with 10 μM OEA, 5 μM Carmofur (CAR) and OEA + Carmofur. β-actin was used as the loading control. Blot signals were quantified by densitometric analysis and reported as % of control (CTR) which was represented by cells without any treatment. Values are the mean ± SEM of five different experiments. * $p < 0.05$; ** $p < 0.005$.

4. Discussion

In the present study, we demonstrated that chronic OEA treatment was able to affect hepatic sphingolipid metabolism, and to significantly decrease TAG level with concurrent reduction of PPARγ and SREBP-1 protein expression.

Lipidomic analysis of the whole hepatic lipid component showed that OEA increased both ceramide and sphingomyelin content. It is particularly noteworthy the increased amount of C24:0ceramide induced by OEA, considering the positive correlation recently reported between the increased plasmatic level of C24:0ceramide and decreased cardiovascular events [57], and the link between C24-ceramide, exogenously administered and the reduction in PPAR-γ expression [58,59]. It is also worth considering the decreased glucosylceramide amount measured in the liver after OEA administration, considering that decreased glucosylceramide synthesis was reported to ameliorate insulin sensitivity and decrease TAG accumulation [60].

It is well known that sphingolipid metabolism is subject to very complex regulatory mechanisms and in this context, sphingosine plays a pivotal role. An increase of sphingosine after OEA treatment is apparently in contrast with inhibitory role of OEA toward ceramidase. However, it must be considered that sphingosine can derive also from sphingosine-1P throughout a phosphatase activity [61], or by a salvage pathway by the breakdown of complex lipids [62]. Overall, our preliminary data open to future lipidomic studies aimed at analyzing quantitatively and comprehensively all these lipid species.

In our study, we observed that OEA-induced a significant decrease of PPARγ expression both in vivo and in vitro. In this latter system the link between OEA and PPARγ was reinforced by using Carmofur, a fatty acid hydrolase inhibitor that can block OEA degradation and, thus, increase OEA level in the cell [56].

A large number of recent studies suggest that PPARγ might represent a relevant clinical target for the treatment of NAFLD [63] and another set of data propose a relationship between sphingolipid biosynthesis and NAFLD [62]. For the first time, we found that OEA decreased the hepatic expression of PPARγ and of its downstream target protein CD36, an effect that in turn, inhibits TAG accumulation.

We overall consider that the hepatic reduction of PPARγ expression was linked to a reduction in TAG content and this observation is in keeping with previous studies demonstrating that a reduced PPARγ expression in the liver is associated with hepatic lipogenesis and TAG content [64].

We confirm our hypothesis by investigating the impact of chronic administration of OEA on the activity and the expression of TAG and fatty acid synthesis enzymes.

The final step in TAG synthesis is catalyzed by DGAT isoforms, DGAT1 and DGAT2 [65] with distinct protein sequences and different biochemical, cellular, and physiological functions [52]. PPARγ is involved in the regulation of both DGAT1 and DGAT2 enzyme expression, and interestingly knockdown of hepatic PPARγ was reported to reduce hepatic lipid accumulation and both DGAT1 and DGAT2 expressions [64–67].

We found that DGAT1 and DGAT2 expression and total DGAT activity decreased in the livers of OEA-treated rats. OEA treatment significantly decreased also the hepatic level of SCD1, a member of the fatty acid desaturase family that catalyses the conversion of stearoyl-CoA to oleoyl-CoA, a major substrate for TAG synthesis [53]. All these data are in line with the lower level of TAG we measured in the liver of OEA-treated rats (Figure 2).

SREBP-1c is a transcription factor that activates the expression of most genes required for hepatic lipogenesis, such as ACC, FAS, and SCD1 [68,69]. Increased levels of SREBP-1c mRNA were demonstrated in livers of several mouse models characterized by insulin resistance and increased rates of hepatic lipogenesis [69].

Li et al. [20] reported that OEA inhibits hepatic de novo lipogenesis throughout a SREBP-1-mediated inhibition of SCD1, and ACC mRNA expression [20]. According to that study, we found that OEA decreases the expression of SREBP-1. Moreover, we observed that OEA decreases ACC, FAS, and SCD1 protein expression, and ACC and FAS activities, thus suggesting a possible link between the decreased level of ACC, FAS, and SCD1 and the down-regulation of SREBP-1.

It is well known that polyunsaturated fatty acids (PUFA) can suppress hepatic SREBP-1 nuclear abundance and the expression of its hepatic target genes [70–72]. Considering that NMR analysis of the hepatic lipid composition highlighted an increased level of PUFA in the liver of OEA-treated rats, we can speculate that the down-regulation of SREBP-1 could be associated with the altered lipidomic profile. About the mechanism responsible for the OEA-induced increase in hepatic PUFA, we hypothesize to be a mere consequence of the suppressed de novo lipogenesis that enriches the cell in saturated fatty acids.

Moreover, it should be also considered that NMR analysis showed a significantly increased level of PC in OEA-treated rats (Figure 1c): interestingly, a study reported that a chronically low PC level can drive TAG production regulating SREBP-1 processing [73]. Thus, according to our results, we cannot exclude a similar mechanism occurring also in the liver of OEA-treated animals.

Taken together, our results propose a dual role played by OEA on reducing hepatic TAG level: by recruiting the SREBP-1 system OEA might regulate key proteins involved in fatty acid synthesis, whereas by affecting PPARγ, it might influence fatty acid uptake, binding, and transport.

While the OEA-induced SREBP-1 down-regulation may be almost expected and essentially linked to changes in fatty acid profile, or probable modification in the OEA-induced insulin signaling [74], the PPARγ effect is rather new and deserves deeper investigation.

5. Conclusions

Taken together, our results describe the antiadipogenic functioning of OEA in the liver, which involves not only the *de novo* synthesis of fatty acids and the sphingolipid metabolism but also the transport of fatty acids into the liver and the synthesis of triacylglycerols. These findings add significant and important insights into the molecular mechanism of OEA on hepatic adipogenesis, suggesting an important role of PPARγ in regulating such process and further support the protective role of OEA in fatty-associated liver diseases.

Supplementary Materials: The following are available online at https://www.mdpi.com/2072-6643/13/2/394/s1, Table S1. MS/MS operating conditions. Multiple reaction monitoring (MRM) functions and settings for detection of sphingolipids are shown. Sa: Sphinganine (d18:0); d 17:0 Sa: Sphinganine (d17:0); So: Sphingosine (d18:1); d 17:1 So: Sphingosine (d17:1); Sa1P: Sphinganine-1-phosphate (d18:0); d 17:0 Sa1P: Sphinganine-1-phosphate (d17:0); So1P: Sphingosine-1-phosphate (d18:1); d 17:1 So1P: Sphingosine-1-phosphate (d17:1); C16Cer: C16:0 Ceramide (d18:1/16:0); C17Cer: C17 Ceramide (d18:1/17:0); C18Cer: C18 Ceramide (d18:1/18:0); C16dHCer: C16 Dihydroceramide (d18:0/16:0); C16GlcCer: C16 Glucosyl(ß) Ceramide (d18:1/16:0); C17GlcCer: C17 Glucosyl(ß) Ceramide (d18:1/17:0); C22Cer: C22 Ceramide (d18:1/22:0); C24Cer: C24 Ceramide (d18:1/24:0); C24dHCer: C24 Dihydroceramide (d18:0/24:0). Figure S1: Rats daily administered with OEA 10 mg kg^{-1} i.p. (black squares) showed a significantly lower body weight gain with respect to VEH treated rats (open gray squares). Such difference was significant starting from the fourth day of treatment and lasted for the rest of the treatment period. Statistical analysis was performed by two way ANOVA for repeated measures, followed by Bonferroni's post hoc test for multiple comparisons. Data are expressed as means $\pm$ SEM; $^{\circ\circ}$ $p < 0.01$, $^{\circ\circ\circ}$ $p < 0.001$ vs. respective VEH; N = 12; Figure S2: ^{1}H NMR spectra of lipid extracts with identified NMR signals (region A: 3.00–0.65 ppm, typical of fatty acids and sterol methyl and methylene resonances; region B: 5.00–3.00 ppm for phospholipids head groups and glycerol backbone proton resonances; region C: 6.00–5.00 ppm for vinyl protons resonances). Numbers are referred to corresponding signal assignments reported in Table 1; Figure S3: (a) Partial least squares discriminant analysis (PLS-DA) score plot and (b) cross validation details (10-fold), in terms of R^2 and Q^2 for OEA and VEH samples.

Author Contributions: Conceptualization, A.M.G. and S.G.; Methodology, D.V., S.G., A.R., M.F., S.L., P.D.B., I.C., and S.V.; Software, M.F., D.V. and L.D.C.; Validation, F.P.F.; Formal Analysis, A.M.G. and L.D.C.; Investigation, A.M.G.; Resources, S.G.; Data Curation, A.M.G.; Writing—Original Draft Preparation, A.M.G.; Writing—Review & Editing, S.G., D.V. and A.R.; Supervision, F.P.F., A.M.G. and S.G.; Funding Acquisition, S.G. All authors have read and agreed to the published version of the manuscript.

Funding: This research was supported by grants (PRIN 2017XZMBYX_004 to SG) of the Italian Ministry of Education, University and Research.

Institutional Review Board Statement: The study was conducted according to the guidelines of the Declaration of Helsinki, in accordance with the European directive 2010/63/UE governing animal welfare, with the Italian Ministry of Health guidelines for the care and use of laboratory animals and upon approval of the local Veterinary Office and of the Italian Ministry of Health (D.lgs 116/92—Art. 7—Comunicazione del 18/06/2013).

Informed Consent Statement: Not applicable.

Conflicts of Interest: The authors declare no conflict of interest.

References

1. Romano, A.; Tempesta, B.; Provensi, G.; Passani, M.B.; Gaetani, S. Central mechanisms mediating the hypophagic effects of oleoylethanolamide and N-acylphosphatidylethanolamines: Different lipid signals? *Front. Pharmacol.* **2015**, *6*. [CrossRef]
2. Piomelli, D. A fatty gut feeling. *Trends Endocrinol. Metab.* **2013**, *24*, 332–341. [CrossRef]

3. De Fonseca, F.R.; Navarro, M.; Gómez, R.; Escuredo, L.; Nava, F.; Fu, J.; Murillo-Rodríguez, E.; Giuffrida, A.; LoVerme, J.; Gaetani, S.; et al. An anorexic lipid mediator regulated by feeding. *Nature* **2001**, *414*, 209–212. [CrossRef]

4. Petersen, G.; Sørensen, C.; Schmid, P.C.; Artmann, A.; Tang-Christensen, M.; Hansen, S.H.; Larsen, P.J.; Schmid, H.H.; Hansen, H.S. Intestinal levels of anandamide and oleoylethanolamide in food-deprived rats are regulated through their precursors. *Biochim. Biophys. Acta (BBA) Mol. Cell Biol. Lipids* **2006**, *1761*, 143–150. [CrossRef] [PubMed]

5. Fu, J.; Astarita, G.; Gaetani, S.; Kim, J.; Cravatt, B.F.; Mackie, K.; Piomelli, D. Food Intake Regulates Oleoylethanolamide Formation and Degradation in the Proximal Small Intestine. *J. Biol. Chem.* **2007**, *282*, 1518–1528. [CrossRef] [PubMed]

6. Schwartz, G.J.; Fu, J.; Astarita, G.; Li, X.; Gaetani, S.; Campolongo, P.; Cuomo, V.; Piomelli, D. The Lipid Messenger OEA Links Dietary Fat Intake to Satiety. *Cell Metab.* **2008**, *8*, 281–288. [CrossRef] [PubMed]

7. Guijarro, A.; Fu, J.; Astarita, G.; Piomelli, D. CD36 gene deletion decreases oleoylethanolamide levels in small intestine of free-feeding mice. *Pharmacol. Res.* **2010**, *61*, 27–33. [CrossRef] [PubMed]

8. Sun, Y.; Alexander, S.; Garle, M.; Gibson, C.L.; Hewitt, K.; Murphy, S.P.; Kendall, D.; Bennett, A. Cannabinoid activation of PPARα; a novel neuroprotective mechanism. *Br. J. Pharmacol.* **2007**, *152*, 734–743. [CrossRef]

9. Sayd, A.; Antón, M.; Alén, F.; Caso, J.R.; Pavón, F.J.; Leza, J.C.; de Fonseca, F.R.; García-Bueno, B.; Orio, L. Systemic Administration of Oleoylethanolamide Protects from Neuroinflammation and Anhedonia Induced by LPS in Rats. *Int. J. Neuropsychopharmacol.* **2015**, *18*, pyu111. [CrossRef]

10. Antón, M.; Alén, F.; de Heras, R.G.; Serrano, A.; Pavón, F.J.; Leza, J.C.; García-Bueno, B.; de Fonseca, F.R.; Orio, L. Oleoylethanolamide prevents neuroimmune HMGB1/TLR4/NF-kB danger signaling in rat frontal cortex and depressive-like behavior induced by ethanol binge administration. *Addict. Biol.* **2016**, *22*, 724–741. [CrossRef]

11. Campolongo, P.; Roozendaal, B.; Trezza, V.; Cuomo, V.; Astarita, G.; Fu, J.; McGaugh, J.L.; Piomelli, D. Fat-induced satiety factor oleoylethanolamide enhances memory consolidation. *Proc. Natl. Acad. Sci. USA* **2009**, *106*, 8027–8031. [CrossRef] [PubMed]

12. Van Kooten, M.J.; Veldhuizen, M.G.; de Araujo, I.E.; O'Malley, S.S.; Small, D.M. Fatty acid amide supplementation decreases impulsivity in young adult heavy drinkers. *Physiol. Behav.* **2016**, *155*, 131–140. [CrossRef] [PubMed]

13. Costa, A.; Cristiano, C.; Cassano, T.; Gallelli, C.A.; Gaetani, S.; Ghelardini, C.; Blandina, P.; Calignano, A.; Passani, M.B.; Provensi, G. Histamine-deficient mice do not respond to the antidepressant-like effects of oleoylethanolamide. *Neuropharmacology* **2018**, *135*, 234–241. [CrossRef] [PubMed]

14. Jin, P.; Yu, H.-L.; Lan, T.; Zhang, F.; Quan, Z.-S. Antidepressant-like effects of oleoylethanolamide in a mouse model of chronic unpredictable mild stress. *Pharmacol. Biochem. Behav.* **2015**, *133*, 146–154. [CrossRef] [PubMed]

15. Gaetani, S.; Romano, A.; Provensi, G.; Ricca, V.; Lutz, T.; Passani, M.B. Eating disorders: From bench to bedside and back. *J. Neurochem.* **2016**, *139*, 691–699. [CrossRef] [PubMed]

16. Romano, A.; di Bonaventura, M.M.; Gallelli, C.A.; Koczwara, J.B.; Smeets, D.; Giusepponi, M.E.; de Ceglia, M.; Friuli, M.; di Bonaventura, E.M.; Scuderi, C.; et al. Oleoylethanolamide decreases frustration stress-induced binge-like eating in female rats: A novel potential treatment for binge eating disorder. *Neuropsychopharmacology* **2020**, *45*, 1931–1941. [CrossRef]

17. Fu, J.; Oveisi, F.; Gaetani, S.; Lin, E.; Piomelli, D. Oleoylethanolamide, an endogenous PPAR-α agonist, lowers body weight and hyperlipidemia in obese rats. *Neuropharmacology* **2005**, *48*, 1147–1153. [CrossRef]

18. Thabuis, C.; Destaillats, F.; Lambert, D.M.; Muccioli, G.G.; Maillot, M.; Harach, T.; Tissot-Favre, D.; Martin, J.-C. Lipid transport function is the main target of oral oleoylethanolamide to reduce adiposity in high-fat-fed mice. *J. Lipid Res.* **2011**, *52*, 1373–1382. [CrossRef]

19. Desvergne, B.; Wahli, W. Peroxisome Proliferator-Activated Receptors: Nuclear Control of Metabolism. *Endocr. Rev.* **1999**, *20*, 649–688. [CrossRef]

20. Li, L.; Li, L.; Chen, L.; Lin, X.; Xu, Y.; Ren, J.; Fu, J.; Qiu, Y. Effect of oleoylethanolamide on diet-induced nonalcoholic fatty liver in rats. *J. Pharmacol. Sci.* **2015**, *127*, 244–250. [CrossRef]

21. Di Paola, M.; Bonechi, E.; Provensi, G.; Costa, A.; Clarke, G.; Ballerini, C.; De Filippo, C.; Passani, M.B. Oleoylethanolamide treatment affects gut microbiota composition and the expression of intestinal cytokines in Peyer's patches of mice. *Sci. Rep.* **2018**, *8*, 14881. [CrossRef] [PubMed]

22. Furlong, S.J.; Ridgway, N.D.; Hoskin, D.W. Modulation of ceramide metabolism in T-leukemia cell lines potentiates apoptosis induced by the cationic antimicrobial peptide bovine lactoferricin. *Int. J. Oncol.* **2008**, *32*, 537–544. [CrossRef] [PubMed]

23. Chavez, J.A.; Knotts, T.A.; Wang, L.-P.; Li, G.; Dobrowsky, R.T.; Florant, G.L.; Summers, S.A. A Role for Ceramide, but Not Diacylglycerol, in the Antagonism of Insulin Signal Transduction by Saturated Fatty Acids. *J. Biol. Chem.* **2003**, *278*, 10297–10303. [CrossRef]

24. Chmura, S.J.; Nodzenski, E.; Kharbanda, S.; Pandey, P.; Quintáns, J.; Kufe, D.; Weichselbaum, R.R. Down-Regulation of Ceramide Production Abrogates Ionizing Radiation-Induced CytochromecRelease and Apoptosis. *Mol. Pharmacol.* **2000**, *57*, 792–796. [CrossRef] [PubMed]

25. Lee, Y.K.; Park, J.E.; Lee, M.; Hardwick, J.P. Hepatic lipid homeostasis by peroxisome proliferator-activated receptor gamma 2. *Liver Res.* **2018**, *2*, 209–215. [CrossRef]

26. Zhang, Y.-L.; Hernandez-Ono, A.; Siri, P.; Weisberg, S.; Conlon, D.; Graham, M.J.; Crooke, R.M.; Huang, L.-S.; Ginsberg, H.N. Aberrant Hepatic Expression of PPARγ2 Stimulates Hepatic Lipogenesis in a Mouse Model of Obesity, Insulin Resistance, Dyslipidemia, and Hepatic Steatosis. *J. Biol. Chem.* **2006**, *281*, 37603–37615. [CrossRef]

27. Yamazaki, T.; Shiraishi, S.; Kishimoto, K.; Miura, S.; Ezaki, O. An increase in liver PPARγ2 is an initial event to induce fatty liver in response to a diet high in butter: PPARγ2 knockdown improves fatty liver induced by high-saturated fat. *J. Nutr. Biochem.* **2011**, *22*, 543–553. [CrossRef]

28. Lee, Y.J.; Ko, E.H.; Kim, J.E.; Kim, E.; Lee, H.; Choi, H.; Yu, J.H.; Kim, H.J.; Seong, J.-K.; Kim, K.-S. Nuclear receptor PPAR -regulated monoacylglycerol O-acyltransferase 1 (MGAT1) expression is responsible for the lipid accumulation in diet-induced hepatic steatosis. *Proc. Natl. Acad. Sci. USA* **2012**, *109*, 13656–13661. [CrossRef]

29. Bonen, A.; Parolin, M.L.; Steinberg, G.R.; Calles-Escandon, J.; Tandon, N.N.; Glatz, J.F.C.; Luiken, J.J.F.P.; Heigenhauser, G.J.F.; Dyck, D.J. Triacylglycerol accumulation in human obesity and type 2 diabetes is associated with increased rates of skeletal muscle fatty acid transport and increased sarcolemmal FAT/CD36. *FASEB J.* **2004**, *18*, 1144–1146. [CrossRef]

30. Su, X.; Abumrad, N.A. Cellular fatty acid uptake: A pathway under construction. *Trends Endocrinol. Metab.* **2009**, *20*, 72–77. [CrossRef]

31. Koonen, D.P.; Jacobs, R.L.; Febbraio, M.; Young, M.E.; Soltys, C.-L.M.; Ong, H.; Vance, D.E.; Dyck, J.R. Increased Hepatic CD36 Expression Contributes to Dyslipidemia Associated With Diet-Induced Obesity. *Diabetes* **2007**, *56*, 2863–2871. [CrossRef] [PubMed]

32. Zhou, J.; Febbraio, M.; Wada, T.; Zhai, Y.; Kuruba, R.; He, J.; Lee, J.H.; Khadem, S.; Ren, S.; Li, S.; et al. Hepatic Fatty Acid Transporter Cd36 Is a Common Target of LXR, PXR, and PPARγ in Promoting Steatosis. *Gastroenterology* **2008**, *134*, 556–567. [CrossRef] [PubMed]

33. Hui, S.T.; Parks, B.W.; Org, E.; Norheim, F.; Che, N.; Pan, C.; Castellani, L.W.; Charugundla, S.; Dirks, D.L.; Psychogios, N.; et al. The genetic architecture of NAFLD among inbred strains of mice. *eLife* **2015**, *4*, e05607. [CrossRef] [PubMed]

34. Gaetani, S.; Fu, J.; Cassano, T.; DiPasquale, P.; Romano, A.; Righetti, L.; Cianci, S.; Laconca, L.; Giannini, E.; Scaccianoce, S.; et al. The fat-induced satiety factor oleoylethanolamide suppresses feeding through central release of oxytocin. *J. Neurosci.* **2010**, *30*, 8096–8101. [CrossRef] [PubMed]

35. Romano, A.; Potes, C.S.; Tempesta, B.; Cassano, T.; Cuomo, V.; Lutz, T.; Gaetani, S. Hindbrain noradrenergic input to the hypothalamic PVN mediates the activation of oxytocinergic neurons induced by the satiety factor oleoylethanolamide. *Am. J. Physiol. Metab.* **2013**, *305*, E1266–E1273. [CrossRef]

36. Romano, A.; Gallelli, C.A.; Koczwara, J.B.; Braegger, F.E.; Vitalone, A.; Falchi, M.; di Bonaventura, M.M.; Cifani, C.; Cassano, T.; Lutz, T.A.; et al. Role of the area postrema in the hypophagic effects of oleoylethanolamide. *Pharmacol. Res.* **2017**, *122*, 20–34. [CrossRef]

37. Giuffrida, A.; de Fonseca, F.R.; Piomelli, D. Quantification of Bioactive Acylethanolamides in Rat Plasma by Electrospray Mass Spectrometry. *Anal. Biochem.* **2000**, *280*, 87–93. [CrossRef]

38. Adosraku, R.K.; Choi, G.T.; Constantinou-Kokotos, V.; Anderson, M.M.; Gibbons, W.A. NMR lipid profiles of cells, tissues, and body fluids: Proton NMR analysis of human erythrocyte lipids. *J. Lipid Res.* **1994**, *35*, 1925–1931. [CrossRef]

39. De Castro, F.; Vergaro, V.; Benedetti, M.; Baldassarre, F.; del Coco, L.; Dell'Anna, M.M.; Mastrorilli, P.; Fanizzi, F.; Ciccarella, G. Visible Light-Activated Water-Soluble Platicur Nanocolloids: Photocytotoxicity and Metabolomics Studies in Cancer Cells. *ACS Appl. Bio Mater.* **2020**. [CrossRef]

40. Yeboah, F.A.; Adosraku, R.K.; Nicolaou, A.; Gibbons, W.A. Proton Nuclear Magnetic Resonance Lipid Profiling of Intact Platelet Membranes. *Ann. Clin. Biochem. Int. J. Lab. Med.* **1995**, *32*, 392–398. [CrossRef]

41. Giudetti, A.M.; Guerra, F.; Longo, S.; Beli, R.; Romano, R.; Manganelli, F.; Nolano, M.; Mangini, V.; Santoro, L.; Bucci, C. An altered lipid metabolism characterizes Charcot-Marie-Tooth type 2B peripheral neuropathy. *Biochim. Biophys. Acta (BBA) Mol. Cell Biol. Lipids* **2020**, *1865*, 158805. [CrossRef] [PubMed]

42. Rossi, C.; Cicalini, I.; Zucchelli, M.; di Ioia, M.; Onofrj, M.; Federici, L.; del Boccio, P.; Pieragostino, D. Metabolomic Signature in Sera of Multiple Sclerosis Patients during Pregnancy. *Int. J. Mol. Sci.* **2018**, *19*, 3589. [CrossRef] [PubMed]

43. Giudetti, A.M.; de Domenico, S.; Ragusa, A.; Lunetti, P.; Gaballo, A.; Franck, J.; Simeone, P.; Nicolardi, G.; de Nuccio, F.; Santino, A.; et al. A specific lipid metabolic profile is associated with the epithelial mesenchymal transition program. *Biochim. Biophys. Acta (BBA) Mol. Cell Biol. Lipids* **2019**, *1864*, 344–357. [CrossRef]

44. Giudetti, A.M.; di Bonaventura, M.V.M.; Ferramosca, A.; Longo, S.; di Bonaventura, E.M.; Friuli, M.; Romano, A.; Gaetani, S.; Cifani, C. Brief daily access to cafeteria-style diet impairs hepatic metabolism even in the absence of excessive body weight gain in rats. *FASEB J.* **2020**, *34*, 9358–9371. [CrossRef] [PubMed]

45. Realini, N.; de Solorzano, C.O.; Pagliuca, C.; Pizzirani, D.; Armirotti, A.; Luciani, R.; Costi, M.P.; Bandiera, T.; Piomelli, D. Discovery of highly potent acid ceramidase inhibitors with in vitro tumor chemosensitizing activity. *Sci. Rep.* **2013**, *3*, 1035. [CrossRef]

46. Damiano, F.; de Benedetto, G.E.; Longo, S.; Giannotti, L.; Fico, D.; Siculella, L.; Giudetti, A.M. Decanoic Acid and Not Octanoic Acid Stimulates Fatty Acid Synthesis in U87MG Glioblastoma Cells: A Metabolomics Study. *Front. Neurosci.* **2020**, *14*, 783. [CrossRef]

47. Van den Berg, R.A.; Hoefsloot, H.C.J.; Westerhuis, J.A.; Smilde, A.K.; van der Werf, M.J. Centering, scaling, and transformations: Improving the biological information content of metabolomics data. *BMC Genom.* **2006**, *7*, 142. [CrossRef]

48. Chong, J.; Soufan, O.; Li, C.; Caraus, I.; Li, S.; Bourque, G.; Wishart, D.S.; Xia, J. MetaboAnalyst 4.0: Towards more transparent and integrative metabolomics analysis. *Nucleic Acids Res.* **2018**, *46*, W486–W494. [CrossRef]

49. Trygg, J.; Wold, S. Orthogonal projections to latent structures (O-PLS). *J. Chemom.* **2002**, *16*, 119–128. [CrossRef]

50. Barrilero, R.; Gil, M.; Amigó, N.; Dias, C.B.; Wood, L.G.; Garg, M.L.; Ribalta, J.; Heras, M.; Vinaixa, M.; Correig, X. LipSpin: A New Bioinformatics Tool for Quantitative 1H NMR Lipid Profiling. *Anal. Chem.* **2018**, *90*, 2031–2040. [CrossRef]

51. Baumstark, D.; Kremer, W.; Boettcher, A.; Schreier, C.; Sander, P.; Schmitz, G.; Kirchhoefer, R.; Huber, F.; Kalbitzer, H.R. 1H NMR spectroscopy quantifies visibility of lipoproteins, subclasses, and lipids at varied temperatures and pressures. *J. Lipid Res.* **2019**, *60*, 1516–1534. [CrossRef] [PubMed]

52. Bonzom, P.M.; Nicolaou, A.; Zloh, M.; Baldeo, W.; Gibbons, W.A. NMR lipid profile of Agaricus bisporus. *Phytochemistry* **1999**, *50*, 1311–1321. [CrossRef]

53. Flowers, M.T.; Ntambi, J.M. Role of stearoyl-coenzyme A desaturase in regulating lipid metabolism. *Curr. Opin. Lipidol.* **2008**, *19*, 248–256. [CrossRef] [PubMed]

54. Coburn, C.T.; Hajri, T.; Ibrahimi, A.; Abumrad, N.A. Role of CD36 in Membrane Transport and Utilization of Long-Chain Fatty Acids by Different Tissues. *J. Mol. Neurosci.* **2001**, *16*, 117–122. [CrossRef]

55. Jump, D.B.; Tripathy, S.; Depner, C.M. Fatty Acid–Regulated Transcription Factors in the Liver. *Annu. Rev. Nutr.* **2013**, *33*, 249–269. [CrossRef] [PubMed]

56. Wu, K.; Xiu, Y.; Zhou, P.; Qiu, Y.; Li, Y. A New Use for an Old Drug: Carmofur Attenuates Lipopolysaccharide (LPS)-Induced Acute Lung Injury via Inhibition of FAAH and NAAA Activities. *Front. Pharmacol.* **2019**, *10*, 818. [CrossRef] [PubMed]

57. Kasumov, T.; Li, L.; Li, M.; Gulshan, K.; Kirwan, J.P.; Liu, X.; Previs, S.; Willard, B.; Smith, J.D.; McCullough, A. Ceramide as a Mediator of Non-Alcoholic Fatty Liver Disease and Associated Atherosclerosis. *PLoS ONE* **2015**, *10*, e0126910. [CrossRef]

58. Li, Y.; Dong, J.; Ding, T.; Kuo, M. S.; Cao, G.; Jiang, X.-C.; Li, Z. Sphingomyelin Synthase 2 Activity and Liver Steatosis: An effect of ceramide-mediated peroxisome proliferator-activated receptor γ2 suppression. *Arter. Thromb. Vasc. Biol.* **2013**, *33*, 1513–1520. [CrossRef]

59. Park, W.-J.; Park, J.-W.; Merrill, A.H.; Storch, J.; Pewzner-Jung, Y.; Futerman, A.H. Hepatic fatty acid uptake is regulated by the sphingolipid acyl chain length. *Biochim. Biophys. Acta (BBA) Mol. Cell Biol. Lipids* **2014**, *1841*, 1754–1766. [CrossRef]

60. Zhao, H.; Przybylska, M.; Wu, I.-H.; Zhang, J.; Maniatis, P.; Pacheco, J.; Piepenhagen, P.; Copeland, D.; Arbeeny, C.; Shayman, J.A.; et al. Inhibiting glycosphingolipid synthesis ameliorates hepatic steatosis in obese mice. *Hepatology* **2009**, *50*, 85–93. [CrossRef]

61. Mao, C.; Obeid, L.M. Ceramidases: Regulators of cellular responses mediated by ceramide, sphingosine, and sphingosine-1-phosphate. *Biochim. Biophys. Acta (BBA) Mol. Cell Biol. Lipids* **2008**, *1781*, 424–434. [CrossRef] [PubMed]

62. Pagadala, M.; Kasumov, T.; McCullough, A.J.; Zein, N.N.; Kirwan, J.P. Role of ceramides in nonalcoholic fatty liver disease. *Trends Endocrinol. Metab.* **2012**, *23*, 365–371. [CrossRef] [PubMed]

63. Fougerat, A.; Montagner, A.; Loiseau, N.; Guillou, H.; Wahli, W. Peroxisome Proliferator-Activated Receptors and Their Novel Ligands as Candidates for the Treatment of Non-Alcoholic Fatty Liver Disease. *Cells* **2020**, *9*, 1638. [CrossRef] [PubMed]

64. Miyazaki, M.; Flowers, M.T.; Sampath, H.; Chu, K.; Otzelberger, C.; Liu, X.; Ntambi, J.M. Hepatic Stearoyl-CoA Desaturase-1 Deficiency Protects Mice from Carbohydrate-Induced Adiposity and Hepatic Steatosis. *Cell Metab.* **2007**, *6*, 484–496. [CrossRef]

65. Zhang, W.; Sun, Q.; Zhong, W.; Sun, X.; Zhou, Z. Hepatic Peroxisome Proliferator-Activated Receptor Gamma Signaling Contributes to Alcohol-Induced Hepatic Steatosis and Inflammation in Mice. *Alcohol Clin. Exp. Res.* **2016**, *40*, 988–999. [CrossRef]

66. Boelsterli, U.A.; Bedoucha, M. Toxicological consequences of altered peroxisome proliferator-activated receptor γ (PPARγ) expression in the liver: Insights from models of obesity and type 2 diabetes. *Biochem. Pharmacol.* **2002**, *63*, 1–10. [CrossRef]

67. Bensinger, S.J.; Tontonoz, P. Integration of metabolism and inflammation by lipid-activated nuclear receptors. *Nature* **2008**, *454*, 470–477. [CrossRef]

68. Horton, J.D.; Goldstein, J.L.; Brown, M.S. SREBPs: Activators of the complete program of cholesterol and fatty acid synthesis in the liver. *J. Clin. Investig.* **2002**, *109*, 1125–1131. [CrossRef]

69. Liang, G.; Yang, J.; Horton, J.D.; Hammer, R.E.; Goldstein, J.L.; Brown, M.S. Diminished Hepatic Response to Fasting/Refeeding and Liver X Receptor Agonists in Mice with Selective Deficiency of Sterol Regulatory Element-binding Protein-1c. *J. Biol. Chem.* **2002**, *277*, 9520–9528. [CrossRef]

70. Worgall, T.S.; Sturley, S.L.; Seo, T.; Osborne, T.F.; Deckelbaum, R.J. Polyunsaturated Fatty Acids Decrease Expression of Promoters with Sterol Regulatory Elements by Decreasing Levels of Mature Sterol Regulatory Element-binding Protein. *J. Biol. Chem.* **1998**, *273*, 25537–25540. [CrossRef]

71. Mater, M.K.; Thelen, A.P.; Pan, D.A.; Jump, D.B. Sterol Response Element-binding Protein 1c (SREBP1c) Is Involved in the Polyunsaturated Fatty Acid Suppression of Hepatic S14 Gene Transcription. *J. Biol. Chem.* **1999**, *274*, 32725–32732. [CrossRef]

72. Xu, J.; Nakamura, M.T.; Cho, H.P.; Clarke, S.D. Sterol Regulatory Element Binding Protein-1 Expression Is Suppressed by Dietary Polyunsaturated Fatty Acids: A mechanism for the coordinate suppression of lipogenic genes by polyunsaturated fats. *J. Biol. Chem.* **1999**, *274*, 23577–23583. [CrossRef] [PubMed]

73. Walker, A.K.; Jacobs, R.L.; Watts, J.L.; Rottiers, V.; Jiang, K.; Finnegan, D.M.; Shioda, T.; Hansen, M.; Yang, F.; Niebergall, L.J.; et al. A Conserved SREBP-1/Phosphatidylcholine Feedback Circuit Regulates Lipogenesis in Metazoans. *Cell* **2011**, *147*, 840–852. [CrossRef] [PubMed]

74. De Ubago, M.M.; García-Oya, I.; Pérez-Pérez, A.; Canfrán-Duque, A.; Quintana-Portillo, R.; de Fonseca, F.R.; González-Yanes, C.; Sánchez-Margalet, V. Oleoylethanolamide, a natural ligand for PPAR-alpha, inhibits insulin receptor signalling in HTC rat hepatoma cells. *Biochim. Biophys. Acta (BBA) Mol. Cell Biol. Lipids* **2009**, *1791*, 740–745. [CrossRef] [PubMed]

 nutrients

Article

Dampened Muscle mTORC1 Response Following Ingestion of High-Quality Plant-Based Protein and Insect Protein Compared to Whey

Gommaar D'Hulst [1,2,*], Evi Masschelein [1] and Katrien De Bock [1]

[1] Laboratory of Exercise and Health, Department of Health Sciences and Technology, Swiss Federal Institute of Technology (ETH) Zurich, 8603 Zurich, Switzerland; evi-masschelein@ethz.ch (E.M.); katrien-debock@ethz.ch (K.D.B.)

[2] Laboratory of Regenerative and Movement Biology, Department of Health Sciences and Technology, Swiss Federal Institute of Technology (ETH) Zurich, 8093 Zurich, Switzerland

* Correspondence: dhulstg@ethz.ch; Tel.: +41-44-655-73-86

Abstract: Increased amino acid availability acutely stimulates protein synthesis partially via activation of mechanistic target of rapamycin complex 1 (mTORC1). Plant-and insect-based protein sources matched for total protein and/or leucine to animal proteins induce a lower postprandial rise in amino acids, but their effects on mTOR activation in muscle are unknown. C57BL/6J mice were gavaged with different protein solutions: whey, a pea–rice protein mix matched for total protein or leucine content to whey, worm protein matched for total protein, or saline. Blood was drawn 30, 60, 105 and 150 min after gavage and muscle samples were harvested 60 min and 150 min after gavage to measure key components of the mTORC1 pathway. Ingestion of plant-based proteins induced a lower rise in blood leucine compared to whey, which coincided with a dampened mTORC1 activation, both acutely and 150 min after administration. Matching total leucine content to whey did not rescue the reduced rise in plasma amino acids, nor the lower increase in mTORC1 compared to whey. Insect protein elicits a similar activation of downstream mTORC1 kinases as plant-based proteins, despite lower postprandial aminoacidemia. The mTORC1 response following ingestion of high-quality plant-based and insect proteins is dampened compared to whey in mouse skeletal muscle.

Keywords: muscle; mTORC1; plant-based protein; whey; insect; muscle protein synthesis

Citation: D'Hulst, G.; Masschelein, E.; De Bock, K. Dampened Muscle mTORC1 Response Following Ingestion of High-Quality Plant-Based Protein and Insect Protein Compared to Whey. *Nutrients* **2021**, *13*, 1396. https://doi.org/10.3390/nu13051396

Academic Editor: Anna M. Giudetti

Received: 5 March 2021
Accepted: 14 April 2021
Published: 21 April 2021

Publisher's Note: MDPI stays neutral with regard to jurisdictional claims in published maps and institutional affiliations.

1. Introduction

Maintenance of skeletal muscle mass is key for metabolic health throughout life [1]. Total muscle mass is generally determined by basal rates of protein synthesis and the ability to stimulate protein synthesis after the intake of protein [2,3]. The enhancement of protein synthesis is primarily induced by the activation of mechanistic target of rapamycin complex 1 (mTORC1)-signalling [4–7]. Growth factors, energetic stress, and muscular contractions control mTORC1 directly or via its upstream inhibitor tuberous sclerosis complex 1/2 (TSC1/2) [8]. Amino acids (AA) regulate mTORC1 via alternative pathways involving Rag GTPases, which recruit mTORC1 to the lysosomal membrane [9–11]. How TSC2-dependent signals regulate mTORC1 in skeletal muscle has been intensely studied, but how fluctuations of amino acids, and in particular fluctuations in the essential amino acid leucine, regulate mTORC1, and thus skeletal muscle mass in vivo is much less well understood.

Our world's population is projected to reach 10 billion by 2050 [12] and protein demand is expected to double by then [13]. In the future, we will no longer be able to cover the animal-based protein demand at rates we are consuming now [14]. Plant-based proteins currently account for 30–50% of total dietary protein intake in many countries [15], with numbers exceeding 60% in less developed countries [16]. More research is emerging on how plant-based proteins affect muscle health [17–19] and muscle remodelling, especially in combination with resistance training [20–23]. Considerable efforts have been made recently to understand how non-animal-based proteins affect the acute protein synthetic response [24–27], but studies assessing the direct effects on acute mTORC1 signalling are lacking.

A rapid postprandial rise in essential amino acids (EAA) concentrations modulates the increase in mTORC1 activity [28,29] after protein ingestion. As animal-based protein sources are rapidly digested and contain high concentrations of EAA, in particular leucine, it is believed that animal-based proteins are more anabolic than plant-based sources [17]. The latter contain several anutritious molecules such as tannins, trypsin inhibitors, and phytates that slow digestion [30]. Moreover, plant-based proteins result in larger portions of splanchnic nitrogen retention/oxidation and hence lower AA appearance in the blood [31]. Nevertheless, innovations in food processing have solved many issues regarding plant-based protein digestibility. Recent advancements in production of plant-based protein concentrates, isolates, and hydrolysates and the potential of mixing different plant sources have been put forward as a strategy to match EAA content of animal-derived protein sources [22], but these hypotheses are yet to be tested. In addition to plant-based proteins, insect-based protein mixtures might be another promising strategy to meet the rapidly growing protein demand. Insect protein has been reported to induce a similar postprandial increase in essential AA (EAA) as soy, but inferior to whey [32]. Similar to vegan proteins, the effects of insect protein isolates on mTORC1 activity remain unexplored.

Hence, the aim of this study is to assess AA appearance in the blood and subsequent mTORC1 signalling acutely after administration of whey or of more sustainable, high quality protein sources based on plants (pea and rice isolate) and insects (buffalo worm protein). We hypothesise that whey protein would promote superior blood concentrations of EAA and branched-chain amino acids (BCAA) compared to total protein-matched vegan proteins or insect isolates, mirrored by superior acute mTORC1 activation. Additionally, we hypothesise to observe a similar EAA and mTORC1 response to whey when a vegan protein mix is leucine-matched to whey.

2. Materials and Methods

2.1. Animals

All experiments were performed on male C57BL/6J mice. Mice were housed in individually ventilated cages (3–4 littermates per cage) at standard housing conditions (22 °C, 12 h light/dark cycle, dark phase starting at 7 pm), with ad libitum access to chow (KlibaNafag, diet #3436 and diet #3437) and water. Health status of all mouse lines was regularly monitored according to FELASA guidelines. All animal procedures were approved by the Veterinary office of the Canton of Zürich (license nr ZH137/2020).

2.2. Experimental Procedures

An overview of the experimental procedures can be found in Figure 1. A number of 8–12-week old male C57BL/6J mice were fasted for 4–5 h from the beginning of the light cycle. Subsequently, mice were gavaged with saline (0.9% NaCl), Whey (myprotein, Northwitch, UK), a pea–rice Vegan Mix (UniProt, Kaltenkirchen, Germany), Vegan Mix++ leucine-matched to whey (Uniprot) and pulverized buffalo Worm protein (Protifarm, Ermelo, The Netherlands) (Table 1). Blood was drawn 30, 60, 105 and 150 min after gavage, mice were sacrificed 60 min and 150 min after gavage for harvesting muscle samples. Each AA-containing condition contained the same amount of total protein per mixture (4.57 g·kg^{-1}), except Vegan Mix++ (5.48 g·kg^{-1}). The Vegan Mix++ was leucine matched to whey (0.475 g leucine·kg^{-1}), hence it contained 20% more total protein. The concentrations chosen correspond to ~40% of total daily leucine intake and have been shown to induce a robust increase in mTORC1 signaling and MPS in rodents [29,33]. AA content of each amino acid mixture was internally assessed by the respective company where the protein source was obtained. Nitrogen to protein ratio (N:P) for whey, vegan mix and worm protein is 6.38, 5.17–5.44 and 5.41, respectively. Values are obtained from previous studies [34–37]. An overview of AA mixture specifications is provided Table 1.

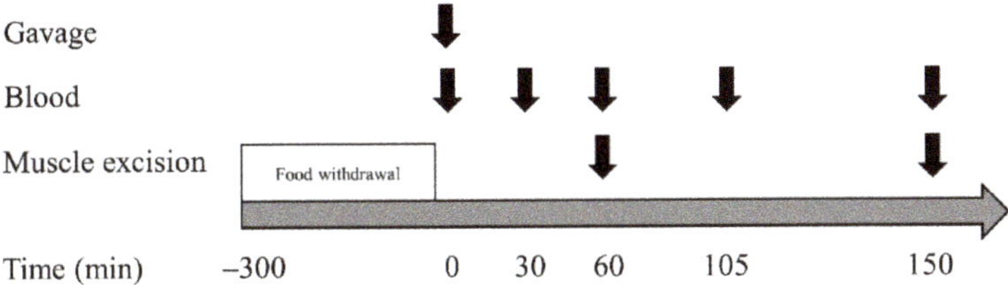

Figure 1. Experimental overview.

Table 1. Amino Acid (AA) mixture specifications.

Composition (g/100g Mix)	Whey		Vegan Mix		Worm Protein	
Protein	80.0		78.0		61.7	
Fat	6.7		0.9		24.3	
Carbohydrate	5.0		3.2		8.0	
AA profile (g/100g)	100 g AA	100 g mix	100 g AA	100 g mix	100 g AA	100 g mix
Lysine	9.24	7.39	6.60	5.15	7.08	4.05
Histidine	1.89	1.51	2.40	1.87	3.58	2.05
Methionine	2.01	1.61	1.60	1.25	1.50	0.86
Phenylanine	3.10	2.48	5.30	4.13	4.72	2.70
Threonine	6.68	5.34	3.40	2.65	4.37	2.50
leucine	10.43	8.34	8.30	6.47	7.20	4.12
Isoleucine	6.31	5.05	4.30	3.35	4.61	2.64
Valine	5.80	4.64	5.30	4.13	6.08	3.48
Alanine	4.93	3.94	4.40	3.43	7.08	4.05
Arginine	2.38	1.90	8.60	6.71	5.80	3.32
Aspartic Acid	10.74	8.59	11.20	8.74	9.67	5.53
Glycine	1.75	1.40	4.10	3.20	4.95	2.83
Glutamic Acid	17.73	14.18	19.10	14.90	12.50	7.15
Cystine	1.87	1.50	1.80	1.40	0.94	0.54
Proline	6.27	5.02	4.00	3.12	6.33	3.62
Serine	4.55	3.64	5.00	3.90	4.53	2.59
Tyrosine	2.88	2.30	3.60	2.81	7.80	4.46
Tryptophan	1.45	1.16	1.10	0.86	1.26	0.72
Amount gavaged (g·kg^{-1})	Saline	Whey	Vegan Mix	Vegan Mix++	Worm	
Total protein	0	4.57	4.57	5.48	4.57	
Leucine	0	0.48	0.38	0.48	0.30	

2.3. Sample Collection

Mice were anesthetized using Ketamine/Xylazine at 115 µg·g^{-1} and 13 µg·g^{-1} body weight respectively via intraperitoneal injection. The depth of anaesthesia was confirmed by testing pedal withdrawal reflex before tissue collection. Subsequently, the m. gastrocnemius (GAS), m. tibialis anterior (TA), m. soleus (SOL), m. plantaris (PLT) were dissected and snap frozen.

2.4. Protein Extraction and Western Blot

Sample Preparation: Between 10 and 25 mg of muscle sample (m. tibialis anterior, TA) was homogenized in ice cold lysis buffer (1:10, *w/v*) (50 mM Tris-HCl pH 7.0, 270 mM sucrose, 5 mM EGTA, 1 mM EDTA, 1 mM sodium orthovanadate, 50 mM glycerophosphate, 5 mM sodium pyrophosphate, 50 mM sodium fluoride, 1 mM DTT, 0.1% Triton-X 100 and 10% protease inhibitor) (20 µL per 1.8–2.5 g of tissue sample) using an OMNI-THq Tissue homogenizer (OMNI International, Kennesaw, GA, USA) for 20 s until a consistent homogenate was formed. Samples were centrifuged at 4 °C at 10,000× g for 10 min and the supernatant with proteins collected. Protein concentration was determined using the DC assay protein method to equalize the amount of protein. Samples were prepared 3:4 with 4× laemmli buffer containing 10% 2-mercaptoethanol and heated at 95° for 5 min. An amount of 20–40 µg of total protein was loaded in a 15-well pre-casted gradient gel (Bio-rad, #456–8086). After electrophoresis, a picture of the gel was taken under UV-light to determine protein loading using stain-free technology. Proteins were transferred via semi-dry transfer onto a polyvinylidene fluoride membrane (Bio-rad, #170–4156) and subsequently blocked for 1 h at room temperature with 5% milk in TBS- Tween. Membranes were incubated overnight at 4 °C with primary antibodies from Cell Signaling Technology; pS6K1^{Thr389} (#9206), pS6$^{Ser235/236}$ (#2211), pmTOR^{Ser2448} (#5536), peEF2^{Thr56} (#2331) and p4E-BP1^{Ser65} (#9451). The appropriate secondary antibodies (1:5000) for anti-rabbit and anti-mouse IgG HRP-linked antibodies (Cell signalling, #7074) were used for chemiluminescent detection of proteins. Membranes were scanned with a chemidoc imaging system (Bio-rad) and quantified using Image lab software (Bio-rad).

2.5. Amino Acid Determination from Blood

A total of 5 nmol of stable isotope labelled amino acids was added (Cambridge Isotope Laboratories) to 10 µL of serum. Proteins were precipitated by adding 9 volumes ice cold methanol at −20 °C. The supernatant was dried and the amino acids were re-constituted in 100 µL 0.1 % acetic acid. Samples (5 µL) were subjected to liquid chromatography coupled multiple reaction monitoring mass spectrometry (LC-MRM-MS). The acquired data were integrated and analysed using the open source software tool Skyline [38]. Area under the Curve (AUC) was calculated using the linear trapezoidal method.

2.6. Statistical and Data Analyses

Results are presented as mean with standard error of the mean (SEM) bars. Additional individual data points are provided for the protein signalling data. Data were subjected to a one-way analysis of variance (ANOVA) (protein signalling) or two-way ANOVA (plasma AA) to generate a p value and post hoc tests were performed using Tukey's post hoc test using Graphpad Prism to compare between groups. All data passed the Shapiro–Wilk test for normality and sample sizes ($n = 5$–7) were calculated a priori via power calculations (1-β: 0.8) using G*Power statistical software. Significance was set at $p < 0.05$. Exact n numbers are provided in the figure and table legends.

3. Results

3.1. Amino Acid Concentrations in Plasma

The postprandial rise in EAA, and in particular BCAA, is essential for mTORC1 activation and subsequent induction of muscle protein synthesis (MPS) in skeletal muscle. BCAA appearance in blood plasma is shown in Figure 2 and Table 2. Plasma leucine increased 30, 60 and 105 min after whey gavage (Figure 2A) and returned to baseline levels after 150 min. Vegan Mix and Vegan Mix++ induced a dampened plasma leucine response compared to whey, with no differences between conditions. Worm protein induced the lowest leucine response, which was lower in comparison to both Vegan Mix, Vegan Mix++ as well as whey (Figure 2A and Table 2). All this resulted in a reduced AUC for both plant-based mixtures and worm protein compared to whey (Figure 2B). The other two BCAA, isoleucine and valine, followed near identical patterns (Figure 2C–F). Essential amino acids (AA) methionine and threonine also displayed a higher, more prolonged peak after whey supplementation compared to Vegan Mix, Vegan Mix++ and worm protein (Table 2). Other amino acids were only marginally affected by any of the AA mixtures (for details, see Table 2).

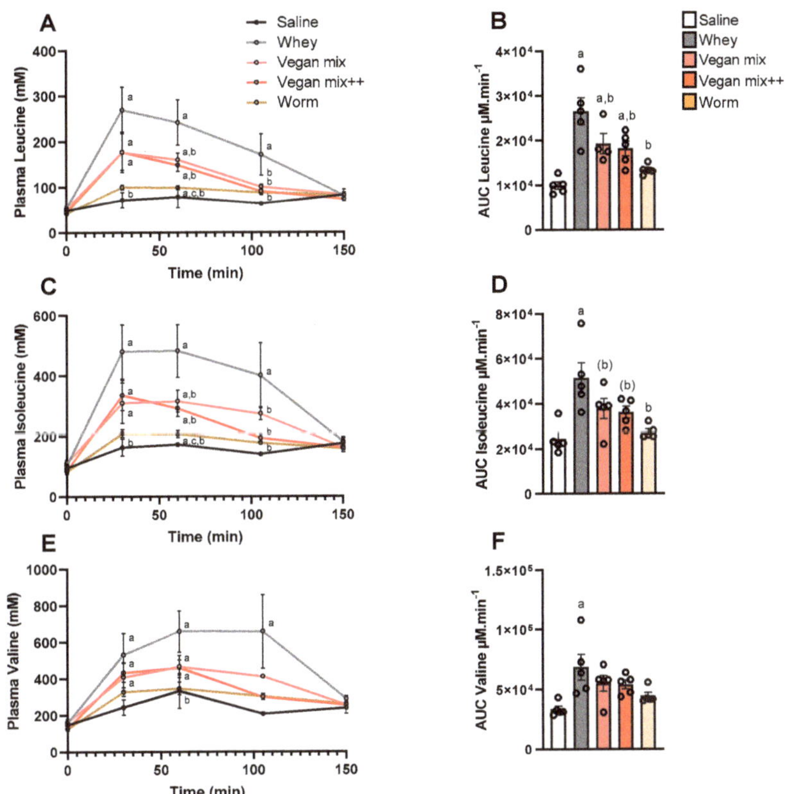

Figure 2. Increase in plasma BCAA upon gavage of different AA mixtures. (**A**) Plasma leucine. (**B**) Leucine area under the curve (AUC). (**C**) Plasma isoleucine. (**D**) Isoleucine AUC. (**E**) Plasma Valine. (**F**) Valine AUC. Data is presented as mean $\pm$ SEM, $n = 5$. [a] $p < 0.05$ vs. saline, [b] $p < 0.05$ vs. whey, [(b)] $p < 0.10$ vs. whey. Vegan mix [++] annotates the leucine-matched (to whey) vegan mix.

3.2. Higher Acute mTORC1 Response with Whey Compared to Plant- and Insect-Based Protein

Blood analysis showed that BCAAs peaked 30–60 min after gavage. Hence, to evaluate whether the higher BCAA response 60 min after whey supplementation would also lead to a higher mTORC1 response in skeletal muscle, we measured the phosphorylation of critical downstream mTORC1 targets ribosomal protein S6 kinase (pS6K1^{Thr389}), ribosomal protein S6 (pS6$^{ser235/236}$) and eukaryotic elongation factor 2 (peEF2^{Thr56}) exactly 60 min after gavage. All AA mixtures strongly increased pS6K1 and pS6, but the effect was larger with whey ($\sim$8-fold vs. saline) compared to plant-based and insect-based protein ($\sim$5-fold), with no differences between both vegan mix conditions and worm (Figure 3A–C,E). Phosphorylation of mTOR at Ser2448 showed an identical pattern (Figure 3C,E). peEF2 remained unaffected by supplementation (Figure 3D,E).

Table 2. Amino acid (AA) in blood plasma after administration of different AA-mixtures.

0 min

	Saline	Whey	Vegan Mix	Vegan Mix++	Worm
Essential AA (µM) $n = 5$					
Lysine	229.4 ± 22.5	253.2 ± 15.5	267.4 ± 20.2	211.7 ± 19.1	216.9 ± 15.9
Histidine	47.9 ± 4.6	59.2 ± 5.6	69.0 ± 5.3	56.3 ± 4.7	51.6 ± 3.3
Methionine	56.7 ± 7.7	53.7 ± 3.7	61.7 ± 7.0	46.8 ± 4.8	44.6 ± 6.7
Phenylalanine	114.7 ± 6.7	139.8 ± 4.2	132.7 ± 16.2	116.8 ± 9.9	115.1 ± 6.1
Threonine	129.3 ± 19.7	136.3 ± 9.5	143.5 ± 22.3	115.4 ± 14.2	118.8 ± 13.2
Leucine	48.4 ± 6.5	53.1 ± 1.4	52.4 ± 6.4	44.7 ± 4.9	42.4 ± 3.3 [b]
Isoleucine	96.5 ± 10.4	104.8 ± 2.6	112.3 ± 9.3	88.6 ± 9.3 [b]	85.1 ± 9.3 [b]
Valine	147.1 ± 16.5	159.3 ± 5.8	164.6 ± 14.7	130.2 ± 13.2	128.5 ± 10.3
Non-essential AA (µM)					
Alanine	205. ± 23.4	213.2 ± 15.6	228.8 ± 29.0	170.0 ± 18.8	197.6 ± 18.6
Arginine	85.9 ± 10.4	90.9 ± 7.2	94.6 ± 7.0	69.4 ± 5.3	84.8 ± 4.5
Aspartate	70.2 ± 14.7	81.0 ± 10.8	82.7 ± 15.1	72.2 ± 8.6	78.9 ± 20.7
Cysteine	1.1 ± 0.4	0.5 ± 0.2	0.7 ± 0.3	0.4 ± 0.2	0.9 ± 0.5
Glutamine	84.2 ± 12.6	98.0 ± 19.2	95.8 ± 10.7	88.5 ± 7.9	71.8 ± 11.7
Glycine	1.1 ± 0.4	0.5 ± 0.2	0.7 ± 0.3	0.4 ± 0.2	0.9 ± 0.5
Proline	71.2 ± 6.6	76.4 ± 2.9	83.1 ± 12.5	62.6 ± 5.8	67.6 ± 5.6
Serine	80.9 ± 10.2	98.5 ± 12.6	92.9 ± 14.7	82.9 ± 11.8	88.3 ± 15.2
Tyrosine	60.2 ± 8.0	63.2 ± 7.0	67.3 ± 11.4	56.2 ± 8.7	53.4 ± 7.9

30 min

	Saline	Whey	Vegan Mix	Vegan Mix++	Worm
Lysine	396.5 ± 65.3	840.6 ± 167.3 [a]	635.8 ± 122.0 [a]	643.1 ± 88.5 [a]	477.8 ± 45.2 [bd]
Histidine	65.5 ± 5.0	75.5 ± 12.5	78.2 ± 4.2	80.4 ± 10.7	74.9 ± 4.2
Methionine	64.1 ± 3.1	124.4 ± 29.1 [a]	83.9 ± 15.2	78.8 ± 15.3 [b]	71.9 ± 5.6
Phenylalanine	173.4 ± 15.2	231.4 ± 28.2 [a]	231.0 ± 13.4 [a]	246.5 ± 7.8 [a]	214.1 ± 10.2 [ad]
Threonine	183.3 ± 8.7	376.3 ± 83.2 [a]	287.8 ± 53.1 [a]	268.3 ± 49.1 [a]	217.4 ± 22.4 [b]
Leucine	71.6 ± 16.7	269.2 ± 50.5 [a]	177.0 ± 44.5 [a]	177.5 ± 40.0 [a]	99.9 ± 4.9 [b]
Isoleucine	163.3 ± 28.1	479.7 ± 89.9 [a]	310.3 ± 67.1 [a]	335.9 ± 49.4 [a]	210.4 ± 5.1 [b]
Valine	243.5 ± 42.1 [b]	531.2 ± 116 [a]	409.9 ± 79.5 [a]	433.1 ± 51.5 [a]	328.2 ± 23.7
Alanine	263.8 ± 19.7	364.7 ± 73.9	350.1 ± 35.2	336.9 ± 43.3	305.9 ± 21.0
Arginine	106.4 ± 29.8	144.4 ± 21.9	172.3 ± 29.4 [a]	168.2 ± 7.8	127.7 ± 15.1 [d]
Aspartate	68.0 ± 11.5	130.0 ± 41.6 [a]	84.1 ± 15.0	106.7 ± 18.6	65.4 ± 11.2
Cysteine	2.4 ± 1.4 [a]	1.3 ± 0.6	1.0 ± 0.5	1.3 ± 0.3	2.2 ± 1.1
Glutamine	75.8 ± 11.3	159.1 ± 75.2 [a]	125.8 ± 21.8 [a]	126.6 ± 16.0 [a]	88.9 ± 11.2
Glycine	2.4 ± 1.4 [a]	1.3 ± 0.6	1.0 ± 0.5	1.3 ± 0.3	2.2 ± 1.1
Proline	103.7 ± 12.8	176.0 ± 32.4	147.3 ± 15.9	153.4 ± 24.5	149.1 ± 11.9
Serine	98.6 ± 4.9	127.6 ± 18.7 [a]	122.6 ± 6.5 [a]	126.6 ± 15.3 [a]	106.6 ± 5.3
Tyrosine	119.0 ± 20.9	231.5 ± 44.6 [a]	192.6 ± 31.9 [a]	209.9 ± 16.4 [a]	176.1 ± 11.0 [a]

60 min

	Saline	Whey	Vegan Mix	Vegan Mix++	Worm
Lysine	505.7 ± 128.6	938.9 ± 201.5 [a]	664.9 ± 70.3 [a]	663.0 ± 61.8 [a]	446.0 ± 41.8 [bcd]
Histidine	94.8 ± 17.0	95.7 ± 17.4	98.9 ± 2.4	96.0 ± 13.5	79.0 ± 3.5 [c]
Methionine	64.0 ± 1.9	128.8 ± 32.0 [a]	66.3 ± 10.1 [b]	56.9 ± 8.9 [b]	70.1 ± 6.0 [b]
Phenylalanine	193.1 ± 9.0	226.6 ± 30.7	241.3 ± 9.9 [a]	259.8 ± 11.5 [a]	217.1 ± 5.9 [cd]
Threonine	232.0 ± 29.3	455.9 ± 96.4 [a]	294.1 ± 22.9	266.9 ± 29.2 [b]	215.5 ± 22.4 [bc]
Leucine	78.5 ± 23.7	241.8 ± 49.5 [a]	160.6 ± 14.2 [ab]	148.8 ± 13.9 [ab]	99.3 ± 4.5 [bcd]
Isoleucine	172.9 ± 87.0	482.1 ± 36.2 [a]	316.3 ± 27.9 [ab]	293.0 ± 12.3 [ab]	208.0 ± 10.9 [bcd]
Valine	333.8 ± 94.7	658.7 ± 113 [a]	467.3 ± 60.6 [a]	461.4 ± 41.4 [a]	347.3 ± 34.3 [b]
Alanine	372.5 ± 59.3	412.6 ± 71.8	374.1 ± 18.1	389.0 ± 48.3	325.9 ± 21.9
Arginine	113.2 ± 26.1	155.1 ± 37.3	152.2 ± 26.2	176.3 ± 25.9	109.5 ± 8.2 [d]
Aspartate	93.4 ± 33.8	133.7 ± 35.8 [a]	95.4 ± 8.1	106.6 ± 18.9	81.4 ± 13.3
Cysteine	1.4 ± 0.6	4.9 ± 3.3 [a]	1.7 ± 0.4	1.5 ± 0.3	1.5 ± 0.3
Glutamine	137.2 ± 42.0	201.3 ± 67.9 [a]	130.2 ± 11.7 [b]	162.9 ± 30.4 [b]	113.2 ± 32.3
Glycine	1.4 ± 0.6	4.9 ± 3.3 [a]	1.7 ± 0.4	1.5 ± 0.3	1.5 ± 0.3
Proline	148.8 ± 26.9	176.2 ± 60.2	150.2 ± 6.8	163.6 ± 23.4	141.0 ± 7.7
Serine	161.4 ± 30.5	154.1 ± 49.6	143.6 ± 11.4	161.6 ± 26.8	130.3 ± 19.1
Tyrosine	130.6 ± 44.6	265.8 ± 63.7	189.6 ± 32.8	196.7 ± 19.2	175.2 ± 12.0

105 min

	Saline	Whey	Vegan Mix	Vegan Mix++	Worm
Lysine	301.4 ± 45.2	1014.1 ± 452.1	580.2 ± 7.1	410.1 ± 34.0	435.2 ± 50.1
Histidine	72.0 ± 6.0	125.3 ± 18.6	90.1 ± 8.0	76.8 ± 8.1	84.4 ± 9.0
Methionine	44.5 ± 5.0	119.3 ± 30.0 [a]	62.1 ± 3.7 [b]	44.4 ± 0.2 [b]	56.2 ± 6.0 [b]
Phenylalanine	178.7 ± 2.0	245.8 ± 83.6	234.1 ± 8.8	210.6 ± 4.9	226.6 ± 7.8
Threonine	145.0 ± 19.5	479.1 ± 176.2	259.8 ± 14.6	187.7 ± 7.6	192.3 ± 8.0
Leucine	63.5 ± 15.5	171.2 ± 46.5 [a]	100.9 ± 5.5 [b]	90.8 ± 8.0 [b]	88.2 ± 4.9 [b]
Isoleucine	141.2 ± 15.2	401.4 ± 107.7 [a]	275.4 ± 21.2 [b]	194.0 ± 13.3 [b]	178.2 ± 15.5 [b]
Valine	208.2 ± 50.1	657.9 ± 202 [a]	411.7 ± 4.8	300.8 ± 16.8	304.0 ± 20.3
Alanine	221.7 ± 20.3	420.3 ± 170.3	357.3 ± 67.5	287.2 ± 22.6	306.3 ± 19.2
Arginine	48.7 ± 16.2	121.2 ± 23.5	157.8 ± 39.4	135.5 ± 29.8	104.8 ± 10.2
Aspartate	52.3 ± 25.6	128.2 ± 35.5 [a]	85.5 ± 16.4	97.7 ± 15.9	76.8 ± 17.9
Cysteine	3.0 ± 0.1	6.3 ± 3.8	1.5 ± 0.3	1.5 ± 0.8	1.4 ± 0.5
Glutamine	64.2 ± 18.3	225.1 ± 75.6	189.3 ± 71.0	181.3 ± 42.1	142.0 ± 23.6
Glycine	3.0 ± 0.2	6.3 ± 3.8	1.5 ± 0.3	1.5 ± 0.8	1.4 ± 0.5
Proline	69.2 ± 20.2	192.5 ± 87.5 [a]	137.1 ± 27.1	110.6 ± 11.9	130.4 ± 8.2
Serine	80.4 ± 8.0	201.1 ± 80.5	162.4 ± 49.8	141.6 ± 19.2	132.2 ± 11.2
Tyrosine	62.7 ± 12.1	212.8 ± 134.0	152.5 ± 37.5	145.1 ± 45.2	94.5 ± 1.4

150 min

	Saline	Whey	Vegan Mix	Vegan Mix++	Worm
Lysine	304.5 ± 42.2	399.4 ± 73.5	332.9 ± 25.3	294.9 ± 25.1	308.9 ± 18.0
Histidine	84.1 ± 14.1	73.4 ± 4.3	64.3 ± 4.8	66.2 ± 4.8	65.5 ± 1.9
Methionine	39.4 ± 1.9	39.9 ± 2.6	34.3 ± 2.1	31.5 ± 1.2	37.4 ± 2.0
Phenylalanine	183.9 ± 12.0	170.1 ± 12.9	180.6 ± 8.0	170.2 ± 5.2	178.1 ± 12.1
Threonine	143.6 ± 19.1	157.7 ± 22.8	132.3 ± 5.4	137.1 ± 15.9	122.9 ± 9.4
Leucine	82.6 ± 12.8	81.7 ± 5.8	83.4 ± 5.8	73.5 ± 4.4	82.8 ± 5.3
Isoleucine	176.6 ± 20.4	182.2 ± 9.8	163.8 ± 9.8	162.5 ± 6.3	159.2 ± 13.45
Valine	239.9 ± 32.4	289.9 ± 16.2	256.3 ± 21.0	257.5 ± 6.3	262.6 ± 13.4
Alanine	219.8 ± 29.5	204.6 ± 15.3	199.0 ± 4.5	228.7 ± 28.5	223.0 ± 20.3
Arginine	46.9 ± 14.2	25.9 ± 7.9	35.2 ± 8.2	34.0 ± 7.5	24.4 ± 4.7
Aspartate	23.1 ± 8.9	19.1 ± 4.2	15.2 ± 4.1	17.6 ± 2.0	[illegible]
Cysteine	2.3 ± 0.5	2.0 ± 0.3	3.0 ± 1.1	2.1 ± 0.3	3.3 ± 0.8
Glutamine	44.6 ± 11.0	25.6 ± 2.0	27.7 ± 3.1	27.6 ± 4.6	28.9 ± 3.7
Glycine	2.3 ± 0.5	2.0 ± 0.3	3.0 ± 1.0	2.1 ± 0.3	3.3 ± 0.3
Proline	62.7 ± 12.0	51.2 ± 3.5	52.4 ± 4.6	51.8 ± 4.4	55.3 ± 4.8
Serine	66.2 ± 8.3	61.9 ± 4.5	62.7 ± 11.0	[illegible]	56.2 ± 4.1
Tyrosine	66.2 ± 13.7	62.2 ± 14.0	62.6 ± 8.1	57.8 ± 6.3	76.9 ± 11.2

Data is presented as mean ± SEM, n = 5. [a] $p < 0.05$ vs. saline, [b] $p < 0.05$ vs. whey, [c] $p < 0.10$ vs. Vegan Mix. [d] $p < 0.10$ vs. Vegan Mix++. Vegan mix++ annotates the leucine-matched (to whey) vegan mix.

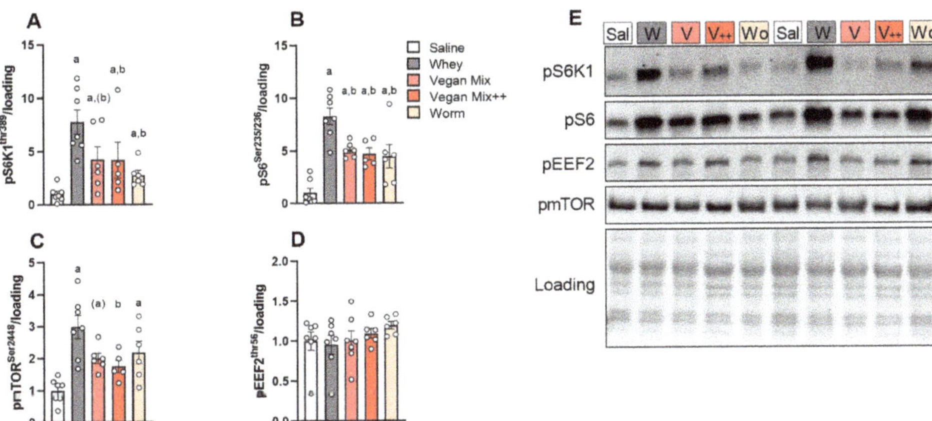

Figure 3. mTORC1 signalling 60 min in *m. Tibialis Anterior* after gavage of different AA mixtures. (**A**) pS6K1^{Thr389}. (**B**) pRPS6$^{Ser235/236}$. (**C**) pmTOR^{Ser2448}. (**D**) pEEF2^{Thr56}. (**E**) Representative blots. Data is presented as mean ± SEM, $n = 6$–7. Sal (Saline), W (Whey), V (Vegan Mix), V++ (Vegan Mix++), Wo (Worm protein). ^{a} $p < 0.05$ vs. saline, ^{b} $p < 0.05$ vs. whey, $^{(a)}$ $p < 0.10$ vs. saline, $^{(b)}$ $p < 0.10$ vs. whey.

3.3. No Prolonged mTORC1 Response with Plant- or Insect-Based Protein

Previous reports in humans have shown a delayed but overall elevated muscle synthetic response with a plant-based source compared to whey, despite a lower overall postprandial amino acid availability with the former [24]. Hence, to evaluate the potential of a delayed mTORC1 response with animal- and/or insect-based protein sources, we assessed critical components of mTORC1 2.5 h after gavage. pS6K1 and pS6 where marginally increased in all AA conditions, but this did not reach statistical significance due to higher variation compared to the acute (60 min after gavage) condition (Figure 4A,B,E). Phosphorylation of 4E-BP1 at Ser65 was increased three-fold in the Vegan mix++ condition (Figure 4C,E, $p < 0.05$). Phospho-mTOR at Ser2448 remained unaffected in all conditions (Figure 4D,E).

4. Discussion

This study assessed changes in AA concentrations in the blood and mTORC1 response after gavage of animal-based proteins (whey), plant-based protein (a pea–rice vegan mix) and insect proteins (pulverized worm). The major findings were that all protein sources robustly increased aminoacidemia and subsequently mTORC1 signalling in mouse skeletal muscle acutely after gavage. However, whey induced a ∼2-fold larger increase in plasma BCAA compared to plant-and insect-based proteins, with a subsequent 1.5-fold increase in downstream mTORC1 signalling. A plant-based protein source that was leucine-matched to whey did not rescue the dampened mTORC1 response. Finally, 2.5 h after gavage, aminoacidemia returned to baseline in all conditions, with no obvious prolonged mTORC1 response with plant-based, nor insect-based proteins.

Ingestion of plant-based proteins results in a lower postprandial muscle synthetic response compared to nitrogen-matched animal-based sources such as whey [26,27], skimmed milk [27] or beef [39] in both resting and post-exercise conditions. Lower anabolic properties of single source plant-based proteins may be attributed to inferior BCAA, in particular, leucine, concentrations compared to most animal proteins [17]. Mixing different plant strains to obtain an amino acid profile similar to those of animal sources might be a promising strategy to enhance protein quality of plant proteins [40–42]. Yet, our data does not confirm this, as the high-quality pea–rice vegan mix used in this study caused a strongly reduced peak in leucine uptake in the blood compared to whey, even when total leucine content was matched (Vegan Mix++). Increased splanchnic extraction/oxidation and subsequent urea synthesis have been shown with plant-based proteins [31] as well as reduced digestion due to the presence of anutritious components such as tannins, trypsin inhibitors and phytates [30]. However, the latter is unlikely an important factor in our study set-up as both vegan mixtures lead to an overall lower AUC for leucine compared to whey. In effect, a delay in AA absorption would have resulted in a similar overall AUC. Interestingly, although the leucine-matched vegan mix (Vegan Mix++) contained 20% more total protein and 25% more leucine than the normal vegan mix, AA appearance was identical, suggesting high splanchnic retention with higher doses of plant-based proteins.

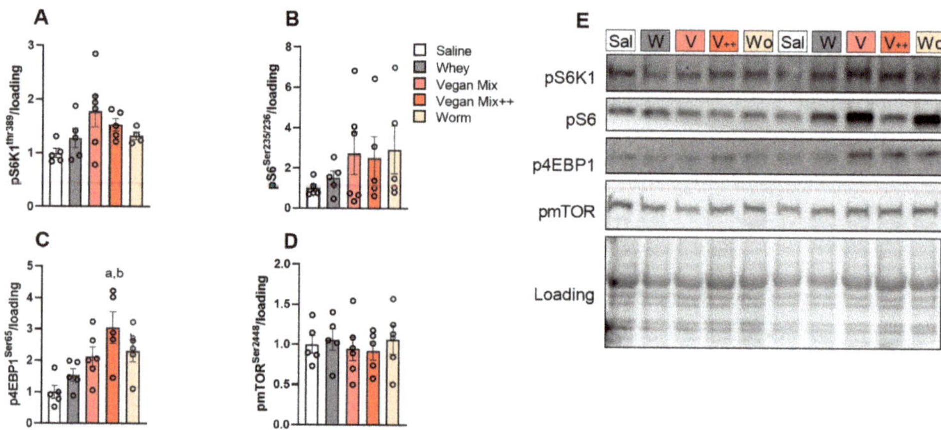

Figure 4. mTORC1 signalling 150 min in *m. Tibialis Anterior* after gavage of different AA mixtures. (**A**) pS6K1Thr389. (**B**) pRPS6Ser235/236. (**C**) p4EBP1Ser65 (**D**) pmTORSer2448. (**E**) Representative blots. Valine AUC. Data is presented as mean ±SEM, n = 5–6. Sal (Saline), W (Whey), V (Vegan Mix), V++ (Vegan Mix++), Wo (Worm). [a] $p < 0.05$ vs. saline, [b] $p < 0.05$ vs. whey.

Next to plant-based protein, other eco-friendly protein options, such as insect proteins, are currently investigated in terms of quality and downstream effects on muscle metabolism. Our data is in line with previous human results showing that insect protein induces blood AA concentrations similar to soy and inferior to whey protein [32]. In fact, we demonstrate that worm protein resulted in a ~30% lower AUC for all BCAA compared to both vegan mixes. It is important to note that although total administered protein was identical to whey and the pea–rice vegan mix, total leucine content was, respectively 38% and 21% lower, which might explain the slightly reduced BCAA blood appearance with insect protein. Similarly, as shown in Table 1, the worm protein contains more carbohydrates and fats than both the vegan mix and whey. The presence of other macronutrients, in particular fats, might have delayed gastric emptying and AA absorption in our study [34]. Future studies should investigate whether an insect source matched for leucine to whey and stripped from excess fats induces a similar leucine response.

BCAAs and in particular leucine trigger protein synthesis via the activation of mTORC1 [4,28,43,44]. Numerous studies have shown potent activation of downstream mTORC1 signalling with whole animal-based protein sources [28,45,46], but data on mTORC1 activation with non-animal-based protein blends is scarce. Recent human data showed a reduced increase in blood leucine with ingestion of fungal derived protein (mycoprotein) compared to milk protein, even though total ingested leucine was matched [23]. Despite lower leucine availability, MPS was higher with mycoprotein in both resting and exercised conditions, while mTOR phosphorylation at ser2448 was unaffected. As no other downstream kinases were assed in this study, it is difficult to draw strong conclusions on how reduced leucine availability in the mycoprotein group affected mTORC1 signalling. Earlier reports have shown a larger increase in *p*-S6K1 in rats fed a meal containing 20% whey compared to 20% soy [47], which is in line with reduced peak in plasma leucine after consumption of various plant-based proteins of which leucine content was matched to whey [42]. Our data supports this as both vegan mixes induced lower rise in BCAA, which resulted in a 1.5-fold lower increase in phosphorylation of various downstream mTORC1 kinases such as *p*-S6K1 and *p*-S6 compared to whey. Plant and insect protein are often described as slow-digesting [32,48] leading to a prolonged increase in protein synthetic rate, despite lower overall postprandial amino acid concentrations, also at later time-points [23,24]. Although we only measured markers of protein synthesis, we did not observe a prolonged mTORC1 response 2.5 h after plant-or insect protein administration, suggesting that mTORC1 activation closely mirrors the rise in plasma leucine, while its downstream effects on protein synthesis might be delayed. Furthermore, the fact that acute (after 60 min) and prolonged (after 2.5 h) mTORC1 activation was similar between the vegan mixes and insect proteins, despite a lower aminoacidemia in the latter, suggests that the origin of ingested proteins can alter mTORC1's sensitivity to leucine, possibly via other micro- or macronutrients [23].

The clinical relevance of our data lies in the nutritional support for alternative protein sources to activate mTORC1 and potentially muscle remodeling. Although the doses used in this study are similar to what is generally used in human studies after accounting for dose translation from rodent to humans [49], caution must be exercised to directly extrapolate our findings to humans. Mice and humans show differential digestion patterns. While mice have an exponential digestion curve, where most of the gastric emptying is finished within the first hour after nutrient ingestion, humans show a more linear digestion pattern [50]. This is a limitation of our study as interspecies digestion patters might have affected blood leucine concentrations at later time-points (e.g., 150 min) after gavage.

5. Conclusions

In conclusion, ingestion of plant-based proteins matched for total protein and leucine induce a lower rise in blood leucine, with a subsequent dampened mTORC1 activation, compared to whey, both acutely and 2.5 h after administration. Insect protein elicits a similar activation of downstream mTORC1 kinases as plant-based proteins, despite lower postprandial aminoacidemia.

Therefore, future research should focus on the mechanisms behind altered mTORC1 sensitivity related to different protein sources. Finally, further methods to increase the ability of plant-based proteins to activate muscle protein synthesis should be explored; these might include improved means to increase digestion and absorption.

Author Contributions: G.D. designed the study, wrote the manuscript and performed all the experiments and data analysis. E.M. helped performing the experiments and edited the manuscript. K.D.B. designed the study and helped drafting and revising the manuscript. All authors have read and agreed to the published version of the manuscript.

Funding: Internal ETH Zurich funding.

Institutional Review Board Statement: The study was conducted according to the guidelines of the Declaration of Helsinki, and approved by the Institutional Review Board (or Ethics Committee) of the Veterinary office of the Canton of Zürich (license nr ZH137/2020).The study was conducted according to the guidelines of the Declaration of Helsinki, and approved by the Veterinary office of the Canton of Zürich, CH under the license nr. ZH137/2020 on 17.11.2020.

Informed Consent Statement: Not applicable.

Data Availability Statement: Not applicable.

Conflicts of Interest: The authors declare to have no conflict of interests. The vegan rice-pea mix was donated by Bertify in April 2020 (https://www.bertify.com/), but the start-up company was not involved in the experimental set-up, nor with the analysis or interpretation of the data. The pulverized worm protein was purchased from Protifarm in April 2020 (https://protifarm.com/), but the company was not involved in the experimental set-up, nor with the data analysis.

References

1. Larsson, L.; Degens, H.; Li, M.; Salviati, L.; Lee, Y.I.; Thompson, W.; Kirkland, J.L.; Sandri, M. Sarcopenia: Aging-Related Loss of Muscle Mass and Function. *Physiol. Rev.* **2019**, *99*, 427–511. [CrossRef] [PubMed]
2. Joanisse, S.; Lim, C.; McKendry, J.; McLeod, J.C.; Stokes, T.; Phillips, S.M. Recent advances in understanding resistance exercise training-induced skeletal muscle hypertrophy in humans. *F1000Research* **2020**, *9*, 141. [CrossRef] [PubMed]
3. Koopman, R.; Van Loon, L.J.C. Aging, exercise, and muscle protein metabolism. *J. Appl. Physiol.* **2009**, *106*, 2040–2048. [CrossRef]
4. D'Hulst, G.; Soro-arnaiz, I.; Masschelein, E.; Veys, K.; Smeuninx, B.; Kim, S.; Deldicque, L.; Blaauw, B.; Breen, L.; Koivunen, P.; et al. PHD1 controls muscle mTORC1 in a hydroxylation-independent manner by stabilizing leucyl tRNA synthetase. *Nat. Commun.* **2020**, *11*, 1–15. [CrossRef] [PubMed]
5. Wang, X.; Proud, C.G. The mTOR Pathway in the Control of Protein Synthesis. *Physiol.* **2006**, *21*, 362–369. [CrossRef]
6. Huo, Y.; Iadevaia, V.; Proud, C.G. Differing effects of rapamycin and mTOR kinase inhibitors on protein synthesis. *Biochem. Soc. Trans.* **2011**, *39*, 446–450. [CrossRef]
7. Drummond, M.J.; Fry, C.S.; Glynn, E.L.; Dreyer, H.C.; Dhanani, S.; Timmerman, K.L.; Volpi, E.; Rasmussen, B.B. Rapamycin administration in humans blocks the contraction-induced increase in skeletal muscle protein synthesis. *J. Physiol.* **2009**, *587*, 1535–1546. [CrossRef]
8. Huang, J.; Manning, B.D. The TSC1–TSC2 complex: A molecular switchboard controlling cell growth. *Biochem. J.* **2008**, *412*, 179–190. [CrossRef]
9. Han, J.M.; Jeong, S.J.; Park, M.C.; Kim, G.; Kwon, N.H.; Kim, H.K.; Ha, S.H.; Ryu, S.H.; Kim, S. Leucyl-tRNA Synthetase Is an Intracellular Leucine Sensor for the mTORC1-Signaling Pathway. *Cell* **2012**, *149*, 410–424. [CrossRef]
10. Wolfson, R.L.; Sabatini, D.M. The Dawn of the Age of Amino Acid Sensors for the mTORC1 Pathway. *Cell Metab.* **2017**, *26*, 301–309. [CrossRef]
11. Wolfson, R.L.; Chantranupong, L.; Saxton, R.A.; Shen, K.; Scaria, S.M.; Cantor, J.R.; Sabatini, D.M. Sestrin2 is a leucine sensor for the mTORC1 pathway. *Science* **2015**, *351*, 43–48. [CrossRef] [PubMed]
12. *World Population Prospects: The 2017 Revision, Key Findings and Advance Tables*; UN Department of Economic and Social Affairs: New York, NY, USA, 2017.
13. Henchion, M.; Hayes, M.; Mullen, A.M.; Fenelon, M.; Tiwari, B. Future Protein Supply and Demand: Strategies and Factors Influencing a Sustainable Equilibrium. *Foods* **2017**, *6*, 53. [CrossRef] [PubMed]
14. FAO. *Mapping Supply and Demand for Animal-Source Foods to 2030, Animal Production and Health Working Paper*; FAO: Rome, Italy, 2011; ISBN 9789251070413.
15. Houston, D.; Nicklas, B.; Ding, J.; Harris, T.B.; Tylavsky, F.; Newman, A.; Sun, L.J.; Sahyoun, N.; Visser, M.; Kritchevsy, S. Dietary protein intake is associated with lean mass change in older, community-dwelling adults: The Health, Aging, and Body Composition (Health ABC). *Am. J. Clin. Nutr.* **2008**, *87*, 150–155. [CrossRef] [PubMed]
16. Statistics Division, FAO. *FAOSTAT Food Balance Sheets*; FAOSTAT: Rome, Italy, 2011.
17. Van Vliet, S.; Burd, N.A.; van Loon, L.J. The Skeletal Muscle Anabolic Response to Plant- versus Animal-based Protein consumption. *J. Nutr.* **2015**, *145*, 1981–1991. [CrossRef] [PubMed]
18. Hackney, K.J.; Trautman, K.; Johnson, N.; McGrath, R.; Stastny, S. Protein and muscle health during aging: Benefits and concerns related to animal-based protein. *Anim. Front.* **2019**, *9*, 12–17. [CrossRef]
19. Benzie, I.F.F.; Wachtel-Galor, S. Vegetarian Diets and Public Health: Biomarker and Redox Connections. *Antioxidants Redox Signal.* **2010**, *13*, 1575–1591. [CrossRef] [PubMed]
20. Joy, J.M.; Lowery, R.P.; Wilson, J.M.; Purpura, M.; De Souza, E.O.; Mc Wilson, S.; Kalman, D.S.; Dudeck, J.E.; Jäger, R. The effects of 8 weeks of whey or rice protein supplementation on body composition and exercise performance. *Nutr. J.* **2013**, *12*, 86. [CrossRef]
21. Haub, M.D.; Wells, A.M.; Tarnopolsky, M.A.; Campbell, W.W. Effect of protein source on resistive-training-induced changes in body composition and muscle size in older men. *Am. J. Clin. Nutr.* **2002**, *76*, 511–517. [CrossRef] [PubMed]
22. Berrazaga, I.; Micard, V.; Gueugneau, M.; Walrand, S. The Role of the Anabolic Properties of Plant- versus Animal-Based Protein Sources in Supporting Muscle Mass Maintenance: A Critical Review. *Nutrients* **2019**, *11*, 1825. [CrossRef]
23. Monteyne, A.J.; Coelho, M.O.C.; Porter, C.; Abdelrahman, D.R.; Jameson, T.S.O.; Jackman, S.R.; Blackwell, J.R.; Finnigan, T.J.A.; Stephens, F.B.; Dirks, M.L.; et al. Mycoprotein ingestion stimulates protein synthesis rates to a greater extent than milk protein in rested and exercised skeletal muscle of healthy young men: A randomized controlled trial. *Am. J. Clin. Nutr.* **2020**, *112*, 318–333. [CrossRef]
24. Gorissen, S.H.M.; Horstman, A.M.H.; Franssen, R.; Crombag, J.J.R.; Langer, H.; Bierau, J.; Respondek, F.; Van Loon, L.J. Ingestion of Wheat Protein Increases In Vivo Muscle Protein Synthesis Rates in Healthy Older Men in a Randomized Trial. *J. Nutr.* **2016**, *146*, 1651–1659. [CrossRef]
25. Wilkinson, S.B.; Tarnopolsky, M.A.; Macdonald, M.J.; Macdonald, J.R.; Armstrong, D.; Phillips, S.M. Consumption of fluid skim milk promotes greater muscle protein accretion after resistance exercise than does consumption of an isonitrogenous and isoenergetic soy-protein beverage. *Am. J. Clin. Nutr.* **2007**, *85*, 1031–1040. [CrossRef] [PubMed]
26. Tang, J.E.; Moore, D.R.; Kujbida, G.W.; Tarnopolsky, M.A.; Phillips, S.M. Ingestion of whey hydrolysate, casein, or soy protein isolate: Effects on mixed muscle protein synthesis at rest and following resistance exercise in young men. *J. Appl. Physiol.* **2009**, *107*, 987–992. [CrossRef]

27. Yang, Y.; Churchward-venne, T.; Burd, N.A.; Breen, L.; Tarnopolsky, M.A.; Phillips, S.M. Greater stimulation of myofibrillar protein synthesis with ingestion of whey protein isolate v. micellar casein at rest and after resistance exercise in elderly men. *Nutr. Metab.* **2012**, *108*, 958–962. [CrossRef]
28. Dijk, F.J.; Van Dijk, M.; Walrand, S.; Van Loon, L.J.C.; Van Norren, K.; Luiking, Y.C. Differential effects of leucine and leucine-enriched whey protein on skeletal muscle protein synthesis in aged mice. *Clin. Nutr. ESPEN* **2018**, *24*, 127–133. [CrossRef]
29. Crozier, S.J.; Kimball, S.R.; Emmert, S.W.; Anthony, J.C.; Jefferson, L.S. Oral Leucine Administration Stimulates Protein Synthesis in Rat Skeletal Muscle. *J. Nutr.* **2005**, *135*, 376–382. [CrossRef] [PubMed]
30. Gilani, G.S.; Xiao, C.W.; Cockell, K.A. Impact of Antinutritional Factors in Food Proteins on the Digestibility of Protein and the Bioavailability of Amino Acids and on Protein Quality. *Br. J. Nutr.* **2012**, *108*, S315–S332. [CrossRef]
31. Bos, C.; Metges, C.C.; Gaudichon, C.; Petzke, K.J.; Pueyo, M.E.; Morens, C.; Everwand, J.; Benamouzig, R.; Tomeé, D. Postprandial Kinetics of Dietary Amino Acids Are the Main Determinant of Their Metabolism after Soy or Milk Protein Ingestion in Humans. *J. Nutr.* **2003**, *133*, 1308–1315. [CrossRef] [PubMed]
32. Vangsoe, M.T.; Thogersen, R.; Bertram, H.C.; Heckmann, L.-H.L.; Hansen, M. Ingestion of Insect Protein Isolate Enhances Blood Amino Acid Concentrations Similar to Soy Protein in A Human Trial. *Nutrients* **2018**, *10*, 1357. [CrossRef]
33. Yoshizawa, F.; Mochizuki, S.; Sugahara, K. Differential Dose Response of mTOR Signaling to Oral Administration of Leucine in Skeletal Muscle and Liver of Rats. *Biosci. Biotechnol. Biochem.* **2013**, *77*, 839–842. [CrossRef] [PubMed]
34. Boulos, S.; Tännler, A.; Nyström, L. Nitrogen-to-Protein Conversion Factors for Edible Insects on the Swiss Market: T. molitor, A. domesticus, and L. migratoria. *Front. Nutr.* **2020**, *7*, 89. [CrossRef] [PubMed]
35. Mariotti, F.; Tomé, D.; Mirand, P. Converting Nitrogen into Protein–Beyond 6.25 and Jones' Factors. *Crit. Rev. Food Sci. Nutr.* **2008**, *2*, 177–184. [CrossRef] [PubMed]
36. Elgar, D.F.; Hill, J.P.; Holroyd, S.E.; Peddie, G.S. Comparison of analytical methods for measuring protein content of whey protein products and investigation of influences on nitrogen conversion factors. *Int. J. Dairy Technol.* **2020**, *73*, 790–794. [CrossRef]
37. Impact Whey Protein—Myprotein—Amino Acid Analysis. Available online: https://www.supplementlabtest.com/tests/impact-whey-protein-myprotein-amino-acid-analysis-carbs-and-sugars-supplementlabtestcom (accessed on 31 August 2015).
38. Adams, K.J.; Pratt, B.; Bose, N.; Dubois, L.G.; St. John-Williams, L.S.; Perrott, K.M.; Ky, K.; Kapahi, P.; Sharma, V.; MacCoss, M.J.; et al. Skyline for Small Molecules: A Unifying Software Package for Quantitative Metabolomics. *J. Proteome Res.* **2020**, *19*, 1447–1458. [CrossRef] [PubMed]
39. Phillips, S.M. Nutrient-rich meat proteins in offsetting age-related muscle loss. *Meat Sci.* **2012**, *92*, 174–178. [CrossRef]
40. Mensa-Wilmot, Y.; Phillips, R.; Hargrove, J. Protein quality evaluation of cowpea-based extrusion cooked cereal/legume weaning mixtures. *Nutr. Res.* **2001**, *21*, 849–857. [CrossRef]
41. Scrimshaw, N.S.; Bressani, R.; Béhar, M.; Viteri, F. Supplementation of Cereal Proteins with Amino Acids: I. Effect of Amino Acid Supplementation of Corn-Masa at High Levels of Protein Intake on the Nitrogen Retention of Young Children. *J. Nutr.* **1958**, *66*, 485–499. [CrossRef]
42. Brennan, J.L.; Keerati-U-Rai, M.; Yin, H.; Daoust, J.; Nonnotte, E.; Quinquis, L.; St-Denis, T.; Bolster, D.R. Differential Responses of Blood Essential Amino Acid Levels Following Ingestion of High-Quality Plant-Based Protein Blends Compared to Whey Protein—A Double-Blind Randomized, Cross-Over, Clinical Trial. *Nutrients* **2019**, *11*, 2987. [CrossRef]
43. Moberg, M.; Apró, W.; Ekblom, B.; Van Hall, G.; Holmberg, H.-C.; Blomstrand, E. Activation of mTORC1 by leucine is potentiated by branched-chain amino acids and even more so by essential amino acids following resistance exercise. *Am. J. Physiol. Cell Physiol.* **2016**, *310*, C874–C884. [CrossRef]
44. Borgenvik, M.; Apró, W.; Blomstrand, E. Intake of branched-chain amino acids influences the levels of MAFbx mRNA and MuRF-1 total protein in resting and exercising human muscle. *Am. J. Physiol. Endocrinol. Metab.* **2012**, *302*, E510–E521. [CrossRef]
45. Moro, T.; Brightwell, C.R.; Velarde, B.; Fry, C.S.; Nakayama, K.; Sanbongi, C.; Volpi, E.; Rasmussen, B.B. Whey Protein Hydrolysate Increases Amino Acid Uptake, mTORC1 Signaling, and Protein Synthesis in Skeletal Muscle of Healthy Young Men in a Randomized Crossover Trial. *J. Nutr.* **2019**, *149*, 1149–1158. [CrossRef]
46. Burd, N.A.; Gorissen, S.H.; Van Vliet, S.; Snijders, T.; van Loon, L.J. Differences in postprandial protein handling after beef compared with milk ingestion during postexercise recovery: A randomized controlled trial. *Am. J. Clin. Nutr.* **2015**, *102*, 828–836. [CrossRef] [PubMed]
47. Márquez-Mota, C.C.; Rodriguez-Gaytan, C.; Adjibade, P.; Mazroui, R.; Gálvez, A.; Granados, O.; Tovar, A.R.; Torres, N. The mTORC1-Signaling Pathway and Hepatic Polyribosome Profile Are Enhanced after the Recovery of a Protein Restricted Diet by a Combination of Soy or Black Bean with Corn Protein. *Nutrients* **2016**, *8*, 573. [CrossRef] [PubMed]
48. Phillips, S.M. A Brief Review of Critical Processes in Exercise-Induced Muscular Hypertrophy. *Sports Med.* **2014**, *44*, 71–77. [CrossRef] [PubMed]
49. Reagan-Shaw, S.; Nihal, M.; Ahmad, N. Dose translation from animal to human studies revisited. *FASEB J.* **2007**, *22*, 659–661. [CrossRef] [PubMed]
50. Schwarz, R.; Kaspar, A.; Seelig, J.; Künnecke, B. Gastrointestinal transit times in mice and humans measured with27Al and19F nuclear magnetic resonance. *Magn. Reson. Med.* **2002**, *48*, 255–261. [CrossRef]

Review

The Accumulation and Molecular Effects of Trimethylamine N-Oxide on Metabolic Tissues: It's Not All Bad

Emily S. Krueger [1], Trevor S. Lloyd [1,2] and Jeffery S. Tessem [1,*]

[1] Department of Nutrition, Dietetics and Food Science, Brigham Young University, Provo, UT 84602, USA; emilys.krueger@gmail.com (E.S.K.); tslloyd@gmail.com (T.S.L.)
[2] Medical Education Program, David Geffen School of Medicine at UCLA, Los Angeles, CA 90095, USA
[*] Correspondence: jeffery_tessem@byu.edu; Tel.: +1-801-422-9082

Abstract: Since elevated serum levels of trimethylamine N-oxide (TMAO) were first associated with increased risk of cardiovascular disease (CVD), TMAO research among chronic diseases has grown exponentially. We now know that serum TMAO accumulation begins with dietary choline metabolism across the microbiome-liver-kidney axis, which is typically dysregulated during pathogenesis. While CVD research links TMAO to atherosclerotic mechanisms in vascular tissue, its molecular effects on metabolic tissues are unclear. Here we report the current standing of TMAO research in metabolic disease contexts across relevant tissues including the liver, kidney, brain, adipose, and muscle. Since poor blood glucose management is a hallmark of metabolic diseases, we also explore the variable TMAO effects on insulin resistance and insulin production. Among metabolic tissues, hepatic TMAO research is the most common, whereas its effects on other tissues including the insulin producing pancreatic β-cells are largely unexplored. Studies on diseases including obesity, diabetes, liver diseases, chronic kidney disease, and cognitive diseases reveal that TMAO effects are unique under pathologic conditions compared to healthy controls. We conclude that molecular TMAO effects are highly context-dependent and call for further research to clarify the deleterious and beneficial molecular effects observed in metabolic disease research.

Keywords: western diet; trimethylamine n-oxide (TMAO); gut microbiome; metabolic tissue function; oxidative stress; metabolic diseases; obesity; diabetes; insulin resistance; insulin production

Citation: Krueger, E.S.; Lloyd, T.S.; Tessem, J.S. The Accumulation and Molecular Effects of Trimethylamine N-Oxide on Metabolic Tissues: It's Not All Bad. *Nutrients* **2021**, *13*, 2873. https://doi.org/10.3390/nu13082873

Academic Editor: Maria D. Mesa

Received: 23 July 2021
Accepted: 19 August 2021
Published: 21 August 2021

Publisher's Note: MDPI stays neutral with regard to jurisdictional claims in published maps and institutional affiliations.

1. Introduction

Trimethylamine N-oxide (TMAO) chemistry has been investigated since the 1890s and TMAO was first reported in human urine samples in 1934 [1–5]. Urine TMAO levels have since been associated with dietary choline consumption and chronic diseases [6–8]. Positive and negative molecular mechanisms of TMAO have been identified in various tissues across many species [9–17]. The 2011 landmark metabolomics study first linked elevated serum TMAO levels to cardiovascular disease (CVD) and TMAO research in the context of chronic diseases has since grown exponentially [8]. We now know that serum TMAO is derived from choline via a gut microbiome metabolite and represents a critical factor for exploring the diet-gut-host effects on health (Figure 1) [8,18–29]. Clinical TMAO effects are most closely tied to atherosclerotic phenotypes, although there is some debate [30,31]. Resent TMAO research has recently expanded to include other chronic metabolic diseases [20,32–47]. Over-nutrition related metabolic phenotypes including insulin resistance, type 2 diabetes (T2D), obesity, metabolic syndrome, and chronic kidney disease (CKD) which are already associated with CVD have been linked to TMAO [39,44,48–68]. Emerging molecular level studies are beginning to elucidate TMAO's effects on various relevant metabolic tissues; however, direct TMAO mechanisms are still unclear. While it is debated whether TMAO plays predominantly positive or negative roles in the metabolic disease contexts, TMAO is generally considered deleterious and strategies to reduce its accumulation are proposed for better CVD treatment [50,69–76]. Here we define how serum TMAO accumulates with a close look at the interaction between the diet, the microbiota, the host liver, and kidney tissues. We then explore the known TMAO effects in metabolic tissues including the liver, kidney, brain, adipose, and muscle. Finally, to further link to metabolic diseases which commonly involve poor blood glucose management, we will review how TMAO affects insulin resistance and insulin secretion. Just as the role of TMAO may differ between CVD and metabolic disease conditions, we hypothesize that it may function differently between healthy and diseased states.

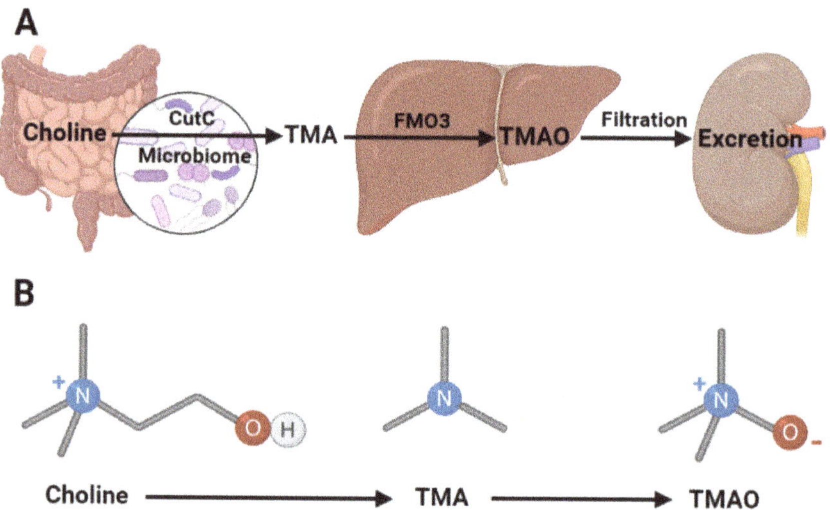

Figure 1. Trimethylamine N-Oxide (TMAO) Accumulation in Serum. (**A**) The microbiome-liver-kidney axis regulates TMAO production and accumulation. Choline and related compounds from dietary animal proteins and fats are metabolized by gut bacteria expressing choline utilization cluster (Cut) genes including *E. coli*. The resultant trimethylamine (TMA) is absorbed by enterocytes and metabolized by hepatic flavin-containing monooxygenase (FMO) enzymes. Serum TMAO is excreted via renal glomerular filtration and uptake by proximal tubular cells through organic cation transport proteins. (**B**) Choline, TMA, and TMAO structures. This figure was created with biorender.com.

2. TMAO Accumulation in Serum

2.1. Intestinal TMA Production

The source of circulating TMAO is the precursor trimethylamine (TMA) produced by the intestinal microbiome metabolism of dietary choline. High levels of quaternary amine-containing semi-essential nutrients such as choline, phosphatidylcholine, carnitine, betaine, and ergothioneine are common in a Western diet containing animal proteins and fats [69,77–80]. These foods and nutrients may or may not be directly linked to CVD independent of TMAO production [8,69,81]. Diets high in plant products, including the Mediterranean, vegetarian, and vegan diets are associated with lower circulating TMAO [50,82–84]. However, even healthy diets containing fish, vegetables, and whole-grain products, which measured high levels of long-chain unsaturated fatty acids can increase serum TMAO levels in patients with at cardiometabolic risk [85]. In patients with obesity, a vegan diet intervention reduced circulating TMAO levels and improved glucose tolerance presumably by reducing intake of the precursor nutrients [50]. Therefore, the first step toward serum TMAO accumulation is the consumption of prerequisite nutrients.

In a Western diet, the abundance of choline-related nutrients can surpass the absorptive capacity of the small intestine and the excess is metabolized by the large intestinal microbiota prior to absorption [86–89]. In carnitine challenged omnivorous and vegetarian subjects, TMAO production increased 10-fold in omnivores [90]. Presumably, the excess carnitine substrate was metabolized by TMA producing bacteria which were disproportionately abundant in the microbiomes of omnivores [71,91]. However, it must be noted that carnitine is not a primary pre-curser of TMA, as discussed below. Aged animal models show higher TMA absorption into the portal vein and eventual TMAO accumulation compared to the young, indicating that changes at the host enterocytes also influences serum TMAO accumulation [92–94].

TMA producing anaerobic bacteria have been identified by screening gastrointestinal isolates and using bioinformatic analysis [95–97]. These studies found that TMA is produced by facultative or obligate anaerobes across 4 phyla including Proteobacteria, Firmicutes, Actinobacteria, and Fusobacteria [98,99]. All species of the genera Desulfosporosinus and Proteus, and most species in the Enterococus, Escherichia, Klebsiella, Conlinsella, Closteridium, and Anaerococcus genera produce TMA [98,99]. These bacteria, including the well-known E. coli, C. bacterium, and C. hathewayi species express the choline utilization cluster (Cut) family of genes [98]. The glycyl radical enzyme homologue choline TMA-lyase (CutC) is activated by CutD and cleaves the C-N bond in choline to produce TMA and acctaldchydc [99–101]. The oxygenase Rieske 2S-2Fe cluster-containing enzymes CntA and CntB have a broader affinity for secondary dietary substrate including carnitine and betaine [97,102]. Although some TMAO research is performed in carnitine supplemented models, these generally report findings contrary to choline feeding studies presumably because carnitine is a poor TMA precursor because it is generally metabolized to γ-butyrobetaine [91,97,103,104]. It must be noted that the gene pair YeaW and YeaX may further modify γ-butyrobetaine to TMA because of its substrate promiscuity [97,105]. Therefore, while TMA can be generated from various dietary quaternary amine-containing semi-essential nutrients, most studies focus on choline metabolism via CutC and CutD. Because TMA-producing species and genes are now well-defined, some studies propose personalized strategies based on the TMA productivity of an individual's microbiome [96]. These strategies include drug and dietary interventions generally aimed at reducing TMAO levels due to their association with CVD and other chronic diseases [78,106–110]. While the common insulin sensitizing drug Metformin decreases gut bacterial TMA production and eventual serum TMAO levels in T2D model mice [111], iodomethylcholine and 3,3-dimethyl-1-butanol directly inhibit the microbiota TMA [78,112]. Although the substrates for the TMA-producing enzymes are largely dietary, it must be noted that the choline moiety is also present in endogenous bile acids which are absorbed and recycled [113,114]. Therefore, the presence of TMA-producing species in the microbiota may be more critical to overall TMAO production and accumulation than the dietary make-up [78]. This conclusion highlights how dietary interventions to reduce meat, milk, and eggs is not always effective at reducing serum TMAO levels. Together these studies establish that microbiota metabolism of choline related nutrients produces the obligatory TMA precursor for TMAO production.

Convincing antibiotic studies further validate the primary role of the intestinal microbiome in TMAO production and accumulation. When the microbiome is intact and diets are supplemented with substrates, circulating TMAO levels increase predictably [36,91]. When the microbiome is absent, as in gnotobiotic or antibiotic treated animals, TMAO levels and its subsequent effects are blocked [42,91,115–117]. In humans, broad-spectrum antibiotics significantly reduce serum TMAO levels which recover after treatment is withdrawn [118]. When gnotobiotic animals are colonized by TMA producing human gastrointestinal isolates, serum TMAO levels increase [95]. These levels also increase when TMAO is supplemented directly [36]. Fish meat containing high levels of TMAO increased serum levels within 15 min of consumption, indicating that TMAO can also be absorbed at the small intestine [19]. In vitro research shows that TMA can be oxidized by reactive oxygen species (ROS) which may play a role during intestinal inflammation [119,120]. Hence, many animal studies use TMAO supplementation to effectively model elevated levels reported in the clinical setting; however, this model is most appropriate for research investigating high fish consumption or inflammatory bowel disease. Other radiolabeled TMAO feeding studies show that TMAO retroconversion to TMA by the microbiome is also possible [21]. This retroconversion is the hallmark of trimethylaminuria, also called fish odor syndrome because this reaction is also present in rotting fish [121–125]. Together, these findings demonstrate that while dietary TMAO can be directly absorbed at the small intestine, microbiome metabolism of choline at the large intestine generates the bulk of TMA levels which are absorbed and delivered to the liver via the portal vein.

2.2. Hepatic TMAO Production

The flavin-containing monooxygenase (FMO) enzyme family, which is highly expressed in the liver, converts TMA to TMAO. TMA is taken up by hepatocytes and oxidized to TMAO by FMO enzymes which function in the same way as cytochrome P450 oxidoreductases [126]. FMO enzymes are located at the endoplasmic reticulum (ER) and oxidize a broad range of neutrophilic substrates with flavin adenine dinucleotide (FAD) and nicotinamide adenine dinucleotide phosphate (NADPH) cofactors [126]. Their capacity to oxidize TMA was established in the 1960s through fish cell biology research and has since been identified as a liver-specific process in many vertebrates [4,5]. Humans have FMO enzymes 1 through 5 with each isoform expressed in a tissue dependent manner. FMO3 and 5 are liver-specific and have the highest overall expression [32,126–129]. FMO3 has a 10-fold higher TMA specific activity than FMO1 in vitro [32,130]. Baseline FMO3 expression is contingent on age and sex such that adults have higher expression than children younger than 6 years old and females have higher levels than males [32,48,131]. Therefore, a common research strategy is to use adult female subjects to investigate FMO3 overexpression and inhibition predictably increases and reduces serum TMAO levels respectively [32]. Serum TMAO levels peak 4 h after nutrient precursor consumption illustrating the stepwise TMAO formation via microbial metabolism, intestinal absorption, and hepatic oxidation [132]. TMAO can be excreted from hepatocytes into the serum by organic cation transporters or it may remain intracellular to influence hepatocyte metabolic functions as will be discussed below [133]. Therefore, after microbiota metabolism of dietary nutrients, FMO3 represents the final step of host mediated TMAO production.

The diet influences TMAO formation not only by providing substrates to the microbiome for TMA production, but also by indirectly regulating FMO3 expression. This control is evident in choline, carnitine, and TMAO supplementation studies, which predictably increased circulating TMAO levels, but also enhanced FMO3 expression [36,91]. This elevated expression may be compensatory due to increased TMA substrate load but may also represent FMO3 expression regulation by TMAO [36]. Mice models of over-nutrition have elevated FMO3 expression [36,93,117,134]. Conversely, one hepatocyte cell culture study showed that free fatty acid treatment inhibited FMO1, 3, and 5 expression [135]. Still, either by providing the TMA precursor or by some other indirect regulating mechanism, the diet alters FMO3 expression.

The diet further affects FMO3 expression through metabolic hormone regulation. Glucagon and corticosteroids, which drive anabolic pathways in the fasted state, induce FMO3 expression up to 14-fold and significantly increase TMAO accumulation [48]. Insulin induces anabolic pathways during the fed state and suppresses FMO3 expression by 60% in primary rat hepatocytes [48]. Inhibition of phosphoinositide 3-kinase (PI3K), a downstream insulin signaling effector, blunted insulin's inhibitory effect [48]. Non-biased metabolomic studies identified TMAO and FMO3 as targets of insulin signaling [48]. In fact, FMO3 expression in mice modelling insulin resistance surpassed expression rates under glucagon stimulation and illustrate that insulin's inhibition of FMO3 eclipses the induction observed under high glucagon conditions [48,136]. Conversely, insulin treatment did not affect FMO1 activity in healthy wildtype rats [137]. This discrepancy in insulin's regulation of TMAO production supports the hypothesis that TMAO may work differently under healthy conditions compared to insulin-resistant conditions. In conclusion, the most convincing studies show that insulin signaling is sufficient to impede FMO3 expression and TMAO production which are induced by glucagon during the fasted state.

Clinical data from insulin resistant patients confirm that metabolic hormones regulate FMO3 expression. Obese patients fasting prior to bariatric surgery had higher FMO3 levels compared to control patients [48]. Of the patients, 79% were diabetic and most were treated with insulin sensitizing drugs. These results strongly confirm that insulin resistance combined with glucagon signaling during fasting drives FMO3 expression. However, unlike the animal studies [48,136], improved insulin signaling due to drug interventions could not overcome these effects. This study speaks to the dysregulated hormonal condition of T2D where insulin resistance may limit an important check on FMO3 expression and allow elevated TMAO production and accumulation [48].

A final known regulator of FMO3 expression is the bile acid-activated farnesoid X receptor (FXR). FXR loss and gain of function experiments link FMO3 transcriptional regulation to lipogenic pathways [32,48]. BA-bound FXR increases FMO3 expression 17-fold and overexpression of FXR induces FMO3 expression in a dose-dependent manner [48]. Studies identified an FXR response element in the FMO3 promoter and FXR regulation was eliminated when this region was mutated [32,48]. These results confirm that FXR directly binds the FMO3 promoter in animals consuming diets ranging from over-nutrition to healthy controls in [32,48]. Together these studies show that hepatic TMAO production by FMO3 is controlled by BA-bound FXR transcriptional regulation along with insulin hormonal regulation.

2.3. Renal TMAO Filtration

Kidney function plays a final role in managing circulating TMAO levels. In healthy subjects, TMAO is efficiently excreted in the urine to maintain serum levels below about 10 μM [60,71,129,138]. Glomerular filtration and uptake by proximal tubular cells through organic cation transport proteins regulates circulating TMAO and prevents excess accumulation [133,139,140]. While small amounts may be secreted as TMA after retroconversion, renal FMO3 and 1 expression accounts for over 96% being excreted as TMAO [128,129,141,142]. Although urine TMAO levels correlate to serum levels and are often reported in TMAO research, the best studies report serum levels which better represents the combined TMAO regulation from the microbiome, liver, and kidney [90]. Indeed, in a case of elevated protein intake resulting in elevated TMAO production, renal excretion of TMAO was unexpectedly decreased due to individualistic factors including these regulatory tissues [143]. Therefore, proper kidney function represents the final check on serum TMAO accumulation.

When kidney function is impaired, as in CKD associated with T2D, elevated TMAO levels range from 20 μM to greater than 100 μM [49,106,144–148]. CKD progression is associated with elevated TMAO levels such that stage 4 and 5 CKD patients had disproportionately elevated serum TMAO levels compared to stage 1 to 3 CKD patients [145,148,149]. Deleterious kidney mechanisms including reduced glomerular filtration rate, fibrosis, and loss of tubular function are associated with elevated serum TMAO [55,150,151]. One study on coronary artery disease patients showed a relationship between TMA levels and glomerular filtration rate [152]. In treating CKD, loop diuretics aggravate the elevated TMAO levels whereas renal transplantation dramatically reduces levels, highlighting that proper renal filtration is sufficient to limit TMAO accumulation [67,153,154]. Together, these data demonstrate that after precursor TMA production by the intestinal microbiome and hepatic conversion to TMAO by FMO3, renal filtration is the final regulator of circulating TMAO levels. Differences in any step of this TMAO production and accumulation may explain the extreme variability in serum TMAO levels observed clinically [155]. Despite this variation, accumulated serum TMAO can affect metabolic functions in tissues throughout the body (Figure 2).

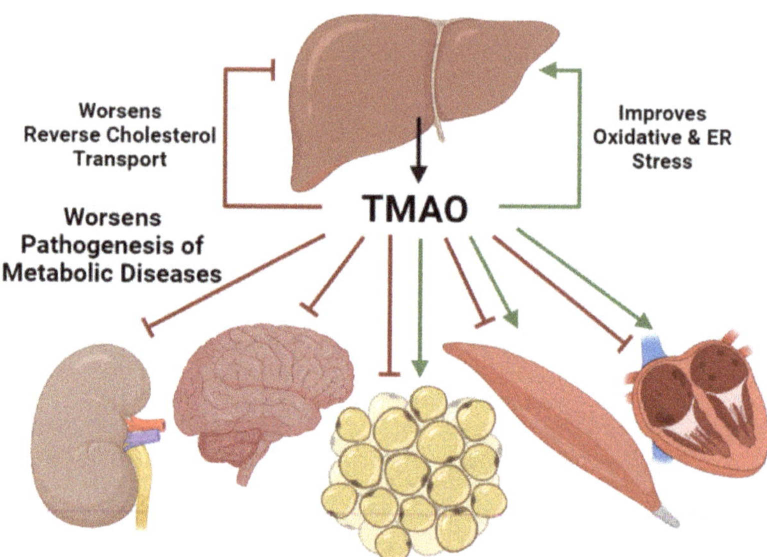

Figure 2. TMAO Effects on Metabolic Tissues. Elevated serum TMAO is generally associated with metabolic diseases and molecular effects are mainly observed in pathogenic and not healthy conditions. Molecular TMAO effects are best defined in hepatocytes where it worsens reverse cholesterol transport and drives non-alcoholic liver disease and gallstone formation. However, TMAO also improves oxidative and endoplasmic reticulum (ER) stress associated with hepatic insulin resistance. While poor renal function drives TMAO accumulation, it in turn aggravates chronic kidney disease. Although direct mechanisms are unclear, TMAO is associated with cognitive diseases including autism spectrum disorder and Alzheimer's disease. Interestingly, TMAO is liked with increased adiposity associated with obesity and type 2 diabetes (T2D), but it also reduces adipocyte ER stress. In cardiovascular disease research, TMAO may increase or decrease cardiac muscle hypertrophy associated with heart failure. In skeletal muscle, TMAO benefits enzyme kinetics but is debated to drive insulin resistance. Therefore, TMAO effects may be positive or negative depending on the context of metabolic disease and molecular mechanisms are not well understood. This figure was created with biorender.com.

3. TMAO Effects on Metabolic Tissues

3.1. TMAO Effects on Liver Function

Because TMAO is produced by hepatic FMO3, potentially higher local concentrations affect liver metabolic functions. Because of the association between elevated serum TMAO levels and various liver diseases, TMAO effects are generally considered deleterious, however TMAO is debated to have beneficial roles in some cellular contexts. Non-alcoholic fatty liver disease (NAFLD) is characterized by hepatic lipid accumulation and is associated with obesity, T2D, and insulin resistance which are also linked to elevated TMAO levels [48,156,157]. NAFLD patients and high fat high cholesterol diet fed animal models demonstrate elevated TMAO compared to healthy controls [134,158]. Hepatic TMAO levels are associated with markers of poor liver function including steatosis, serum bile acid levels, and inflammation in patients stratified by disease severity [159]. Gallstones and primary sclerosing cholangitis involve dysregulated hepatic bile formation and are associated with increased serum TMAO levels [36,160]. Cholangitis patients with serum TMAO levels over 4 µM have shorter liver transplant free survival rates [160]. Although these studies do not demonstrate a mechanistic role for TMAO at the liver, they provide the clinical rational for molecular research on TMAO effects on liver function.

Mouse liver disease models more directly link TMAO to altered hepatic function. In these models, over-nutrition induces disease phenotypes including increased hepatic lipogenesis rates, steatosis, serum aspartate transaminase or alanine transaminase levels, and insulin resistance, which are associated with elevated serum and urine TMAO levels [134,156]. While most studies use female mice, some report TMAO accumulation and aggravated liver disease in male mice which are known to have a lower basal expression of hepatic FMO3 and lower average TMAO production than females [32,48,142]. Despite the potential inhibition from androgen hormones in these male mice, the high fat diet significantly increased TMAO levels and escalated liver disease phenotypes [134]. Interestingly, TMAO supplementation in chow fed male mice did not exhibit altered steatosis scores [134]. Similarly, in healthy elderly female patients taking carnitine supplements, plasma lipid and inflammation markers were not associated with TMAO accumulation, presumably because carnitine is a poor TMA pre-curser as described above [103,104]. These studies on healthy subjects may also demonstrate that TMAO may not influence hepatic functions in a disease-free environment [103,104,134]. In gallstone susceptible mice, a high cholesterol lithogenic diet combined with TMAO supplementation demonstrated increased gallstone formation [36]. Even gallstone resistant mice developed gallstones on the same dietary regimen yielding 7 µM serum TMAO levels [36]. These various liver disease model studies further associate elevated TMAO with poor liver function and motivate a deeper look at TMAO molecular effects.

The liver is a major insulin sensitive tissue assisting with blood glucose management. In healthy conditions, proper insulin signaling upregulates glycogen and lipid synthesis to reduce blood glucose levels. TMAO is however linked to insulin resistant conditions resulting in elevated blood glucose levels. In mice modeling hepatic insulin resistance, TMAO reduction by FMO3 knockdown prevented hyperlipidaemia [48]. TMAO supplemented high fat diet fed mice had decreased glycogen synthesis and FMO3 knockout mice with reduced TMAO levels had increased synthesis which connects TMAO to hepatic insulin resistance [161,162]. Gluconeogenic genes glucose 6-phosphatase (G6pase) and phosphoenolpyruvate carboxykinase (PEPCK) typically suppressed by insulin were concomitantly upregulated in the liver, muscle, and adipose tissue of TMAO supplemented mice [161]. Furthermore, FMO3 overexpression in hepatic cell cultures increased glucose secretion, which is also typically suppressed by insulin [162]. Finally, TMAO induction of insulin resistance was more directly measured by reduced expression of the insulin signaling cascade including insulin receptor substrate 2 (IRS2), PI3K, RAC-β serine/threonine-protein kinase (AKT) and glucose transporter 2 (GLUT2) [161]. These studies demonstrate that TMAO drives T2D associated insulin resistance by downregulating hepatic insulin signaling.

3.1.1. TMAO and Reverse Cholesterol Transport

The liver is essential for lipid metabolism and regulates cholesterol levels. After dietary lipids, including cholesterol, are trafficked as apolipoproteins to deliver substrates to target tissues, the excess returns to the liver through reverse cholesterol transport (RCT). This cholesterol is metabolized to bile acid stored at the gallbladder until secreted for intestinal dietary lipid emulsification, digestion, and absorption. Bile acids are either reabsorbed by enterocytes or excreted in the feces, an important cholesterol efflux route. Impaired RCT is associated with CVD and NAFLD and increased bile cholesterol content is connected to gallstone formation and elevated TMAO levels are linked with both pathogenic pathways [36,51,91,117]. Conversely, intestinal TMAO effects relating to RCT are beneficial and are discussed further below. Rodent FMO3 overexpression elevated plasma cholesterol levels by 20% and FMO3 knockdown reduced plasma and hepatic lipid levels indicative of NAFLD [117,162]. Since TMAO reduction by FMO3 knockout decreased bile cholesterol content, elevated TMAO presumably drives gallstone formation [32]. Most liver-specific findings indicate that altering TMAO levels is sufficient to alter cholesterol flux through RCT via bile acid and cholesterol synthesis. Therefore, these studies set the precedence for using NAFLD and gallstone formation models to investigate the molecular pathways involved in hepatic TMAO effects.

The RCT rate and hepatic cholesterol pool is controlled by cholesterol transporters and synthesis enzymes, which are altered by TMAO. Low-density lipoproteins and high-density lipoproteins are a major source of the hepatocyte cholesterol pool, and their receptors include low-density lipoprotein receptor (LDLR) and scavenger receptor class B type 1 (SRB1). Alternatively, cholesterol and bile acids can be taken up through transporters sodium/taurocholate co-transporting polypeptide (NTCP) and organic-anion-transporting polypeptides (Oatp1 and Oatp4), which are upregulated in hepatocytes cultured with supra-physiological 250 μM TMAO [36]. This upregulation by TMAO is validated in gallstone-susceptible mice with 6 μM serum TMAO and in gallstone patients with 3.3–4 μM serum TMAO [36]. One study on NAFLD patient biopsies also demonstrated no change in NTCP levels [134]. Despite some inconsistency, the evidence supporting that TMAO increases hepatic cholesterol uptake is convincing because results are validated in human subjects with both extremely elevated and slightly elevated TMAO levels [36].

Hydroxymethylglutaryl-CoA synthase 1 (HMGCS1) is the key regulated enzyme for endogenous cholesterol synthesis which contributes to deleterious hepatic cholesterol accumulation. Mice with reduced TMAO levels from FMO3 or FMO5 deficiency had downregulated HMGCS1, supporting the hypothesis that TMAO increases cholesterol synthesis [163]. Similarly, triacylglyceride synthesis was increased with FMO3 overexpression and decreased with knockout [162,163]. Furthermore, malic enzyme 1, which produces NADPH used in lipid biosynthesis, was decreased in FMO5 knockout mice [163]. Increased hepatic cholesterol can be further metabolized to BA. Bile acid synthesis is primarily regulated by cholesterol α-hydroxylases (Cyp7a1 and Cyp27a1). Male wildtype high fat diet fed mice and NAFLD patients present elevated serum TMAO and increased Cyp7a1 expression [134]. Cyp7a1 levels are also elevated in female apolipoprotein E knockout mice on a chow diet with choline supplementation [164]. These studies expand the lipogenic role for TMAO to include the production of endogenous lipids cholesterol, triacylglycerides, and bile acids. Therefore, TMAO increases the hepatic cholesterol pool by upregulating lipid uptake and synthesis which validates the aggravated hepatic lipid accumulation observed clinically.

Cell culture, animal, and clinical research shows that TMAO alters canalicular cholesterol and bile acid transport to promote gallstone formation. The ATP binding cassette transporters (Mrp2, BSEP, ABCG5, and ABCG8) are located at the cholesterol-rich apical cell membrane on hepatocytes. Their sterol and bile acid substrates are stored in the gallbladder until they are secreted to aid in intestinal lipid digestion and may eventually be excreted in the stool. ABCG5 and ABCG8 primarily transport sterols, including cholesterol, whereas BSEP transports bile salts. Expression of these transporters control the cholesterol to bile acid ratio at the gallbladder which when increased can gallstone formation. Hepatocytes cultured with TMAO had increased ABCG5 and ABCG8 levels, while TMAO reduction by FMO3 knockout decreased expression [36]. Data from genetic or over-nutrition induced gallstone formation models support this finding [36,117]. Similarly, patients with gallstones had increased ABCG5 and ABCG8 levels and decreased BSEP levels [36]. FMO3 loss or gain-of-function experiments in cholesterol fed mice demonstrated that TMAO induces ABCG5 and ABCG8 expression, and presumably increases bile cholesterol content to drive gallstone formation [117]. Together, these data illustrate that TMAO induces canalicular cholesterol transport in preference over BAs and promotes gallstone formation.

In contrast to the deleterious TMAO effects on hepatocytes, TMAO may trigger interesting beneficial changes at the intestines related to RCT and cholesterol efflux [51]. Dietary cholesterol and endogenous bile acids in the lumen of the intestine are absorbed by enterocytes, packaged into apolipoproteins, and delivered to the lymphatic system. The transporter niemann-pick C1-like 1 (Npc1l1) is critical for cholesterol absorption into enterocytes. While there is some debate, TMAO generally reduces Npc1l1 which presumably combats high blood cholesterol levels associated with CVD and CKD [117,162]. TMAO supplementation in high cholesterol diet fed mice reduced Npc1l1 levels and cholesterol absorption rates by 26% [91,165]. Conversely, cholesterol may be transported out of enterocytes and into the lumen via ABCG5 and ABCG8 for fecal excretion. As shown in hepatocytes, TMAO supplementation increased enterocyte ABCG5 and ABCG8 expression and decreased TMAO from FMO3 knockout blunted expression [91,162,165]. While in hepatocytes this elevated expression increased bile cholesterol content and promoted gallstone formation, in enterocytes this elevated expression may benefits cholesterol efflux [117,166]. Therefore, intestinal TMAO effects could combat chronic diseases associated with cholesterol accumulation including obesity, CVD, NAFLD, and gallstone formation. It must be noted however that since TMAO is produced at the liver and not in the intestine as described above, results on enterocyte changes from TMAO feeding studies my not be physiologically relevant except when high fish consumption or inflamed bowel disease is considered.

A final regulator of hepatic cholesterol metabolism affected by TMAO is FXR which can be bound by various agonist or antagonist BA. Because TMAO influences bile acid formation, it alters the FXR ligand abundance which ultimately impacts its regulation of lipogenic pathways, including cholesterol metabolism and bile acid synthesis. Agonist-bound FXR activates small heterodimer partner (SHP) which inhibits the lipogenesis regulator sterole regulatory element-binding protein 1 (SREBP1c) to reduce cholesterol and bile acid synthesis [167,168]. Conversely, antagonist-bound FXR inhibits SHP, leaving cholesterol and bile acid synthesis active. Most studies show that TMAO drives the agonist-bound FXR inhibition of SREBP1c and the bile acid synthesis enzyme Cyp7a1 [91,117,164]. However, one study using cellular, animal, and clinical data showed that TMAO increased lipogenesis similar to antagonist-bound FXR studies [134]. Because of this confusion, more research is needed to decipher the positive or negative effects that TMAO exerts on FXR regulation via altering its ligand abundance.

In the context of liver cholesterol metabolism during pathogenic conditions, studies generally show that TMAO is deleterious to liver function. In terms of hepatic cholesterol uptake, synthesis, and canalicular transport, TMAO plays a negative role. It aggravates NAFLD by increasing hepatic cholesterol content and exacerbates gallstone formation by elevating biliary cholesterol content. In experiments surrounding bile acid synthesis and bile acid-bound FXR transcriptional activity there are conflicting results. Interestingly, cholesterol absorption and excretion transporters at the enterocyte are regulated by TMAO in a similar manor to hepatocytes, but the phenotype may be considered beneficial. While more research on the RCT system is needed to clarify some details, hepatic molecular studies typically highlight TMAO as an aggravator of poor liver cholesterol metabolism leading to NAFLD and gallstones.

3.1.2. TMAO, Oxidative Stress, and Endoplasmic Reticulum Stress

Contrary to the previous reports that TMAO aggravates liver diseases, some studies show that TMAO may beneficially combat cellular stresses. There is a close relationship between ER stress and oxidative stress exerted by deleterious ROS. Both stresses are clearly implicated in chronic liver diseases [116,169–171]. Although it is well established that TMAO increases oxidative and ER stress levels in arterial tissue, studies in other tissue types including hepatocytes report a surprising beneficial role for TMAO [94,116,172–174]. In fish, TMAO is an important osmoregulator for maintaining proper cell volume during osmotic pressure [175,176]. The promoter region of FMO genes in many fish species have putative osmoregulatory response elements [177]. When accumulated, urea and other ER stressors cause protein denaturation or mis-folding, against which TMAO is protective [175,178–182]. Thus, TMAO is also defined as a protein folding chaperone which reduces ER stress and the resultant unfolded protein response (UPR) leading to apoptosis [16,183]. Recent molecular dynamics studies show that TMAO acts as a surfactant between the folding protein and its aqueous environment by reorganizing the hydrogen bond network to selectively stabilize proteins experiencing structural collapse [17,182,184,185]. Indeed, many molecular experiments investigating ER stress and the UPR, including those using primary human tissues, use TMAO treatments as experimental controls [186–188]. As a chaperone, TMAO is, therefore, a strong candidate for combating oxidative ER stress and the UPR relevant to many chronic diseases including insulin resistance [189].

Cell culture and animal studies validate this beneficial TMAO effect in hepatocytes modeling various metabolic diseases. Over-nutrition-induced NAFLD model mice supplemented with TMAO had reduced liver damage as measured by serum aspartate transaminase and alanine transaminase levels [165]. TMAO also improved cholesterol metabolism by reducing liver and serum cholesterol and serum low density lipoprotien levels which correlated with downregulated ER stress, UPR, and apoptosis genes [165]. Similarly, palmitate induced chronic ER stress associated with hepatic insulin resistance was reversed by short and long-term TMAO treatments [171]. However, one palmitate induced oxidative stress study showed that TMAO reduction by FMO3 knockdown reduced palmitate induced ROS levels by 20% [135]. In high cholesterol diet fed mice, TMAO supplementation reduced ER stress and inflammation [117]. As discussed earlier, this study investigated the RCT in FMO3 knockdown mice and identified a reduction in oxysterol availability. Since oxysterols are a bile acid ligand for the transcription factor liver X receptor (LXR), they measured the bile acid-bound LXR transcriptional suppression of the inflammation and ER stress response in hepatocytes. In the absence of TMAO, inflammation was increased by macrophage-derived proinflammatory cytokines and ER stress genes associated with fatty acid-induced stress. FMO3 overexpression downregulated the relevant deleterious genes [117]. Despite some debate, in vitro and in vivo studies generally demonstrate that TMAO protects against hepatic ER stress and the associated inflammation and apoptosis at the transcriptional level [117,165,171].

One robust study identified that TMAO directly binds an ER stress response protein which may underpin the beneficial TMAO effects described. Using healthy and ER stress animal and hepatocyte models, this study showed that TMAO upregulated UPR and apoptosis genes which were reversed by FMO3 knockdown [136]. Furthermore, radiolabeled TMAO co-precipitated with the protein kinase R-like ER kinase (PERK) indicating direct binding between TMAO and the UPR regulating transcription factor. Seemingly, these findings contrast with the earlier results showing that TMAO reduced ER stress, UPR, and apoptosis gene expression [165]. However, activation of the UPR during acute ER stress by PERK is debatably adaptive [190]. Acute PERK activation reduces ROS and preserves proteostasis, redox homeostasis, and mitochondrial function [191–194]. Indeed, this study is the first to identify a direct binding effect of TMAO during short-term ER stress conditions [136]. Together with other studies, these findings demonstrate a uniquely beneficial role for TMAO in healthy and pathologic hepatocytes [117,165,171].

TMAO effects on ER and oxidative stress are of particular interest to metabolic disease research. Reducing ER stress is sufficient to normalize phenotypes including hyperglycemia and insulin sensitivity by improving insulin action at the liver [195]. These hepatocyte studies elucidate that TMAO reduces ER and oxidative stress and limits the UPR and apoptosis. Other chronic diseases implicated by these cellular stresses include CVD, NAFLD, obesity, T2D, CKD, and cognitive diseases where TMAO effects the various tissues relevant to each disease as discussed below. The molecular studies mentioned here provide evidence through short and long-term studies with low and high TMAO concentrations using in vivo, ex vivo, and in vitro experiments. Therefore, strong evidence supports the hypothesis that, similar to its beneficial effects in fish cell biology, TMAO plays a beneficial oxidative and ER stress-mitigating role in human physiology. Further research will help to elucidate the various health conditions where this beneficial TMAO effect is most relevant to clinical outcomes.

3.2. TMAO Effects on Kidney Function

While TMAO can accumulate due to poor filtration during CKD, some studies suggest that TMAO aggravates CKD in turn. CKD animal models are used to investigate potential mechanisms that underpin TMAO elevation beyond that expected with impaired kidney during CKD [55,147,196]. The TMAO mechanisms related to CKD involve crosstalk between the TMAO producing tissues including the intestinal microbiome, liver, and kidney. Fecal transplants from CKD or healthy patients to antibiotic treated mice altered the microbiome with a bias toward opportunistic TMA producing species [147]. CKD mouse models have upregulated hepatic FMO3 which highlights the liver involvement in CKD [146]. Hepatic FMO3 genetic allelic variants corresponding with elevated TMAO production are also associated with CKD [197]. FMO3 also has higher substrate affinity in CKD animals, suggesting that elevated serum TMAO levels may be dependent on hepatic function independent of the low filtration rates expected during CKD [145,197]. FMO3 kinetic studies utilizing FMO3 inhibitors and activators with CKD or healthy control serum found that CKD serum contains an unknown compound which induces FMO3 [145]. While TMAO is a candidate for this regulatory effect, its involvement was not directly measured [145]. Together these studies allude to potential inter-tissue signals driving TMAO accumulation beyond the levels expected during CKD. But further studies are required to establish kidney-specific molecular actions of TMAO.

Molecular studies are required to elucidate mechanisms linking TMAO and CKD development. Clinical studies categorize TMAO as a uremic toxin which induce pathophysiologic changes implicated in CKD [144,154,198]. Increased TMAO levels at a healthy baseline predicts future CKD development and 5-year mortality risk after controlling for traditional CKD risk factors [199,200]. In over-nutrition animal studies, choline or TMAO supplementation raised serum TMAO levels which corresponded with CKD phenotypes including collagen deposition, tubulointerstitial fibrosis, and kidney injury markers [150,173]. The TMA production inhibitors iodomethylcholine and 3,3-dimetyl-1-butanol reduced TMAO levels in CKD and obesity mice models and improved renal function by suppressing tubulointerstitial fibrosis and collagen deposition [106,173]. While other clinical studies argue no relationship between TMAO and CKD [201], or only report a deleterious association [55], these molecular studies begin to define kidney-specific TMAO effects which aggravate CKD pathogenesis.

Contrary to the deleterious TMAO effects observed in CKD, TMAO shows positive and negative effects on kidney function in the context of CVD. During a 58-week study on rats modeling heart failure, all the rats supplemented with TMAO with about 40μM serum levels survived whereas 3 of the un-supplemented group died from ischemic stroke or lung edema [202]. This increased survival corresponded with increased diuresis. Although protein expression across the renin-angiotensin pathway was altered in the TMAO group, this did not affect sodium or potassium tubular transport. Therefore, the beneficial diuretic effect of TMAO was attributed to the osmotic activity of TMAO which decreased the reabsorption of water [202]. In atherosclerosis model mice, TMAO inhibition protected against poor renal phenotypes and suggests that inhibiting microbiota TMA production is a potential treatment for renal damage during CVD [203]. These mixed TMAO effects possibly depend on the animal model used which highlights how careful selection of animal models is critical to investigating clinically relevant molecular effects of TMAO. Together these studies illustrate that TMAO is generally deleterious for CKD, but its effects on renal damage during CVD are still unclear.

3.3. TMAO Effects on Brain Function

TMAO crosses the blood–brain barrier, making it a candidate for influencing cognitive function and neurological diseases [93,129,204–206]. Research on the gut–brain axis reveals that microbiome alterations accompany changes to the cardio-sympathetic nervous system, the central nervous system, and brain chemistry [207–209]. Many bacteria species and their metabolite products regulate neurotransmitter expression associated with physiological and psychological stress [209–211]. Clinical and animal research on aged individuals highlight a connection between oxidative stress or inflammation and TMAO during age-related microbiome remodeling and cognitive deficiency [93,94,116,205,211–214]. Conversely, TMAO treatment on hippocampal sections induced oxidative and ER stress [93]. Studies on traumatic brain injury demonstrate that TMAO reduced expression of a protective antioxidant enzyme in the hippocampus, whereas treatments for reducing the injury reduced TMAO [214,215]. One study supports fecal transplants as a means of reducing TMA and TMAO production to treat stroke patients [215] Others report no relationship between TMAO and oxidative stress in healthy human or animal subjects and, therefore, highlight that TMAO's deleterious cognitive effects may be dependent on an already diseased condition [214,216–218]. Interestingly, TMAO alleviated neuronal dysfunction by abrogating ER stress in diabetic neuropathy models [219]. These studies further validate that TMAO effects on neuronal tissue are context-dependent.

Cognitive disease studies demonstrate a tandem increase in oxidative stress and TMAO concentration and suggest deleterious TMAO effects at the brain. Autism spectrum disorder is associated with microbiome composition changes, inflammation, and oxidative stress [220–223]. Autism spectrum patients with higher severity scores have higher serum TMAO levels [224]. Animal models show TMAO supplementation disrupts the blood brain barrier by decreased tight junction proteins [181,225]. The heat shock protein 70 is a degradation regulator of proteins damaged by oxidative stress in barrier cells where disruption is implicated in many cognitive diseases [226,227]. Conversely, the protein folding chaperone capacity of TMAO may prevent aggregation of proteins associated with other neurodegenerative diseases including Parkinson's, dementia, or Alzheimer's disease [228–231]. However, it must be noted that the in vitro studies investigating this potential use supra-physiological levels of TMAO ranging from 100 mM to 2M [228,229]. Thus, because the clinical relevance of beneficial TMAO effects on neurodegenerative disease is not well established, reducing TMAO levels seems a more promising strategy for slowing the progression of neurological diseases.

TMAO is a biomarker in Alzheimer's disease and studies identify a molecular link between TMAO, oxidative stress, and poor neuronal health [205]. An Alzheimer's disease computational model ranked TMAO as the top associated metabolite and proposed that it may compound with other genetic and neurological factors during pathogenesis [29]. Nine molecular pathways linked elevated TMAO and Alzheimer's disease, including various neuronal pathways, lipid and protein metabolism, the immune system, and ephrin receptor signaling [29]. Ephrin receptors are a subfamily of receptor tyrosine kinases which are integral to the function of secretory cells, including neurons and pancreatic insulin secreting β-cells [232–234]. Therefore, alterations in the ephrin forward signaling could link TMAO to Alzheimer's disease and diabetes. Downstream, the metabolic regulator protein mechanistic target of rapamycin kinase (mTOR) is inhibited in the neurons of TMAO supplemented aged mice [235]. Following mTOR inhibition, increased oxidative stress and mitochondrial impairments lead to synaptic damage in the hippocampus which culminated in reduced spatial working memory indicative of the aged cognitively deficient phenotype [235]. Another metabolic regulator protein peroxisome proliferator activated receptor α (PPARα) is implicated in AD, T2D, and NAFLD [236]. FMO3 kockout mice had decreased hepatic PPARα expression [162]. Although this study did not investigate brain tissue levels, the hepatic results expand the known TMAO effects on metabolic TFs expression. While direct metabolic TMAO mechanisms are not yet established in brain tissues, these studies demonstrate that TMAO inhibits integral neuronal processes via transcription factor inhibition, oxidative damage, and altered lipid metabolism, which culminates in age-related neurodegenerative disease progression.

3.4. TMAO Effects on Adipose Function

Elevated TMAO increases adiposity associated with various metabolic diseases whereas decreased TMAO triggers beneficial changes in adipose tissue [132,237,238]. In humans and mice, TMAO levels increase with body mass index and visceral adiposity [237,239]. TMAO levels over 8.2 μM predict the metabolic syndrome associated with obesity [237]. Obese mice have elevated TMAO levels [157,240] and TMAO reduction by FMO knockout produces leaner mice [157,163,241]. This decreased adiposity coincided with increased metabolic flexibility in the white adipose tissue which is more typical in brown adipose tissue [157]. Furthermore, FMO3 knockout prevented obesity in insulin resistant high fat diet fed mice and increased brown adipocyte gene expression [157]. FMO5 knockout mice had greater leanness due to a 55% increase in fatty acid oxidation compared to controls [163]. The reduced adiposity in FMO1, 2 and 4 knockout mice was attributed to a futile cycle in triglyceride catabolism and re-esterification [241]. These results show a positive correlation between TMAO levels and adiposity where FMO3 knockout enabled a beneficial energy consuming phenotype.

Data further suggest that elevated TMAO levels enhance adipocyte insulin resistance and inflammation. Insulin signals exocytosis of adipocyte glucose transporter GLUT4 to help regulate blood glucose which is obstructed by TMAO. TMAO supplements in high fat diet fed mice modeling insulin resistance and inflammation demonstrated elevated fasting insulin levels and reduced insulin signaling cascade expression along with increased inflammation markers [161,242–245]. These markers included pro-inflammatory adipokines, which regulate insulin sensitivity in other metabolic tissues and coincide with insulin resistance and chronic low-grade inflammation [244,246,247]. These results indicate that elevated TMAO generally worsens adipocyte insulin resistance and drives inflammation during obesity.

Conversely, TMAO is linked with reduced adipocyte ER stress and improved insulin sensitivity. In one study, the pre-intervention urine TMAO levels predicted obesity through increased weight gain, body mass index, and adiposity [239]. However, after 5 weeks of a high fat diet, increased TMAO correlated with reduced adiposity. These surprising results corresponded with downregulated ER stress, lipid biosynthesis, insulin signaling, and adipocyte differentiation genes [239]. Reduced TMAO levels in FMO1, 2 and 4 knockout mice had 64% reduced GLUT4 expression indicating that TMAO levels are correlated to GLUT4 expression and may benefit insulin sensitivity [241]. Since these effects were not observed in healthy mice, this study supports the hypothesis that a stressed cellular environment elicits TMAO effects that are not apparent in healthy conditions [242]. In contrast to previous studies, these studies illustrate that TMAO can beneficially combat high fat diet-induced adiposity by improving insulin sensitivity and reducing ER stress which was further validated in other tissues including liver, muscle, and pancreatic β-cells [239]. When adipocyte studies using similar high fat diet models are aggregated, the bulk support deleterious TMAO effects, but a few purport beneficial effects. Therefore, further research is needed to clarify TMAO's role in adipocyte function.

3.5. TMAO Effects on Muscle Function

Proper muscle function is vital for metabolic health and TMAO may benefit muscle tissue under cellular stress. The beneficial protein chaperone role for TMAO is investigated in muscle tissues similar to the previous hepatocyte studies. TMAO was first identified as a protein chaperone in fish where it accumulates in muscle tissue and protects cellular functions from the challenges of the deep-sea environment including osmotic pressure [180,248,249]. In animal muscle tissue enzyme kinetic studies, TMAO treatment beneficially increased lactate dehydrogenase (LDH) substrate affinity [250,251]. When urea treatment inhibited rabbit LDH activity, TMAO addition recovered it to control levels [250]. TMAO also increased chicken skeletal muscle myosin ATPase activity [252]. Again, ATPase activity was inhibited by urea and recovered by TMAO [252]. These in vitro studies demonstrate that the TMAO protein chaperone capacity is independent of evolutionary history because its effects on enzyme stability are observed in various species [10–12,14,15]. Because hydrostatic pressure stress is similar between deep-sea conditions and cardiac muscle contraction during heart failure, these beneficial TMAO effects on LDH are clinically relevant. However, in heart failure modeling rats, TMAO supplementation did not affect the tertiary or quaternary structures of LDH [202]. While these in vitro results highlight a potential protective TMAO effect on muscle enzyme activity, future in vivo studies may provide a more convincing connection to clinically relevant cellular conditions.

TMAO effects on cardiac muscle investigated in CVD research report variable results. While most CVD research investigates TMAO effects on atherosclerosis at the vascular tissue, some studies investigate cardiac muscle tissue effects independent of vascular damage which relate to metabolic health. Poor cardiac function phenotypes including altered ventricular wall dimensions, ejection fraction, fractional shortening, and interventricular wall thickness are worsened by TMAO [24,150,253,254]. The key cardiac muscle functions contraction and relaxation are also impaired by TMAO. Left ventricle contraction and relaxation times were prolonged in high fat high carbohydrate diet fed mice [255]. These times were recovered by TMAO reduction via the intestinal TMAO production inhibitor 3,3-dimethyl-1-butanol [255]. Primary rat cardiomyocytes cultured with 20 μM and 100 μM TMAO also had prolonged re-lengthening times and decreased fractional shortening [256]. These findings connect TMAO to heart failure [257]. Conversely, human atrial appendage biopsy tissue cultured with pharmacological concentrations of 300 μM to 3 mM TMAO had increased contractile tension and rate of relaxation [258]. This enhanced contractility was attributed to increased calcium signaling in primary animal cardiac tissue cultured with more physiological 20–100 μM TMAO [256,258]. Furthermore, TMAO supplemented rats accumulated TMAO in cardiac tissue which coincided with improved mitochondrial energy metabolism [259]. Other studies suggest that TMA, but not TMAO is deleterious to cardiac function, however TMA accumulation in cardiac tissue has not been demonstrated [92,260]. Since these studies show that TMAO can depress and enhance contractile function, it is unclear if TMAO is beneficial or harmful to cardiac muscle.

In animal models, TMAO alters cardiacmyocytes in potentially beneficial or deleterious ways. TMAO increases cardiac hypertrophy which may beneficially increase cardiac output, but can aggravate heart failure when combined with fibrosis, collagen deposition, and inflammation [254,261,262]. Cardiac fibroblasts cultured with 10 μM, 50 μM, and 100 μM TMAO had increased proliferation and viability [253]. TMAO and choline supplemented animals had increased heart weight to body weight ratios and cardiomyocyte fibrosys [23,24,150,253,254,263]. TMAO further increases the risk of heart failure by altering fuel utilization in cardiac tissue [264–269]. Cardiac biopsies from high fat high carbohydrate diet fed mice with elevated TMAO levels also showed increased fibrosis and inflammatory markers [255]. TMAO treatments correspond with elevated ROS levels and inflammation in cell culture [35,254,256]. In vascular smooth muscle, TMAO treatments induced ROS accumulation at the mitochondria and increased inflammation while downregulating the endogenous antioxidant defense system [35,270]. Other studies contradict these reports that TMAO worsens cardiac health by reporting no change in TMAO treated cardiomyocytes or smooth muscle cell viability or ROS levels [92,264,271]. Finally, one study observed beneficial effects where cardiomyocytes from TMAO supplemented hypertensive rats showed reduced hypertrophy and fibrosis [272]. Therefore, while many studies support the hypothesis that TMAO drives deleterious cardiac hypertrophy through fibrosis, inflammation, and ROS accumulation, potentially beneficial TMAO effects on cardiomyocytes are also reported.

In the metabolic context, insulin-sensitive muscle tissue function helps to regulate whole-body carbohydrate and lipid energy balance. Various studies highlight that TMAO alters skeletal muscle insulin sensitivity; however they report negative and positive effects. In TMAO supplemented high fat diet fed mice, the muscle tissue insulin signaling cascade was inhibited [161]. Gluconeogenic genes typically suppressed by insulin were upregulated in the muscle and other metabolic tissues of TMAO supplemented mice, indicating a link between TMAO and insulin resistance [161]. TMAO reduction by FMO1, 2, 4, and 5 knockout produced leaner mice with increased exercise capacity and resting energy expenditure by increasing muscle fatty acid oxidation rates which presumably links elevated TMAO levels to insulin resistance [163,241]. By contrast, in over-nutrition induced insulin resistant monkeys, elevated serum TMAO levels corresponded with skeletal muscle hyperlipidaemia indicating that insulin induced lipogenesis was functional [158]. TMAO enhanced insulin triggered glycogen accumulation in primary rat cardiomyocytes [256]. Despite some debate, the majority of data on insulin sensitivity show that TMAO drives muscle carbohydrate and lipid metabolism associated with insulin resistance and heart failure. However, because TMAO muscle tissue effects include beneficial changes to enzyme activity and contractile function, a dominant role for TMAO is debated and requires further research.

4. TMAO Effects on Blood Glucose Management

Poor blood glucose management is a hallmark of metabolic diseases. The hormone regulators insulin and glucagon manage metabolic functions across the body to manage blood glucose levels. Since TMAO does not alter glucagon levels, research focuses on its effects on insulin secretion and signaling (Figure 3) [162]. Situated within the pancreatic islets of Langerhans, β-cells match insulin secretion to elevated blood glucose levels to trigger glucose uptake into responsive target tissues including hepatocytes, adipocytes, and muscle tissues as discussed [273–275]. In T2D, elevated blood glucose stems from β-cell glucose intolerance and target tissue insulin resistance. Therefore, these mechanisms are investigated using obese patients and high fat diet fed animals which link elevated serum TMAO to metabolic diseases including obesity, gestational diabetes, and T2D [157,158,162,237,242]. In over-nutrition-induced obese mice, TMAO reduction by FMO3 knockdown reduced, and overexpression increased, body weight and fat pad adiposity [162]. This connection between TMAO and obesity is also observed clinically [50,237,238]. A vegan diet intervention in obese glucose intolerant patients reduced TMAO levels and improved postprandial blood glucose levels [50]. While these studies link TMAO to poor blood glucose management typical in metabolic diseases, a closer look at target tissue insulin resistance and β-cell glucose intolerance is necessary to identify the underlying TMAO mechanisms.

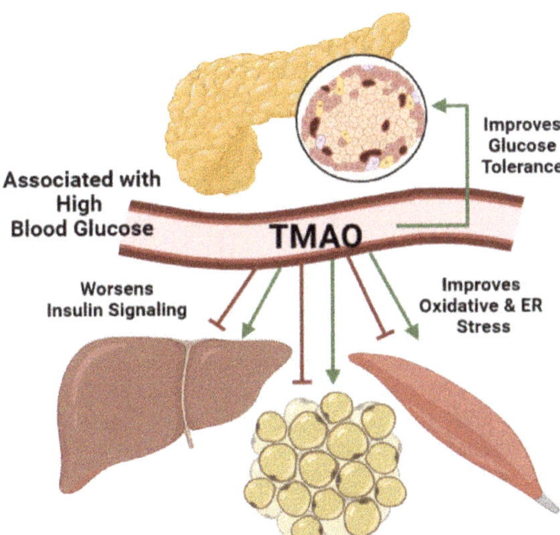

Figure 3. TMAO Effects on Blood Glucose Management. Elevated serum TMAO levels are associated with elevated blood glucose levels, a hallmark of metabolic diseases. Insulin resistance at target tissues or β-cell glucose intolerance can drive this phenotype. While associative studies generally link TMAO to worsened insulin resistance, the molecular evidence in hepatocytes, adipocytes, and skeletal muscles is divided. One study investigates TMAO effects on β-cell containing pancreatic islets and report improved glucose tolerance which beneficially lowers blood glucose in the T2D condition. This figure was created with biorender.com.

Clinical and animal studies demonstrate varying TMAO effects on insulin resistance. Although insulin inhibits FMO3 expression [48,135,136], TMAO accumulation under insulin resistant conditions may alter insulin signaling in turn. As described earlier, associative studies generally link TMAO to worsened insulin resistance phenotypes [134,156,161,163,241,242]. In obese patients, increased TMAO levels are associated with elevated fasting blood glucose and insulin levels and increased insulin resistance as assessed by homeostatic model assessment for insulin resistance (HOMA-IR) [237]. When obese patients were treated with exercise and a hypocaloric diet, TMAO levels decreased by 30% which correlated with improved insulin sensitivity and glucose disposition rates [238]. This connection between clinical insulin resistance and TMAO is matched by high fat diet fed animals [158,161]. Reduced TMAO levels in FMO3 knockdown mice reversed insulin resistance as measured by blood glucose and insulin levels [48,162]. Indeed, metabolomic studies in obese and insulin resistant mice identify TMAO and FMO3 as markers of altered metabolism [217]. Therefore, TMAO is generally associated with worsened insulin resistance and elevated blood glucose levels.

However, when TMAO effects on insulin signaling are investigated more closely in hepatocytes, adipocytes, and skeletal muscle, the molecular evidence is divided. While positive and negative TMAO effects on insulin signaling are reported in these tissues, the molecular mechanisms are best described in hepatic studies. Interestingly, TMAO downregulates the hepatic insulin signaling cascade [161,162], but beneficially reduces oxidative and ER stress associated with insulin resistance [117,136,165,171,195]. Other studies report no relationship between TMAO and insulin. Reducing TMAO levels by FMO3 and FMO5 knockdown in mice did not affect insulin resistance, nor did hepatic FMO3 knockdown alter AKT expression [135]. Diet-induced TMAO levels in healthy elderly subjects were not associated with blood glucose, insulin, HOMA-IR or hemoglobin A1c (HbA1c) measurements [143]. This result shows that TMAO may either not be causally related to poor blood glucose management or that TMAO effects differ in healthy versus diseased conditions. Therefore, while TMAO is generally associated with worsened insulin resistance and elevated blood glucose levels, there is molecular evidence for beneficial and deleterious TMAO effects on insulin signaling at target tissues.

Whole-body blood glucose management also hinges on pancreatic β-cell glucose tolerance and insulin secretion. As a nutrient sensing tissue, β-cells couple insulin secretion to elevated blood glucose levels. Impeded β-cell function, measured by in vivo glucose tolerance tests or the in vitro glucose stimulated insulin secretion assay, is a hallmark of diabetes. In high fat diet fed mice, TMAO supplementation increased serum levels to 17 μM and worsened glucose tolerance test results while TMAO reduction by FMO3 knockdown improved performance [48,161,242]. Although these studies highlight a connection between TMAO and blood glucose mismanagement, they do not directly measure β-cell function. Only one study on high fat diet-fed mice with a continuous TMAO intravenous pump suggests that TMAO ameliorated blood glucose levels via improved β-cell glucose tolerance and function [239]. Compared to vehicle controls, TMAO treated mice had better blood glucose and insulin levels during a glucose tolerance test [239]. Interestingly, TMAO showed no effect on glucose tolerance in healthy conditions, again supporting these pathogenic conditions elucidated a unique TMAO effect not mirrored in healthy contexts [239,242]. While this study approximated insulin secretion function in T2D mice, future β-cell culture experiments are needed to measure direct TMAO effects. Therefore, while TMAO has variable effects on blood glucose management such that it is generally linked to insulin resistance at the target tissue while it may benefit β-cell function under T2D conditions.

5. Conclusions

Although TMAO research in the human health context is young, we know that changes across the diet–microbiota–liver–kidney axis leads to serum accumulation, and that TMAO metabolic effects are context dependent. In CVD research, studies generally assert that TMAO is a deleterious dietary gut microbiome metabolite biomarker, although its molecular effects on vascular and cardiac tissue are mixed. TMAO's role in tissues relating to metabolic diseases are further divided and may differ from its role in CVD [57]. While some data demonstrate that it is deleterious, others support beneficial TMAO effects. TMAO effects also differ between healthy and diseased conditions, such that metabolic stress seems to be a prerequisite for observable TMAO functions. Using such pathologic models, TMAO effects are best defined for hepatocytes while future research is needed to clarify mechanisms involving pancreatic β-cells. Negative TMAO effects are mainly observed in patients and animal models of NAFLD, gallstone formation, CKD, cognitive diseases, and T2D by aggravating insulin resistance and impairing cellular functions. The positive TMAO effects generally involve anti-oxidative or anti-inflammatory effects observed at the tissue level, especially in hepatocytes, adipocytes, muscle tissue, and pancreatic β-cells under stress from over-nutrition models. When considering its effects on blood glucose management, it is unclear if it improves or worsens insulin resistance overall. This debate between positive and negative TMAO effects calls for more research to better connect direct cellular mechanisms to clinical outcomes. It must be noted that the studies discussed here investigated TMAO effects via various research models. Studies induced elevated TMAO levels by altering the dietary substrate load, the microbiome TMA production, or hepatic FMO3 TMAO production. Other studies measure elevated TMAO levels in animal models using dietary or genetically mutant animals originally established to model metabolic diseases. Therefore, these models could alter metabolic tissues in other more direct ways that may overshadow or compound TMAO effects. Since interpreting causality in these studies in fraught with concerns, future TMAO research should of course consider clinical relevance of the research design closely. Since TMAO is an indicator of excess dietary consumption of choline and related compounds, future studies should continue to explore TMAO effects within the context of over-nutrition associated metabolic diseases contributing to our global health challenge.

Author Contributions: Conceptualization, writing—original draft preparation, review, and editing, E.S.K., T.S.L. and J.S.T.; supervision, project administration, and funding acquisition, J.S.T. All authors have read and agreed to the published version of the manuscript.

Funding: This research was funded by The Beatson Foundation [2019-003], NIH NIDDK [R15DK1248 3501A1] and USDA-AFRI [2020-67017-30846].

Acknowledgments: We thank Andrew Neilson, Jason Kenealey, Chad Hancock, and members of the Tessem lab for review and editing support.

Conflicts of Interest: The authors declare no conflict of interest.

References

1. Anderlini, F. Index of authors' names. *J. Chem. Soc. Abstr.* **1894**, *66*, B493–B539. [CrossRef]
2. Dunstan, W.R.; Goulding, E. XCVII.—The action of hydrogen peroxide on secondary and tertiary aliphatic amines. Formation of alkylated hydroxylamines and oxamines. *J. Chem. Soc. Trans.* **1899**, *75*, 1004–1011. [CrossRef]
3. Lintzel, W. Trimethylaminoxyd im Menschlichen Harn. *Klin. Wochenschr.* **1934**, *13*, 304. [CrossRef]
4. Baker, J.; Chaykin, S. The biosynthesis of trimethylamine-N-oxide. *Biochim. Biophys. Acta* **1960**, *41*, 548–550. [CrossRef]
5. Baker, J.R.; Struempler, A.; Chaykin, S. A comparative study of trimethylamine-N-oxide biosynthesis. *Biochim. Biophys. Acta* **1963**, *71*, 58–64. [CrossRef] [PubMed]
6. de la Huerga, J.; Popper, H. Urinary excretion of choline metabolites following choline administration in normals and patients with hepatobiliary diseases. *J. Clin. Investig.* **1951**, *30*, 463–470. [CrossRef] [PubMed]
7. Tarr, H.L.A. The fate of trimethylamine oxide and trimethylamine in man. *J. Fish. Res. Board Can.* **1941**, *5b*, 211–216. [CrossRef]
8. Wang, Z.; Klipfell, E.; Bennett, B.J.; Koeth, R.; Levison, B.S.; Dugar, B.; Feldstein, A.E.; Britt, E.B.; Fu, X.; Chung, Y.M.; et al. Gut flora metabolism of phosphatidylcholine promotes cardiovascular disease. *Nature* **2011**, *472*, 57–63. [CrossRef]
9. Kuffel, A.; Zielkiewicz, J. The hydrogen bond network structure within the hydration shell around simple osmolytes: Urea, tetramethylurea, and trimethylamine-N-oxide, investigated using both a fixed charge and a polarizable water model. *J. Chem. Phys.* **2010**, *133*, 035102. [CrossRef]
10. Bennion, B.J.; Daggett, V. Counteraction of urea-induced protein denaturation by trimethylamine N-oxide: A chemical chaperone at atomic resolution. *Proc. Natl. Acad. Sci. USA* **2004**, *101*, 6433–6438. [CrossRef]

11. Bennion, B.J.; DeMarco, M.L.; Daggett, V. Preventing misfolding of the prion protein by trimethylamine N-oxide. *Biochemistry* **2004**, *43*, 12955–12963. [CrossRef] [PubMed]

12. Hu, C.Y.; Lynch, G.C.; Kokubo, H.; Pettitt, B.M. Trimethylamine N-oxide influence on the backbone of proteins: An oligoglycine model. *Proteins* **2010**, *78*, 695–704. [CrossRef] [PubMed]

13. Mulhern, M.L.; Madson, C.J.; Kador, P.F.; Randazzo, J.; Shinohara, T. Cellular osmolytes reduce lens epithelial cell death and alleviate cataract formation in galactosemic rats. *Mol. Vis.* **2007**, *13*, 1397–1405. [PubMed]

14. Arakawa, T.; Timasheff, S.N. The stabilization of proteins by osmolytes. *Biophys. J.* **1985**, *47*, 411–414. [CrossRef]

15. Mashino, T.; Fridovich, I. Effects of urea and trimethylamine-N-oxide on enzyme activity and stability. *Arch. Biochem. Biophys.* **1987**, *258*, 356–360. [CrossRef]

16. Pincus, D.L.; Hyeon, C.; Thirumalai, D. Effects of trimethylamine N-oxide (TMAO) and crowding agents on the stability of RNA hairpins. *J. Am. Chem. Soc.* **2008**, *130*, 7364–7372. [CrossRef]

17. Rezus, Y.L.; Bakker, H.J. Destabilization of the hydrogen-bond structure of water by the osmolyte trimethylamine N-oxide. *J. Phys. Chem. B* **2009**, *113*, 4038–4044. [CrossRef]

18. Chhibber-Goel, J.; Singhal, V.; Parakh, N.; Bhargava, B.; Sharma, A. The metabolite trimethylamine-N-Oxide is an emergent biomarker of human health. *Curr. Med. Chem.* **2017**, *24*, 3942–3953. [CrossRef]

19. Cho, C.E.; Taesuwan, S.; Malysheva, O.V.; Bender, E.; Tulchinsky, N.F.; Yan, J.; Sutter, J.L.; Caudill, M.A. Trimethylamine-N-oxide (TMAO) response to animal source foods varies among healthy young men and is influenced by their gut microbiota composition: A randomized controlled trial. *Mol. Nutr. Food Res.* **2017**, *61*, 1600324. [CrossRef]

20. Heianza, Y.; Ma, W.; DiDonato, J.A.; Sun, Q.; Rimm, E.B.; Hu, F.B.; Rexrode, K.M.; Manson, J.E.; Qi, L. Long-term changes in gut microbial metabolite trimethylamine N-Oxide and coronary heart disease risk. *J. Am. Coll. Cardiol.* **2020**, *75*, 763–772. [CrossRef]

21. Hoyles, L.; Jimenez-Pranteda, M.L.; Chilloux, J.; Brial, F.; Myridakis, A.; Aranias, T.; Magnan, C.; Gibson, G.R.; Sanderson, J.D.; Nicholson, J.K.; et al. Metabolic retroconversion of trimethylamine N-oxide and the gut microbiota. *Microbiome* **2018**, *6*, 73. [CrossRef]

22. Moraes, C.; Fouque, D.; Amaral, A.C.; Mafra, D. Trimethylamine N-Oxide from gut microbiota in chronic kidney disease patients: Focus on diet. *J. Ren. Nutr.* **2015**, *25*, 459–465. [CrossRef]

23. Nagatomo, Y.; Tang, W.H.W. Intersections between microbiome and heart failure: Revisiting the gut hypothesis. *J. Card. Fail.* **2015**, *21*, 973–980. [CrossRef] [PubMed]

24. Organ, C.L.; Otsuka, H.; Bhushan, S.; Wang, Z.; Bradley, J.; Trivedi, R.; Polhemus, D.J.; Tang, W.H.; Wu, Y.; Hazen, S.L.; et al. Choline diet and its gut microbe-derived metabolite, trimethylamine N-Oxide, exacerbate pressure overload-induced heart failure. *Circ. Heart Fail.* **2016**, *9*, e002314. [CrossRef] [PubMed]

25. Servillo, L.; D'Onofrio, N.; Giovane, A.; Casale, R.; Cautela, D.; Castaldo, D.; Iannaccone, F.; Neglia, G.; Campanile, G.; Balestrieri, M.L. Ruminant meat and milk contain delta-valerobetaine, another precursor of trimethylamine N-oxide (TMAO) like gamma-butyrobetaine. *Food Chem.* **2018**, *260*, 193–199. [CrossRef]

26. Tang, W.H.; Hazen, S.L. Microbiome, trimethylamine N-oxide, and cardiometabolic disease. *Transl. Res.* **2017**, *179*, 108–115. [CrossRef] [PubMed]

27. Wang, F.; Xu, J.; Jakovlic, I.; Wang, W.M.; Zhao, Y.H. Dietary betaine reduces liver lipid accumulation via improvement of bile acid and trimethylamine-N-oxide metabolism in blunt-snout bream. *Food Funct.* **2019**, *10*, 6675–6689. [CrossRef] [PubMed]

28. Wang, Z.; Bergeron, N.; Levison, B.S.; Li, X.S.; Chiu, S.; Jia, X.; Koeth, R.A.; Li, L.; Wu, Y.; Tang, W.H.W.; et al. Impact of chronic dietary red meat, white meat, or non-meat protein on trimethylamine N-oxide metabolism and renal excretion in healthy men and women. *Eur. Heart J.* **2019**, *40*, 583–594. [CrossRef] [PubMed]

29. Xu, R.; Wang, Q. Towards understanding brain-gut-microbiome connections in Alzheimer's disease. *BMC Syst. Biol.* **2016**, *10*, 63. [CrossRef]

30. Aldana-Hernandez, P.; Leonard, K.A.; Zhao, Y.Y.; Curtis, J.M.; Field, C.J.; Jacobs, R.L. Dietary choline or trimethylamine N-oxide supplementation does not influence atherosclerosis development in Ldlr-/-and Apoe-/-male mice. *J. Nutr.* **2020**, *150*, 249–255. [CrossRef]

31. Lindskog Jonsson, A.; Caesar, R.; Akrami, R.; Reinhardt, C.; Fak Hallenius, F.; Boren, J.; Backhed, F. Impact of gut microbiota and diet on the development of atherosclerosis in Apoe(-/-) mice. *Arterioscler. Thromb. Vasc. Biol.* **2018**, *38*, 2318–2326. [CrossRef] [PubMed]

32. Bennett, B.J.; de Aguiar Vallim, T.Q.; Wang, Z.; Shih, D.M.; Meng, Y.; Gregory, J.; Allayee, H.; Lee, R.; Graham, M.; Crooke, R.; et al. Trimethylamine-N-oxide, a metabolite associated with atherosclerosis, exhibits complex genetic and dietary regulation. *Cell Metab.* **2013**, *17*, 49–60. [CrossRef]

33. Boini, K.M.; Hussain, T.; Li, P.L.; Koka, S. Trimethylamine-N-Oxide instigates NLRP3 inflammasome activation and endothelial dysfunction. *Cell Physiol. Biochem.* **2017**, *44*, 152–162. [CrossRef] [PubMed]

34. Bordoni, L.; Sawicka, A.K.; Szarmach, A.; Winklewski, P.J.; Olek, R.A.; Gabbianelli, R. A pilot study on the effects of l-carnitine and trimethylamine-N-Oxide on platelet mitochondrial DNA methylation and CVD biomarkers in aged women. *Int. J. Mol. Sci.* **2020**, *21*, 1047. [CrossRef]

35. Chen, M.L.; Zhu, X.H.; Ran, L.; Lang, H.D.; Yi, L.; Mi, M.T. Trimethylamine-N-Oxide induces vascular inflammation by activating the NLRP3 inflammasome through the SIRT3-SOD2-mtROS signaling pathway. *J. Am. Heart Assoc.* **2017**, *6*, e006347. [CrossRef] [PubMed]

36. Chen, Y.; Weng, Z.; Liu, Q.; Shao, W.; Guo, W.; Chen, C.; Jiao, L.; Wang, Q.; Lu, Q.; Sun, H.; et al. FMO3 and its metabolite TMAO contribute to the formation of gallstones. *Biochim. Biophys. Acta Mol. Basis Dis.* **2019**, *1865*, 2576–2585. [CrossRef] [PubMed]
37. Cheng, X.; Qiu, X.; Liu, Y.; Yuan, C.; Yang, X. Trimethylamine N-oxide promotes tissue factor expression and activity in vascular endothelial cells: A new link between trimethylamine N-oxide and atherosclerotic thrombosis. *Thromb. Res.* **2019**, *177*, 110–116. [CrossRef]
38. Collins, H.L.; Drazul-Schrader, D.; Sulpizio, A.C.; Koster, P.D.; Williamson, Y.; Adelman, S.J.; Owen, K.; Sanli, T.; Bellamine, A. L-Carnitine intake and high trimethylamine N-oxide plasma levels correlate with low aortic lesions in ApoE(-/-) transgenic mice expressing CETP. *Atherosclerosis* **2016**, *244*, 29–37. [CrossRef] [PubMed]
39. Croyal, M.; Saulnier, P.J.; Aguesse, A.; Gand, E.; Ragot, S.; Roussel, R.; Halimi, J.M.; Ducrocq, G.; Cariou, B.; Montaigne, D.; et al. Plasma trimethylamine N-Oxide and risk of cardiovascular events in patients with type 2 diabetes. *J. Clin. Endocrinol. Metab.* **2020**, *105*, dgaa188. [CrossRef] [PubMed]
40. Ding, L.; Chang, M.; Guo, Y.; Zhang, L.; Xue, C.; Yanagita, T.; Zhang, T.; Wang, Y. Trimethylamine-N-oxide (TMAO)-induced atherosclerosis is associated with bile acid metabolism. *Lipids Health Dis.* **2018**, *17*, 286. [CrossRef]
41. Fu, Q.; Zhao, M.; Wang, D.; Hu, H.; Guo, C.; Chen, W.; Li, Q.; Zheng, L.; Chen, B. Coronary plaque characterization assessed by optical coherence tomography and plasma trimethylamine-N-oxide levels in patients with coronary artery disease. *Am. J. Cardiol.* **2016**, *118*, 1311–1315. [CrossRef]
42. Haghikia, A.; Li, X.S.; Liman, T.G.; Bledau, N.; Schmidt, D.; Zimmermann, F.; Krankel, N.; Widera, C.; Sonnenschein, K.; Haghikia, A.; et al. Gut microbiota dependent Trimethylamine N-Oxide predicts risk of cardiovascular events in patients with stroke and is related to proinflammatory monocytes. *Arterioscler. Thromb. Vasc. Biol.* **2018**, *38*, 2225–2235. [CrossRef]
43. Hou, L.; Zhang, Y.; Zheng, D.; Shi, H.; Zou, C.; Zhang, H.; Lu, Z.; Du, H. Increasing trimethylamine N-oxide levels as a predictor of early neurological deterioration in patients with acute ischemic stroke. *Neurol. Res.* **2020**, *42*, 153–158. [CrossRef] [PubMed]
44. Kanitsoraphan, C.; Rattanawong, P.; Charoensri, S.; Senthong, V. Trimethylamine N-Oxide and risk of cardiovascular disease and mortality. *Curr. Nutr. Rep.* **2018**, *7*, 207–213. [CrossRef] [PubMed]
45. Ma, G.; Pan, B.; Chen, Y.; Guo, C.; Zhao, M.; Zheng, L.; Chen, B. Trimethylamine N-oxide in atherogenesis: Impairing endothelial self-repair capacity and enhancing monocyte adhesion. *Biosci. Rep.* **2017**, *37*, BSR20160244. [CrossRef] [PubMed]
46. Liu, Y.; Dai, M. Trimethylamine N-Oxide generated by the gut microbiota is associated with vascular inflammation: New insights into atherosclerosis. *Mediat. Inflamm.* **2020**, *2020*, 4634172. [CrossRef] [PubMed]
47. Mafune, A.; Iwamoto, T.; Tsutsumi, Y.; Nakashima, A.; Yamamoto, I.; Yokoyama, K.; Yokoo, T.; Urashima, M. Associations among serum trimethylamine-N-oxide (TMAO) levels, kidney function and infarcted coronary artery number in patients undergoing cardiovascular surgery: A cross-sectional study. *Clin. Exp. Nephrol.* **2016**, *20*, 731–739. [CrossRef]
48. Miao, J.; Ling, A.V.; Manthena, P.V.; Gearing, M.E.; Graham, M.J.; Crooke, R.M.; Croce, K.J.; Esquejo, R.M.; Clish, C.B.; Vicent, D.; et al. Flavin-containing monooxygenase 3 as a potential player in diabetes-associated atherosclerosis. *Nat. Commun.* **2015**, *6*, 6498. [CrossRef]
49. Al-Obaide, M.A.I.; Singh, R.; Datta, P.; Rewers-Felkins, K.A.; Salguero, M.V.; Al-Obaidi, I.; Kottapalli, K.R.; Vasylyeva, T.L. Gut microbiota-dependent Trimethylamine-N-Oxide and serum biomarkers in patients with T2DM and advanced CKD. *J. Clin. Med.* **2017**, *6*, 86. [CrossRef]
50. Argyridou, S.; Davies, M.J.; Biddle, G.J.H.; Bernieh, D.; Suzuki, T.; Dawkins, N.P.; Rowlands, A.V.; Khunti, K.; Smith, A.C.; Yates, T. Evaluation of an 8-week vegan diet on plasma Trimethylamine-N-Oxide and postchallenge glucose in adults with dysglycemia or obesity. *J. Nutr.* **2021**, *151*, 1844–1853. [CrossRef]
51. Canyelles, M.; Tondo, M.; Cedo, L.; Farras, M.; Escola-Gil, J.C.; Blanco-Vaca, F. Trimethylamine N-Oxide: A link among diet, gut microbiota, gene regulation of liver and intestine cholesterol homeostasis and HDL function. *Int. J. Mol. Sci.* **2018**, *19*, 3228. [CrossRef]
52. Dambrova, M.; Latkovskis, G.; Kuka, J.; Strele, I.; Konrade, I.; Grinberga, S.; Hartmane, D.; Pugovics, O.; Erglis, A.; Liepinsh, E. Diabetes is Associated with Higher Trimethylamine N-oxide Plasma Levels. *Exp. Clin. Endocrinol. Diabetes* **2016**, *124*, 251–256. [CrossRef]
53. Dong, Z.; Liang, Z.; Guo, M.; Hu, S.; Shen, Z.; Hai, X. The association between plasma levels of Trimethylamine N-Oxide and the risk of coronary heart disease in chinese patients with or without type 2 diabetes mellitus. *Dis. Markers* **2018**, *2018*, 1578320. [CrossRef] [PubMed]
54. Garcia, E.; Oste, M.C.J.; Bennett, D.W.; Jeyarajah, E.J.; Shalaurova, I.; Gruppen, E.G.; Hazen, S.L.; Otvos, J.D.; Bakker, S.J.L.; Dullaart, R.P.F.; et al. High betaine, a Trimethylamine N-Oxide related metabolite, is prospectively associated with low future risk of type 2 diabetes mellitus in the PREVEND study. *J. Clin. Med.* **2019**, *8*, 1813. [CrossRef] [PubMed]
55. Koppe, L.; Fouque, D.; Soulage, C.O. Metabolic abnormalities in diabetes and kidney disease: Role of uremic toxins. *Curr. Diab. Rep.* **2018**, *18*, 97. [CrossRef] [PubMed]
56. Leustean, A.M.; Ciocoiu, M.; Sava, A.; Costea, C.F.; Floria, M.; Tarniceriu, C.C.; Tanase, D.M. Implications of the intestinal microbiota in diagnosing the progression of diabetes and the presence of cardiovascular complications. *J. Diabetes Res.* **2018**, *2018*, 5205126. [CrossRef] [PubMed]
57. Lever, M.; George, P.M.; Slow, S.; Bellamy, D.; Young, J.M.; Ho, M.; McEntyre, C.J.; Elmslie, J.L.; Atkinson, W.; Molyneux, S.L.; et al. Betaine and Trimethylamine-N-Oxide as predictors of cardiovascular outcomes show different patterns in diabetes mellitus: An observational study. *PLoS ONE* **2014**, *9*, e114969. [CrossRef]

58. McEntyre, C.J.; Lever, M.; Chambers, S.T.; George, P.M.; Slow, S.; Elmslie, J.L.; Florkowski, C.M.; Lunt, H.; Krebs, J.D. Variation of betaine, N,N-dimethylglycine, choline, glycerophosphorylcholine, taurine and trimethylamine-N-oxide in the plasma and urine of overweight people with type 2 diabetes over a two-year period. *Ann. Clin. Biochem* **2015**, *52*, 352–360. [CrossRef] [PubMed]

59. Mente, A.; Chalcraft, K.; Ak, H.; Davis, A.D.; Lonn, E.; Miller, R.; Potter, M.A.; Yusuf, S.; Anand, S.S.; McQueen, M.J. The relationship between Trimethylamine-N-Oxide and prevalent cardiovascular disease in a multiethnic population living in Canada. *Can. J. Cardiol.* **2015**, *31*, 1189–1194. [CrossRef] [PubMed]

60. Mueller, D.M.; Allenspach, M.; Othman, A.; Saely, C.H.; Muendlein, A.; Vonbank, A.; Drexel, H.; von Eckardstein, A. Plasma levels of trimethylamine-N-oxide are confounded by impaired kidney function and poor metabolic control. *Atherosclerosis* **2015**, *243*, 638–644. [CrossRef] [PubMed]

61. Oellgaard, J.; Winther, S.A.; Hansen, T.S.; Rossing, P.; von Scholten, B.J. Trimethylamine N-oxide (TMAO) as a new potential therapeutic target for insulin resistance and cancer. *Curr. Pharm. Des.* **2017**, *23*, 3699–3712. [CrossRef] [PubMed]

62. Ottiger, M.; Nickler, M.; Steuer, C.; Bernasconi, L.; Huber, A.; Christ-Crain, M.; Henzen, C.; Hoess, C.; Thomann, R.; Zimmerli, W.; et al. Gut, microbiota-dependent trimethylamine-N-oxide is associated with long-term all-cause mortality in patients with exacerbated chronic obstructive pulmonary disease. *Nutrition* **2018**, *45*, 135–141.e1. [CrossRef] [PubMed]

63. Papandreou, C.; Bullo, M.; Zheng, Y.; Ruiz-Canela, M.; Yu, E.; Guasch-Ferre, M.; Toledo, E.; Clish, C.; Corella, D.; Estruch, R.; et al. Plasma trimethylamine-N-oxide and related metabolites are associated with type 2 diabetes risk in the prevencion con dieta mediterranea (PREDIMED) trial. *Am. J. Clin. Nutr.* **2018**, *108*, 163–173. [CrossRef] [PubMed]

64. Randrianarisoa, E.; Lehn-Stefan, A.; Wang, X.; Hoene, M.; Peter, A.; Heinzmann, S.S.; Zhao, X.; Konigsrainer, I.; Konigsrainer, A.; Balletshofer, B.; et al. Relationship of serum Trimethylamine N-Oxide (TMAO) levels with early atherosclerosis in humans. *Sci. Rep.* **2016**, *6*, 26745. [CrossRef] [PubMed]

65. Shan, Z.; Sun, T.; Huang, H.; Chen, S.; Chen, L.; Luo, C.; Yang, W.; Yang, X.; Yao, P.; Cheng, J.; et al. Association between microbiota-dependent metabolite trimethylamine-N-oxide and type 2 diabetes. *Am. J. Clin. Nutr.* **2017**, *106*, 888–894. [CrossRef]

66. Tang, W.H.; Wang, Z.; Li, X.S.; Fan, Y.; Li, D.S.; Wu, Y.; Hazen, S.L. Increased Trimethylamine N-Oxide portends high mortality risk independent of glycemic control in patients with type 2 diabetes mellitus. *Clin. Chem.* **2017**, *63*, 297–306. [CrossRef]

67. Winther, S.A.; Ollgaard, J.C.; Tofte, N.; Tarnow, L.; Wang, Z.; Ahluwalia, T.S.; Jorsal, A.; Theilade, S.; Parving, H.H.; Hansen, T.W.; et al. Utility of plasma concentration of Trimethylamine N-Oxide in predicting cardiovascular and renal complications in individuals with type 1 diabetes. *Diabetes Care* **2019**, *42*, 1512–1520. [CrossRef]

68. Zhuang, R.; Ge, X.; Han, L.; Yu, P.; Gong, X.; Meng, Q.; Zhang, Y.; Fan, H.; Zheng, L.; Liu, Z.; et al. Gut microbe-generated metabolite trimethylamine N-oxide and the risk of diabetes: A systematic review and dose-response meta-analysis. *Obes. Rev.* **2019**, *20*, 883–894. [CrossRef]

69. Velasquez, M.T.; Ramezani, A.; Manal, A.; Raj, D.S. Trimethylamine N-Oxide: The good, the bad and the unknown. *Toxins* **2016**, *8*, 326. [CrossRef]

70. Angiletta, C.J.; Griffin, L.E.; Steele, C.N.; Baer, D.J.; Novotny, J.A.; Davy, K.P.; Neilson, A.P. Impact of short-term flavanol supplementation on fasting plasma trimethylamine N-oxide concentrations in obese adults. *Food Funct.* **2018**, *9*, 5350–5361. [CrossRef]

71. Cho, C.E.; Caudill, M.A. Trimethylamine-N-Oxide: Friend, foe, or simply caught in the cross-fire? *Trends Endocrinol. Metab.* **2017**, *28*, 121–130. [CrossRef]

72. Nowinski, A.; Ufnal, M. Trimethylamine N-oxide: A harmful, protective or diagnostic marker in lifestyle diseases? *Nutrition* **2018**, *46*, 7–12. [CrossRef] [PubMed]

73. Papandreou, C.; More, M.; Bellamine, A. Trimethylamine N-Oxide in relation to cardiometabolic health-cause or effect? *Nutrients* **2020**, *12*, 1330. [CrossRef] [PubMed]

74. Washburn, R.L.; Cox, J.E.; Muhlestein, J.B.; May, H.T.; Carlquist, J.F.; Le, V.T.; Anderson, J.L.; Horne, B.D. Pilot study of novel intermittent fasting effects on metabolomic and Trimethylamine N-oxide changes during 24-hour water-only fasting in the FEELGOOD trial. *Nutrients* **2019**, *11*, 246. [CrossRef]

75. Baugh, M.E.; Steele, C.N.; Angiletta, C.J.; Mitchell, C.M.; Neilson, A.P.; Davy, B.M.; Hulver, M.W.; Davy, K.P. Inulin supplementation does not reduce plasma Trimethylamine N-Oxide concentrations in individuals at risk for type 2 diabetes. *Nutrients* **2018**, *10*, 793. [CrossRef] [PubMed]

76. Smits, L.P.; Kootte, R.S.; Levin, E.; Prodan, A.; Fuentes, S.; Zoetendal, E.G.; Wang, Z.; Levison, B.S.; Cleophas, M.C.P.; Kemper, E.M.; et al. Effect of vegan fecal microbiota transplantation on carnitine- and choline-derived Trimethylamine-N-Oxide production and vascular inflammation in patients with metabolic syndrome. *J. Am. Heart Assoc.* **2018**, *7*, e008342. [CrossRef]

77. Svingen, G.F.; Schartum-Hansen, H.; Pedersen, E.R.; Ueland, P.M.; Tell, G.S.; Mellgren, G.; Njolstad, P.R.; Seifert, R.; Strand, E.; Karlsson, T.; et al. Prospective associations of systemic and urinary choline metabolites with incident type 2 diabetes. *Clin. Chem.* **2016**, *62*, 755–765. [CrossRef]

78. Roberts, A.B.; Gu, X.; Buffa, J.A.; Hurd, A.G.; Wang, Z.; Zhu, W.; Gupta, N.; Skye, S.M.; Cody, D.B.; Levison, B.S.; et al. Development of a gut microbe-targeted nonlethal therapeutic to inhibit thrombosis potential. *Nat. Med.* **2018**, *24*, 1407–1417. [CrossRef]

79. Zhang, A.Q.; Mitchell, S.C.; Smith, R.L. Dietary precursors of trimethylamine in man: A pilot study. *Food Chem. Toxicol.* **1999**, *37*, 515–520. [CrossRef]

80. Muramatsu, H.; Matsuo, H.; Okada, N.; Ueda, M.; Yamamoto, H.; Kato, S.; Nagata, S. Characterization of ergothionase from Burkholderia sp. HME13 and its application to enzymatic quantification of ergothioneine. *Appl. Microbiol. Biotechnol.* **2013**, *97*, 5389–5400. [CrossRef]

81. Meyer, K.A.; Shea, J.W. Dietary choline and betaine and risk of CVD: A systematic review and meta-analysis of prospective studies. *Nutrients* **2017**, *9*, 711. [CrossRef]

82. Tindall, A.M.; Petersen, K.S.; Kris-Etherton, P.M. Dietary patterns affect the gut microbiome-the link to risk of cardiometabolic diseases. *J. Nutr.* **2018**, *148*, 1402–1407. [CrossRef] [PubMed]

83. Kaiser, J.; van Daalen, K.R.; Thayyil, A.; Cocco, M.; Caputo, D.; Oliver-Williams, C. A Systematic review of the association between vegan diets and risk of cardiovascular disease. *J. Nutr.* **2021**, *151*, 1539–1552. [CrossRef] [PubMed]

84. De Filippis, F.; Pellegrini, N.; Vannini, L.; Jeffery, I.B.; La Storia, A.; Laghi, L.; Serrazanetti, D.I.; Di Cagno, R.; Ferrocino, I.; Lazzi, C.; et al. High-level adherence to a Mediterranean diet beneficially impacts the gut microbiota and associated metabolome. *Gut* **2016**, *65*, 1812–1821. [CrossRef]

85. Costabile, G.; Vetrani, C.; Bozzetto, L.; Giacco, R.; Bresciani, L.; Del Rio, D.; Vitale, M.; Della Pepa, G.; Brighenti, F.; Riccardi, G.; et al. Plasma TMAO increase after healthy diets: Results from 2 randomized controlled trials with dietary fish, polyphenols, and whole-grain cereals. *Am. J. Clin. Nutr.* **2021**. [CrossRef] [PubMed]

86. Zeisel, S.H.; Wishnok, J.S.; Blusztajn, J.K. Formation of methylamines from ingested choline and lecithin. *J. Pharm. Exp. Ther.* **1983**, *225*, 320–324.

87. Rebouche, C.J.; Chenard, C.A. Metabolic fate of dietary carnitine in human adults: Identification and quantification of urinary and fecal metabolites. *J. Nutr.* **1991**, *121*, 539–546. [CrossRef]

88. Al-Waiz, M.; Mitchell, S.C.; Idle, J.R.; Smith, R.L. The metabolism of 14C-labelled trimethylamine and its N-oxide in man. *Xenobiotica* **1987**, *17*, 551–558. [CrossRef]

89. Ufnal, M.; Zadlo, A.; Ostaszewski, R. TMAO: A small molecule of great expectations. *Nutrition* **2015**, *31*, 1317–1323. [CrossRef] [PubMed]

90. Wu, W.K.; Chen, C.C.; Liu, P.Y.; Panyod, S.; Liao, B.Y.; Chen, P.C.; Kao, H.L.; Kuo, H.C.; Kuo, C.H.; Chiu, T.H.T.; et al. Identification of TMAO-producer phenotype and host-diet-gut dysbiosis by carnitine challenge test in human and germ-free mice. *Gut* **2019**, *68*, 1439–1449. [CrossRef]

91. Koeth, R.A.; Wang, Z.; Levison, B.S.; Buffa, J.A.; Org, E.; Sheehy, B.T.; Britt, E.B.; Fu, X.; Wu, Y.; Li, L.; et al. Intestinal microbiota metabolism of L-carnitine, a nutrient in red meat, promotes atherosclerosis. *Nat. Med.* **2013**, *19*, 576–585. [CrossRef] [PubMed]

92. Jaworska, K.; Konop, M.; Hutsch, T.; Perlejewski, K.; Radkowski, M.; Grochowska, M.; Bielak-Zmijewska, A.; Mosieniak, G.; Sikora, E.; Ufnal, M. Trimethylamine but not Trimethylamine Oxide increases with age in rat plasma and affects smooth muscle cells viability. *J. Gerontol. A Biol. Sci. Med. Sci.* **2020**, *75*, 1276–1283. [CrossRef]

93. Govindarajulu, M.; Pinky, P.D.; Steinke, I.; Bloemer, J.; Ramesh, S.; Kariharan, T.; Rella, R.T.; Bhattacharya, S.; Dhanasekaran, M.; Suppiramaniam, V.; et al. Gut metabolite TMAO induces synaptic plasticity deficits by promoting endoplasmic reticulum stress. *Front. Mol. Neurosci.* **2020**, *13*, 138. [CrossRef] [PubMed]

94. Li, T.; Chen, Y.; Gua, C.; Li, X. Elevated circulating Trimethylamine N-Oxide levels contribute to endothelial dysfunction in aged rats through vascular inflammation and oxidative stress. *Front. Physiol.* **2017**, *8*, 350. [CrossRef] [PubMed]

95. Romano, K.A.; Vivas, E.I.; Amador-Noguez, D.; Rey, F.E. Intestinal microbiota composition modulates choline bioavailability from diet and accumulation of the proatherogenic metabolite trimethylamine-N-oxide. *mBio* **2015**, *6*, e02481. [CrossRef] [PubMed]

96. Wu, W.K.; Panyod, S.; Liu, P.Y.; Chen, C.C.; Kao, H.L.; Chuang, H.L.; Chen, Y.H.; Zou, H.B.; Kuo, H.C.; Kuo, C.H.; et al. Characterization of TMAO productivity from carnitine challenge facilitates personalized nutrition and microbiome signatures discovery. *Microbiome* **2020**, *8*, 162. [CrossRef]

97. Falony, G.; Vieira-Silva, S.; Raes, J. Microbiology meets big data: The case of gut microbiota-derived trimethylamine. *Annu. Rev. Microbiol.* **2015**, *69*, 305–321. [CrossRef]

98. Martinez-del Campo, A.; Bodea, S.; Hamer, H.A.; Marks, J.A.; Haiser, H.J.; Turnbaugh, P.J.; Balskus, E.P. Characterization and detection of a widely distributed gene cluster that predicts anaerobic choline utilization by human gut bacteria. *mBio* **2015**, *6*, e00042. [CrossRef]

99. Kalnins, G.; Kuka, J.; Grinberga, S.; Makrecka-Kuka, M.; Liepinsh, E.; Dambrova, M.; Tars, K. Structure and function of CutC choline lyase from human microbiota bacterium klebsiella pneumoniae. *J. Biol. Chem.* **2015**, *290*, 21732–21740. [CrossRef]

100. Craciun, S.; Marks, J.A.; Balskus, E.P. Characterization of choline trimethylamine-lyase expands the chemistry of glycyl radical enzymes. *ACS Chem. Biol.* **2014**, *9*, 1408–1413. [CrossRef] [PubMed]

101. Craciun, S.; Balskus, E.P. Microbial conversion of choline to trimethylamine requires a glycyl radical enzyme. *Proc. Natl. Acad. Sci. USA* **2012**, *109*, 21307–21312. [CrossRef] [PubMed]

102. Kalnins, G.; Sevostjanovs, E.; Hartmane, D.; Grinberga, S.; Tars, K. CntA oxygenase substrate profile comparison and oxygen dependency of TMA production in Providencia rettgeri. *J. Basic Microbiol.* **2018**, *58*, 52–59. [CrossRef] [PubMed]

103. Samulak, J.J.; Sawicka, A.K.; Hartmane, D.; Grinberga, S.; Pugovics, O.; Lysiak-Szydlowska, W.; Olek, R.A. L-carnitine supplementation increases Trimethylamine-N-Oxide but not markers of atherosclerosis in healthy aged women. *Ann. Nutr. Metab.* **2019**, *74*, 11–17. [CrossRef]

104. Samulak, J.J.; Sawicka, A.K.; Samborowska, E.; Olek, R.A. Plasma Trimethylamine-N-oxide following cessation of l-carnitine supplementation in healthy aged women. *Nutrients* **2019**, *11*, 1322. [CrossRef]

105. Koeth, R.A.; Levison, B.S.; Culley, M.K.; Buffa, J.A.; Wang, Z.; Gregory, J.C.; Org, E.; Wu, Y.; Li, L.; Smith, J.D.; et al. Gamma-Butyrobetaine is a proatherogenic intermediate in gut microbial metabolism of L-carnitine to TMAO. *Cell Metab.* **2014**, *20*, 799–812. [CrossRef] [PubMed]

106. Gupta, N.; Buffa, J.A.; Roberts, A.B.; Sangwan, N.; Skye, S.M.; Li, L.; Ho, K.J.; Varga, J.; DiDonato, J.A.; Tang, W.H.W.; et al. Targeted inhibition of gut microbial Trimethylamine N-Oxide production reduces renal tubulointerstitial fibrosis and functional impairment in a murine model of chronic kidney disease. *Arter. Thromb. Vasc. Biol.* **2020**, *40*, 1239–1255. [CrossRef]

107. Fortino, M.; Marino, T.; Russo, N.; Sicilia, E. Mechanistic investigation of trimethylamine-N-oxide reduction catalysed by biomimetic molybdenum enzyme models. *Phys. Chem. Chem. Phys.* **2016**, *18*, 8428–8436. [CrossRef]

108. Kruk, M.; Lee, J.S. Inhibition of escherichia coli Trimethylamine-N-oxide reductase by food preservatives (1). *J. Food Prot.* **1982**, *45*, 241–243. [CrossRef]

109. Chen, P.Y.; Li, S.; Koh, Y.C.; Wu, J.C.; Yang, M.J.; Ho, C.T.; Pan, M.H. Oolong tea extract and citrus peel polymethoxyflavones reduce transformation of l-carnitine to Trimethylamine-N-Oxide and decrease vascular inflammation in l-carnitine feeding mice. *J. Agric. Food Chem.* **2019**, *67*, 7869–7879. [CrossRef]

110. He, Z.; Hao, W.; Kwek, E.; Lei, L.; Liu, J.; Zhu, H.; Ma, K.Y.; Zhao, Y.; Ho, H.M.; He, W.S.; et al. Fish oil is more potent than flaxseed oil in modulating gut microbiota and reducing Trimethylamine-N-oxide-exacerbated atherogenesis. *J. Agric. Food Chem.* **2019**, *67*, 13635–13647. [CrossRef] [PubMed]

111. Kuka, J.; Videja, M.; Makrecka-Kuka, M.; Liepins, J.; Grinberga, S.; Sevostjanovs, E.; Vilks, K.; Liepinsh, E.; Dambrova, M. Metformin decreases bacterial trimethylamine production and trimethylamine N-oxide levels in db/db mice. *Sci. Rep.* **2020**, *10*, 14555. [CrossRef]

112. Wang, Z.; Roberts, A.B.; Buffa, J.A.; Levison, B.S.; Zhu, W.; Org, E.; Gu, X.; Huang, Y.; Zamanian-Daryoush, M.; Culley, M.K.; et al. Non-lethal inhibition of gut microbial trimethylamine production for the treatment of atherosclerosis. *Cell* **2015**, *163*, 1585–1595. [CrossRef] [PubMed]

113. Boyer, J.L. Bile formation and secretion. *Compr. Physiol.* **2013**, *3*, 1035–1078. [CrossRef] [PubMed]

114. Phillips, G.B. The lipid composition of human bile. *Biochim. Biophys. Acta* **1960**, *41*, 361–363. [CrossRef]

115. al-Waiz, M.; Mikov, M.; Mitchell, S.C.; Smith, R.L. The exogenous origin of trimethylamine in the mouse. *Metabolism* **1992**, *41*, 135–136. [CrossRef]

116. Brunt, V.E.; Gioscia-Ryan, R.A.; Richey, J.J.; Zigler, M.C.; Cuevas, L.M.; Gonzalez, A.; Vazquez-Baeza, Y.; Battson, M.L.; Smithson, A.T.; Gilley, A.D.; et al. Suppression of the gut microbiome ameliorates age-related arterial dysfunction and oxidative stress in mice. *J. Physiol.* **2019**, *597*, 2361–2378. [CrossRef]

117. Warrier, M.; Shih, D.M.; Burrows, A.C.; Ferguson, D.; Gromovsky, A.D.; Brown, A.L.; Marshall, S.; McDaniel, A.; Schugar, R.C.; Wang, Z.; et al. The TMAO-generating enzyme flavin monooxygenase 3 is a central regulator of cholesterol balance. *Cell Rep.* **2015**, *10*, 326–338. [CrossRef] [PubMed]

118. Tang, W.H.; Wang, Z.; Levison, B.S.; Koeth, R.A.; Britt, E.B.; Fu, X.; Wu, Y.; Hazen, S.L. Intestinal microbial metabolism of phosphatidylcholine and cardiovascular risk. *N. Engl. J. Med.* **2013**, *368*, 1575–1584. [CrossRef] [PubMed]

119. Winter, S.E.; Lopez, C.A.; Baumler, A.J. The dynamics of gut-associated microbial communities during inflammation. *EMBO Rep.* **2013**, *14*, 319–327. [CrossRef] [PubMed]

120. Balagam, B.; Richardson, D.E. The mechanism of carbon dioxide catalysis in the hydrogen peroxide N-oxidation of amines. *Inorg. Chem.* **2008**, *47*, 1173–1178. [CrossRef] [PubMed]

121. Fraser-Andrews, E.A.; Manning, N.J.; Ashton, G.H.; Eldridge, P.; McGrath, J.; Menage Hdu, P. Fish odour syndrome with features of both primary and secondary trimethylaminuria. *Clin. Exp. Dermatol* **2003**, *28*, 203–205. [CrossRef]

122. Al-Waiz, M.; Ayesh, R.; Mitchell, S.C.; Idle, J.R.; Smith, R.L. Trimethylaminuria ('fish-odour syndrome'): A study of an affected family. *Clin. Sci.* **1988**, *74*, 231–236. [CrossRef]

123. Al-Waiz, M.; Ayesh, R.; Mitchell, S.C.; Idle, J.R.; Smith, R.L. Trimethylaminuria (fish-odour syndrome): An inborn error of oxidative metabolism. *Lancet* **1987**, *1*, 634–635. [CrossRef]

124. D'Angelo, R.; Esposito, T.; Calabro, M.; Rinaldi, C.; Robledo, R.; Varriale, B.; Sidoti, A. FMO3 allelic variants in Sicilian and Sardinian populations: Trimethylaminuria and absence of fish-like body odor. *Gene* **2013**, *515*, 410–415. [CrossRef]

125. Humbert, J.A.; Hammond, K.B.; Hathaway, W.E. Trimethylaminuria: The fish-odour syndrome. *Lancet* **1970**, *2*, 770–771. [CrossRef]

126. Krueger, S.K.; Williams, D.E. Mammalian flavin-containing monooxygenases: Structure/function, genetic polymorphisms and role in drug metabolism. *Pharm. Ther.* **2005**, *106*, 357–387. [CrossRef] [PubMed]

127. Cherrington, N.J.; Cao, Y.; Cherrington, J.W.; Rose, R.L.; Hodgson, E. Physiological factors affecting protein expression of flavin-containing monooxygenases 1, 3 and 5. *Xenobiotica* **1998**, *28*, 673–682. [CrossRef]

128. Zhang, J.; Cashman, J.R. Quantitative analysis of FMO gene mRNA levels in human tissues. *Drug Metab. Dispos.* **2006**, *34*, 19–26. [CrossRef] [PubMed]

129. Vernetti, L.; Gough, A.; Baetz, N.; Blutt, S.; Broughman, J.R.; Brown, J.A.; Foulke-Abel, J.; Hasan, N.; In, J.; Kelly, E.; et al. Functional coupling of human microphysiology systems: Intestine, liver, kidney proximal tubule, blood-brain barrier and skeletal muscle. *Sci. Rep.* **2017**, *7*, 42296. [CrossRef] [PubMed]

130. Falls, J.G.; Blake, B.L.; Cao, Y.; Levi, P.E.; Hodgson, E. Gender differences in hepatic expression of flavin-containing monooxygenase isoforms (FMO1, FMO3, and FMO5) in mice. *J. Biochem. Toxicol.* **1995**, *10*, 171–177. [CrossRef]

131. Xu, M.; Bhatt, D.K.; Yeung, C.K.; Claw, K.G.; Chaudhry, A.S.; Gaedigk, A.; Pearce, R.E.; Broeckel, U.; Gaedigk, R.; Nickerson, D.A.; et al. Genetic and nongenetic factors associated with protein abundance of flavin-containing monooxygenase 3 in human liver. *J. Pharm. Exp.* **2017**, *363*, 265–274. [CrossRef]

132. Schugar, R.C.; Willard, B.; Wang, Z.; Brown, J.M. Postprandial gut microbiota-driven choline metabolism links dietary cues to adipose tissue dysfunction. *Adipocyte* **2018**, *7*, 49–56. [CrossRef]

133. Miyake, T.; Mizuno, T.; Mochizuki, T.; Kimura, M.; Matsuki, S.; Irie, S.; Ieiri, I.; Maeda, K.; Kusuhara, H. Involvement of organic cation transporters in the kinetics of Trimethylamine N-oxide. *J. Pharm. Sci.* **2017**, *106*, 2542–2550. [CrossRef]

134. Tan, X.; Liu, Y.; Long, J.; Chen, S.; Liao, G.; Wu, S.; Li, C.; Wang, L.; Ling, W.; Zhu, H. Trimethylamine N-Oxide aggravates liver steatosis through modulation of bile acid metabolism and inhibition of farnesoid x receptor signaling in nonalcoholic fatty liver disease. *Mol. Nutr. Food Res.* **2019**, *63*, e1900257. [CrossRef]

135. Liao, B.M.; McManus, S.A.; Hughes, W.E.; Schmitz-Peiffer, C. Flavin-containing monooxygenase 3 reduces endoplasmic reticulum stress in lipid-treated hepatocytes. *Mol. Endocrinol.* **2016**, *30*, 417–428. [CrossRef] [PubMed]

136. Chen, S.; Henderson, A.; Petriello, M.C.; Romano, K.A.; Gearing, M.; Miao, J.; Schell, M.; Sandoval-Espinola, W.J.; Tao, J.; Sha, B.; et al. Trimethylamine N-Oxide binds and activates PERK to promote metabolic dysfunction. *Cell Metab.* **2019**, *30*, 1141–1151 e1145. [CrossRef] [PubMed]

137. Borbas, T.; Benko, B.; Dalmadi, B.; Szabo, I.; Tihanyi, K. Insulin in flavin-containing monooxygenase regulation. Flavin-containing monooxygenase and cytochrome P450 activities in experimental diabetes. *Eur. J. Pharm. Sci.* **2006**, *28*, 51–58. [CrossRef] [PubMed]

138. Zhu, W.; Wang, Z.; Tang, W.H.W.; Hazen, S.L. Gut microbe-generated Trimethylamine N-Oxide from dietary choline is prothrombotic in subjects. *Circulation* **2017**, *135*, 1671–1673. [CrossRef] [PubMed]

139. Gessner, A.; Konig, J.; Fromm, M.F. Contribution of multidrug and toxin extrusion protein 1 (MATE1) to renal secretion of trimethylamine-N-oxide (TMAO). *Sci. Rep.* **2018**, *8*, 6659. [CrossRef]

140. Teft, W.A.; Morse, B.L.; Leake, B.F.; Wilson, A.; Mansell, S.E.; Hegele, R.A.; Ho, R.H.; Kim, R.B. Identification and characterization of Trimethylamine-N-oxide uptake and efflux transporters. *Mol. Pharm.* **2017**, *14*, 310–318. [CrossRef]

141. Krause, R.J.; Lash, L.H.; Elfarra, A.A. Human kidney flavin-containing monooxygenases and their potential roles in cysteine s-conjugate metabolism and nephrotoxicity. *J. Pharm. Exp. Ther.* **2003**, *304*, 185–191. [CrossRef]

142. Ripp, S.L.; Itagaki, K.; Philpot, R.M.; Elfarra, A.A. Species and sex differences in expression of flavin-containing monooxygenase form 3 in liver and kidney microsomes. *Drug Metab. Dispos.* **1999**, *27*, 46–52.

143. Mitchell, S.M.; Milan, A.M.; Mitchell, C.J.; Gillies, N.A.; D'Souza, R.F.; Zeng, N.; Ramzan, F.; Sharma, P.; Knowles, S.O.; Roy, N.C.; et al. Protein intake at twice the RDA in older men increases circulatory concentrations of the microbiome metabolite Trimethylamine-N-Oxide (TMAO). *Nutrients* **2019**, *11*, 2207. [CrossRef]

144. Tang, W.H.W.; Wang, Z.; Kennedy, D.J.; Wu, Y.; Buffa, J.A.; Agatisa-Boyle, B.; Li, X.M.S.; Levison, B.S.; Hazen, S.L. Gut microbiota-dependent Trimethylamine N-Oxide (TMAO) pathway contributes to both development of renal insufficiency and mortality risk in chronic kidney disease. *Circ. Res.* **2015**, *116*, 448–455. [CrossRef] [PubMed]

145. Prokopienko, A.J.; West, R.E., 3rd; Schrum, D.P.; Stubbs, J.R.; Leblond, F.A.; Pichette, V.; Nolin, T.D. Metabolic activation of flavin monooxygenase-mediated Trimethylamine-N-Oxide formation in experimental kidney disease. *Sci. Rep.* **2019**, *9*, 15901. [CrossRef] [PubMed]

146. Johnson, C.; Prokopienko, A.J.; West, R.E., 3rd; Nolin, T.D.; Stubbs, J.R. Decreased kidney function is associated with enhanced hepatic flavin monooxygenase activity and increased circulating Trimethylamine N-Oxide concentrations in mice. *Drug Metab. Dispos.* **2018**, *46*, 1304–1309. [CrossRef] [PubMed]

147. Xu, K.Y.; Xia, G.H.; Lu, J.Q.; Chen, M.X.; Zhen, X.; Wang, S.; You, C.; Nie, J.; Zhou, H.W.; Yin, J. Impaired renal function and dysbiosis of gut microbiota contribute to increased trimethylamine-N-oxide in chronic kidney disease patients. *Sci. Rep.* **2017**, *7*, 1445. [CrossRef] [PubMed]

148. Stubbs, J.R.; House, J.A.; Ocque, A.J.; Zhang, S.; Johnson, C.; Kimber, C.; Schmidt, K.; Gupta, A.; Wetmore, J.B.; Nolin, T.D.; et al. Serum Trimethylamine-N-Oxide is elevated in CKD and correlates with coronary atherosclerosis burden. *J. Am. Soc. Nephrol.* **2016**, *27*, 305–313. [CrossRef]

149. Kim, R.B.; Morse, B.L.; Djurdjev, O.; Tang, M.; Muirhead, N.; Barrett, B.; Holmes, D.T.; Madore, F.; Clase, C.M.; Rigatto, C.; et al. Advanced chronic kidney disease populations have elevated trimethylamine N-oxide levels associated with increased cardiovascular events. *Kidney Int.* **2016**, *89*, 1144–1152. [CrossRef]

150. Organ, C.L.; Li, Z.; Sharp, T.E., 3rd; Polhemus, D.J.; Gupta, N.; Goodchild, T.T.; Tang, W.H.W.; Hazen, S.L.; Lefer, D.J. Nonlethal inhibition of gut microbial Trimethylamine N-oxide production improves cardiac function and remodeling in a murine model of heart failure. *J. Am. Heart Assoc.* **2020**, *9*, e016223. [CrossRef]

151. Bordoni, L.; Samulak, J.J.; Sawicka, A.K.; Pelikant-Malecka, I.; Radulska, A.; Lewicki, L.; Kalinowski, L.; Gabbianelli, R.; Olek, R.A. Trimethylamine N-oxide and the reverse cholesterol transport in cardiovascular disease: A cross-sectional study. *Sci. Rep.* **2020**, *10*, 18675. [CrossRef]

152. Bordoni, L.; Petracci, I.; Pelikant-Malecka, I.; Radulska, A.; Piangerelli, M.; Samulak, J.J.; Lewicki, L.; Kalinowski, L.; Gabbianelli, R.; Olek, R.A. Mitochondrial DNA copy number and trimethylamine levels in the blood: New insights on cardiovascular disease biomarkers. *FASEB J.* **2021**, *35*, e21694. [CrossRef] [PubMed]

153. Latkovskis, G.; Makarova, E.; Mazule, M.; Bondare, L.; Hartmane, D.; Cirule, H.; Grinberga, S.; Erglis, A.; Liepinsh, E.; Dambrova, M. Loop diuretics decrease the renal elimination rate and increase the plasma levels of trimethylamine-N-oxide. *Br. J. Clin. Pharm.* **2018**, *84*, 2634–2644. [CrossRef] [PubMed]
154. Missailidis, C.; Hallqvist, J.; Qureshi, A.R.; Barany, P.; Heimburger, O.; Lindholm, B.; Stenvinkel, P.; Bergman, P. Serum Trimethylamine-N-Oxide is strongly related to renal function and predicts outcome in chronic kidney disease. *PLoS ONE* **2016**, *11*, e0141738. [CrossRef]
155. Kuhn, T.; Rohrmann, S.; Sookthai, D.; Johnson, T.; Katzke, V.; Kaaks, R.; von Eckardstein, A.; Muller, D. Intra-individual variation of plasma trimethylamine-N-oxide (TMAO), betaine and choline over 1 year. *Clin. Chem. Lab. Med.* **2017**, *55*, 261–268. [CrossRef]
156. Dumas, M.E.; Barton, R.H.; Toye, A.; Cloarec, O.; Blancher, C.; Rothwell, A.; Fearnside, J.; Tatoud, R.; Blanc, V.; Lindon, J.C.; et al. Metabolic profiling reveals a contribution of gut microbiota to fatty liver phenotype in insulin-resistant mice. *Proc. Natl. Acad. Sci. USA* **2006**, *103*, 12511–12516. [CrossRef]
157. Schugar, R.C.; Shih, D.M.; Warrier, M.; Helsley, R.N.; Burrows, A.; Ferguson, D.; Brown, A.L.; Gromovsky, A.D.; Heine, M.; Chatterjee, A.; et al. The TMAO-producing enzyme flavin-containing monooxygenase 3 regulates obesity and the beiging of white adipose tissue. *Cell Rep.* **2017**, *19*, 2451–2461. [CrossRef]
158. Li, X.; Chen, Y.; Liu, J.; Yang, G.; Zhao, J.; Liao, G.; Shi, M.; Yuan, Y.; He, S.; Lu, Y.; et al. Serum metabolic variables associated with impaired glucose tolerance induced by high-fat-high-cholesterol diet in Macaca mulatta. *Exp. Biol. Med.* **2012**, *237*, 1310–1321. [CrossRef]
159. Chen, Y.M.; Liu, Y.; Zhou, R.F.; Chen, X.L.; Wang, C.; Tan, X.Y.; Wang, L.J.; Zheng, R.D.; Zhang, H.W.; Ling, W.H.; et al. Associations of gut-flora-dependent metabolite trimethylamine-N-oxide, betaine and choline with non-alcoholic fatty liver disease in adults. *Sci. Rep.* **2016**, *6*, 19076. [CrossRef] [PubMed]
160. Kummen, M.; Vesterhus, M.; Troseid, M.; Moum, B.; Svardal, A.; Boberg, K.M.; Aukrust, P.; Karlsen, T.H.; Berge, R.K.; Hov, J.R. Elevated trimethylamine-N-oxide (TMAO) is associated with poor prognosis in primary sclerosing cholangitis patients with normal liver function. *United Eur. Gastroenterol. J.* **2017**, *5*, 532–541. [CrossRef] [PubMed]
161. Gao, X.; Xu, J.; Jiang, C.; Zhang, Y.; Xue, Y.; Li, Z.; Wang, J.; Xue, C.; Wang, Y. Fish oil ameliorates trimethylamine N-oxide-exacerbated glucose intolerance in high-fat diet-fed mice. *Food Funct.* **2015**, *6*, 1117–1125. [CrossRef]
162. Shih, D.M.; Wang, Z.; Lee, R.; Meng, Y.; Che, N.; Charugundla, S.; Qi, H.; Wu, J.; Pan, C.; Brown, J.M.; et al. Flavin containing monooxygenase 3 exerts broad effects on glucose and lipid metabolism and atherosclerosis. *J. Lipid Res.* **2015**, *56*, 22–37. [CrossRef]
163. Gonzalez Malagon, S.G.; Melidoni, A.N.; Hernandez, D.; Omar, B.A.; Houseman, L.; Veeravalli, S.; Scott, F.; Varshavi, D.; Everett, J.; Tsuchiya, Y.; et al. The phenotype of a knockout mouse identifies flavin-containing monooxygenase 5 (FMO5) as a regulator of metabolic ageing. *Biochem. Pharm.* **2015**, *96*, 267–277. [CrossRef]
164. Chen, M.L.; Yi, L.; Zhang, Y.; Zhou, X.; Ran, L.; Yang, J.; Zhu, J.D.; Zhang, Q.Y.; Mi, M.T. Resveratrol attenuates Trimethylamine-N-Oxide (TMAO)-induced atherosclerosis by regulating TMAO synthesis and bile acid metabolism via remodeling of the gut microbiota. *mBio* **2016**, *7*, e02210–e02215. [CrossRef]
165. Zhao, Z.H.; Xin, F.Z.; Zhou, D.; Xue, Y.Q.; Liu, X.L.; Yang, R.X.; Pan, Q.; Fan, J.G. Trimethylamine N-oxide attenuates high-fat high-cholesterol diet-induced steatohepatitis by reducing hepatic cholesterol overload in rats. *World J. Gastroenterol.* **2019**, *25*, 2450–2462. [CrossRef] [PubMed]
166. Chen, H.; Peng, L.; Perez de Nanclares, M.; Trudeau, M.P.; Yao, D.; Cheng, Z.; Urriola, P.E.; Mydland, L.T.; Shurson, G.C.; Overland, M.; et al. Identification of sinapine-derived choline from a rapeseed diet as a source of serum Trimethylamine N-Oxide in pigs. *J. Agric. Food Chem.* **2019**, *67*, 7748–7754. [CrossRef] [PubMed]
167. Watanabe, M.; Houten, S.M.; Wang, L.; Moschetta, A.; Mangelsdorf, D.J.; Heyman, R.A.; Moore, D.D.; Auwerx, J. Bile acids lower triglyceride levels via a pathway involving FXR, SHP, and SREBP-1c. *J. Clin. Investig.* **2004**, *113*, 1408–1418. [CrossRef] [PubMed]
168. Matsukuma, K.E.; Bennett, M.K.; Huang, J.; Wang, L.; Gil, G.; Osborne, T.F. Coordinated control of bile acids and lipogenesis through FXR-dependent regulation of fatty acid synthase. *J. Lipid Res.* **2006**, *47*, 2754–2761. [CrossRef] [PubMed]
169. Touyz, R.M.; Rios, F.J.; Alves-Lopes, R.; Neves, K.B.; Camargo, L.L.; Montezano, A.C. Oxidative stress: A unifying paradigm in hypertension. *Can. J. Cardiol.* **2020**, *36*, 659–670. [CrossRef] [PubMed]
170. Ufnal, M.; Jazwiec, R.; Dadlez, M.; Drapala, A.; Sikora, M.; Skrzypecki, J. Trimethylamine-N-oxide: A carnitine-derived metabolite that prolongs the hypertensive effect of angiotensin II in rats. *Can. J. Cardiol.* **2014**, *30*, 1700–1705. [CrossRef]
171. Achard, C.S.; Laybutt, D.R. Lipid-induced endoplasmic reticulum stress in liver cells results in two distinct outcomes: Adaptation with enhanced insulin signaling or insulin resistance. *Endocrinology* **2012**, *153*, 2164–2177. [CrossRef]
172. Brunt, V.E.; Gioscia-Ryan, R.A.; Casso, A.G.; VanDongen, N.S.; Ziemba, B.P.; Sapinsley, Z.J.; Richey, J.J.; Zigler, M.C.; Neilson, A.P.; Davy, K.P.; et al. Trimethylamine-N-Oxide promotes age-related vascular oxidative stress and endothelial dysfunction in mice and healthy humans. *Hypertension* **2020**, *76*, 101–112. [CrossRef] [PubMed]
173. Sun, X.; Jiao, X.; Ma, Y.; Liu, Y.; Zhang, L.; He, Y.; Chen, Y. Trimethylamine N-oxide induces inflammation and endothelial dysfunction in human umbilical vein endothelial cells via activating ROS-TXNIP-NLRP3 inflammasome. *Biochem. Biophys. Res. Commun.* **2016**, *481*, 63–70. [CrossRef] [PubMed]
174. Ke, Y.; Li, D.; Zhao, M.; Liu, C.; Liu, J.; Zeng, A.; Shi, X.; Cheng, S.; Pan, B.; Zheng, L.; et al. Gut flora-dependent metabolite Trimethylamine-N-oxide accelerates endothelial cell senescence and vascular aging through oxidative stress. *Free Radic. Biol. Med.* **2018**, *116*, 88–100. [CrossRef]

175. Wang, A.; Bolen, D.W. A naturally occurring protective system in urea-rich cells: Mechanism of osmolyte protection of proteins against urea denaturation. *Biochemistry* **1997**, *36*, 9101–9108. [CrossRef] [PubMed]

176. Parkin, K.L.; Hultin, H.O. Characterization of trimethylamine-N-oxide (TMAO) demethylase activity from fish muscle microsomes. *J. Biochem.* **1986**, *100*, 77–86. [CrossRef]

177. Rodriguez-Fuentes, G.; Aparicio-Fabre, R.; Li, Q.; Schlenk, D. Osmotic regulation of a novel flavin-containing monooxygenase in primary cultured cells from rainbow trout (Oncorhynchus mykiss). *Drug Metab. Dispos.* **2008**, *36*, 1212–1217. [CrossRef]

178. Sackett, D.L. Natural osmolyte trimethylamine N-oxide stimulates tubulin polymerization and reverses urea inhibition. *Am. J. Physiol.* **1997**, *273*, R669–R676. [CrossRef]

179. Song, J.L.; Chuang, D.T. Natural osmolyte trimethylamine N-oxide corrects assembly defects of mutant branched-chain alpha-ketoacid decarboxylase in maple syrup urine disease. *J. Biol. Chem.* **2001**, *276*, 40241–40246. [CrossRef]

180. Larsen, B.K.; Schlenk, D. Effect of salinity on flavin-containing monooxygenase expression and activity in rainbow trout (Oncorhynchus mykiss). *J. Comp. Physiol. B* **2001**, *171*, 421–429. [CrossRef]

181. Villalobos, A.R.; Renfro, J.L. Trimethylamine oxide suppresses stress-induced alteration of organic anion transport in choroid plexus. *J. Exp. Biol.* **2007**, *210*, 541–552. [CrossRef]

182. Yancey, P.H.; Siebenaller, J.F. Co-evolution of proteins and solutions: Protein adaptation versus cytoprotective micromolecules and their roles in marine organisms. *J. Exp. Biol* **2015**, *218*, 1880–1896. [CrossRef]

183. Ganguly, P.; Boserman, P.; van der Vegt, N.F.A.; Shea, J.E. Trimethylamine N-oxide counteracts urea denaturation by inhibiting protein-urea preferential interaction. *J. Am. Chem. Soc.* **2018**, *140*, 483–492. [CrossRef]

184. Liao, Y.T.; Manson, A.C.; DeLyser, M.R.; Noid, W.G.; Cremer, P.S. Trimethylamine N-oxide stabilizes proteins via a distinct mechanism compared with betaine and glycine. *Proc. Natl. Acad. Sci. USA* **2017**, *114*, 2479–2484. [CrossRef]

185. Mondal, J.; Stirnemann, G.; Berne, B.J. When does trimethylamine N-oxide fold a polymer chain and urea unfold it? *J. Phys. Chem. B* **2013**, *117*, 8723–8732. [CrossRef] [PubMed]

186. Shepshelovich, J.; Goldstein-Magal, L.; Globerson, A.; Yen, P.M.; Rotman-Pikielny, P.; Hirschberg, K. Protein synthesis inhibitors and the chemical chaperone TMAO reverse endoplasmic reticulum perturbation induced by overexpression of the iodide transporter pendrin. *J. Cell Sci.* **2005**, *118*, 1577–1586. [CrossRef] [PubMed]

187. Chan, J.Y.; Luzuriaga, J.; Bensellam, M.; Biden, T.J.; Laybutt, D.R. Failure of the adaptive unfolded protein response in islets of obese mice is linked with abnormalities in beta-cell gene expression and progression to diabetes. *Diabetes* **2013**, *62*, 1557–1568. [CrossRef]

188. Huang, C.; Wang, J.J.; Ma, J.H.; Jin, C.; Yu, Q.; Zhang, S.X. Activation of the UPR protects against cigarette smoke-induced RPE apoptosis through up-regulation of Nrf2. *J. Biol. Chem.* **2015**, *290*, 5367–5380. [CrossRef]

189. Petersen, M.C.; Shulman, G.I. Mechanisms of insulin action and insulin resistance. *Physiol. Rev.* **2018**, *98*, 2133–2223. [CrossRef] [PubMed]

190. Romine, I.C.; Wiseman, R.L. PERK signaling regulates extracellular proteostasis of an amyloidogenic protein during endoplasmic reticulum stress. *Sci. Rep.* **2019**, *9*, 410. [CrossRef]

191. Lebeau, J.; Saunders, J.M.; Moraes, V.W.R.; Madhavan, A.; Madrazo, N.; Anthony, M.C.; Wiseman, R.L. The PERK arm of the unfolded protein response regulates mitochondrial morphology during acute endoplasmic reticulum stress. *Cell Rep.* **2018**, *22*, 2827–2836. [CrossRef] [PubMed]

192. Cullinan, S.B.; Diehl, J.A. PERK-dependent activation of Nrf2 contributes to redox homeostasis and cell survival following endoplasmic reticulum stress. *J. Biol. Chem.* **2004**, *279*, 20108–20117. [CrossRef] [PubMed]

193. Harding, H.P.; Zhang, Y.; Zeng, H.; Novoa, I.; Lu, P.D.; Calfon, M.; Sadri, N.; Yun, C.; Popko, B.; Paules, R.; et al. An integrated stress response regulates amino acid metabolism and resistance to oxidative stress. *Mol. Cell* **2003**, *11*, 619–633. [CrossRef]

194. Rainbolt, T.K.; Atanassova, N.; Genereux, J.C.; Wiseman, R.L. Stress-regulated translational attenuation adapts mitochondrial protein import through Tim17A degradation. *Cell Metab.* **2013**, *18*, 908–919. [CrossRef]

195. Ozcan, U.; Yilmaz, E.; Ozcan, L.; Furuhashi, M.; Vaillancourt, E.; Smith, R.O.; Gorgun, C.Z.; Hotamisligil, G.S. Chemical chaperones reduce ER stress and restore glucose homeostasis in a mouse model of type 2 diabetes. *Science* **2006**, *313*, 1137–1140. [CrossRef] [PubMed]

196. Gruppen, E.G.; Garcia, E.; Connelly, M.A.; Jeyarajah, E.J.; Otvos, J.D.; Bakker, S.J.L.; Dullaart, R.P.F. TMAO is Associated with Mortality: Impact of modestly impaired renal function. *Sci. Rep.* **2017**, *7*, 13781. [CrossRef]

197. Robinson-Cohen, C.; Newitt, R.; Shen, D.D.; Rettie, A.E.; Kestenbaum, B.R.; Himmelfarb, J.; Yeung, C.K. Association of FMO3 variants and Trimethylamine N-Oxide concentration, disease progression, and mortality in CKD patients. *PLoS ONE* **2016**, *11*, e0161074. [CrossRef]

198. Pelletier, C.C.; Croyal, M.; Ene, L.; Aguesse, A.; Billon-Crossouard, S.; Krempf, M.; Lemoine, S.; Guebre-Egziabher, F.; Juillard, L.; Soulage, C.O. Elevation of Trimethylamine-N-Oxide in chronic kidney disease: Contribution of decreased glomerular filtration rate. *Toxins* **2019**, *11*, 635. [CrossRef]

199. Rhee, E.P.; Clish, C.B.; Ghorbani, A.; Larson, M.G.; Elmariah, S.; McCabe, E.; Yang, Q.; Cheng, S.; Pierce, K.; Deik, A.; et al. A combined epidemiologic and metabolomic approach improves CKD prediction. *J. Am. Soc. Nephrol.* **2013**, *24*, 1330–1338. [CrossRef]

200. Jia, J.Z.; Dou, P.; Gao, M.; Kong, X.J.; Li, C.W.; Liu, Z.H.; Huang, T. Assessment of causal direction between gut microbiota-dependent metabolites and cardiometabolic health: A bidirectional mendelian randomization analysis. *Diabetes* **2019**, *68*, 1747–1755. [CrossRef]
201. Kaysen, G.A.; Johansen, K.L.; Chertow, G.M.; Dalrymple, L.S.; Kornak, J.; Grimes, B.; Dwyer, T.; Chassy, A.W.; Fiehn, O. Associations of Trimethylamine N-Oxide with nutritional and inflammatory biomarkers and cardiovascular outcomes in patients new to dialysis. *J. Ren. Nutr.* **2015**, *25*, 351–356. [CrossRef]
202. Gawrys-Kopczynska, M.; Konop, M.; Maksymiuk, K.; Kraszewska, K.; Derzsi, L.; Sozanski, K.; Holyst, R.; Pilz, M.; Samborowska, E.; Dobrowolski, L.; et al. TMAO, a seafood-derived molecule, produces diuresis and reduces mortality in heart failure rats. *eLife* **2020**, *9*, e57028. [CrossRef]
203. Liu, J.; Lai, L.; Lin, J.; Zheng, J.; Nie, X.; Zhu, X.; Xue, J.; Liu, T. Ranitidine and finasteride inhibit the synthesis and release of trimethylamine N-oxide and mitigates its cardiovascular and renal damage through modulating gut microbiota. *Int. J. Biol. Sci.* **2020**, *16*, 790–802. [CrossRef] [PubMed]
204. Del Rio, D.; Zimetti, F.; Caffarra, P.; Tassotti, M.; Bernini, F.; Brighenti, F.; Zini, A.; Zanotti, I. The gut microbial metabolite Trimethylamine-N-Oxide is present in human cerebrospinal fluid. *Nutrients* **2017**, *9*, 1053. [CrossRef]
205. Vogt, N.M.; Romano, K.A.; Darst, B.F.; Engelman, C.D.; Johnson, S.C.; Carlsson, C.M.; Asthana, S.; Blennow, K.; Zetterberg, H.; Bendlin, B.B.; et al. The gut microbiota-derived metabolite trimethylamine N-oxide is elevated in Alzheimer's disease. *Alzheimers Res. Ther.* **2018**, *10*, 124. [CrossRef] [PubMed]
206. Yu, L.; Meng, G.; Huang, B.; Zhou, X.; Stavrakis, S.; Wang, M.; Li, X.; Zhou, L.; Wang, Y.; Wang, M.; et al. A potential relationship between gut microbes and atrial fibrillation: Trimethylamine N-oxide, a gut microbe-derived metabolite, facilitates the progression of atrial fibrillation. *Int. J. Cardiol.* **2018**, *255*, 92–98. [CrossRef] [PubMed]
207. Ntranos, A.; Casaccia, P. The microbiome-gut-behavior axis: Crosstalk between the gut microbiome and oligodendrocytes modulates behavioral responses. *Neurotherapeutics* **2018**, *15*, 31–35. [CrossRef] [PubMed]
208. Bercik, P.; Denou, E.; Collins, J.; Jackson, W.; Lu, J.; Jury, J.; Deng, Y.; Blennerhassett, P.; Macri, J.; McCoy, K.D.; et al. The intestinal microbiota affect central levels of brain-derived neurotropic factor and behavior in mice. *Gastroenterology* **2011**, *141*, 599–609, 609.e1–609.e3. [CrossRef] [PubMed]
209. Tran, S.M.; Mohajeri, M.H. The role of gut bacterial metabolites in brain development, aging and disease. *Nutrients* **2021**, *13*, 732. [CrossRef]
210. Bravo, J.A.; Forsythe, P.; Chew, M.V.; Escaravage, E.; Savignac, H.M.; Dinan, T.G.; Bienenstock, J.; Cryan, J.F. Ingestion of Lactobacillus strain regulates emotional behavior and central GABA receptor expression in a mouse via the vagus nerve. *Proc. Natl. Acad. Sci. USA* **2011**, *108*, 16050–16055. [CrossRef]
211. Ahmadi, S.; Wang, S.; Nagpal, R.; Wang, B.; Jain, S.; Razazan, A.; Mishra, S.P.; Zhu, X.; Wang, Z.; Kavanagh, K.; et al. A human-origin probiotic cocktail ameliorates aging-related leaky gut and inflammation via modulating the microbiota/taurine/tight junction axis. *JCI Insight* **2020**, *5*, e132055. [CrossRef]
212. Claesson, M.J.; Jeffery, I.B.; Conde, S.; Power, S.E.; O'Connor, E.M.; Cusack, S.; Harris, H.M.; Coakley, M.; Lakshminarayanan, B.; O'Sullivan, O.; et al. Gut microbiota composition correlates with diet and health in the elderly. *Nature* **2012**, *488*, 178–184. [CrossRef]
213. O'Toole, P.W.; Jeffery, I.B. Gut microbiota and aging. *Science* **2015**, *350*, 1214–1215. [CrossRef]
214. Meng, F.; Li, N.; Li, D.; Song, B.; Li, L. The presence of elevated circulating trimethylamine N-oxide exaggerates postoperative cognitive dysfunction in aged rats. *Behav. Brain Res.* **2019**, *368*, 111902. [CrossRef]
215. Du, D.; Tang, W.; Zhou, C.; Sun, X.; Wei, Z.; Zhong, J.; Huang, Z. Fecal microbiota transplantation is a promising method to restore gut microbiota dysbiosis and relieve neurological deficits after traumatic brain injury. *Oxid. Med. Cell Longev.* **2021**, *2021*, 5816837. [CrossRef]
216. Olek, R.A.; Samulak, J.J.; Sawicka, A.K.; Hartmane, D.; Grinberga, S.; Pugovics, O.; Lysiak-Szydlowska, W. Increased Trimethylamine N-Oxide is not associated with oxidative stress markers in healthy aged women. *Oxid. Med. Cell Longev.* **2019**, *2019*, 6247169. [CrossRef] [PubMed]
217. Schirra, H.J.; Anderson, C.G.; Wilson, W.J.; Kerr, L.; Craik, D.J.; Waters, M.J.; Lichanska, A.M. Altered metabolism of growth hormone receptor mutant mice: A combined NMR metabonomics and microarray study. *PLoS ONE* **2008**, *3*, e2764. [CrossRef]
218. Zhao, L.; Zhang, C.; Cao, G.; Dong, X.; Li, D.; Jiang, L. Higher circulating Trimethylamine N-oxide sensitizes sevoflurane-induced cognitive dysfunction in aged rats probably by downregulating hippocampal methionine sulfoxide reductase, A. *Neurochem. Res.* **2019**, *44*, 2506–2516. [CrossRef] [PubMed]
219. Lupachyk, S.; Watcho, P.; Stavniichuk, R.; Shevalye, H.; Obrosova, I.G. Endoplasmic reticulum stress plays a key role in the pathogenesis of diabetic peripheral neuropathy. *Diabetes* **2013**, *62*, 944–952. [CrossRef] [PubMed]
220. de Theije, C.G.; Wopereis, H.; Ramadan, M.; van Eijndthoven, T.; Lambert, J.; Knol, J.; Garssen, J.; Kraneveld, A.D.; Oozeer, R. Altered gut microbiota and activity in a murine model of autism spectrum disorders. *Brain Behav. Immun.* **2014**, *37*, 197–206. [CrossRef]
221. Li, Q.; Han, Y.; Dy, A.B.C.; Hagerman, R.J. The gut microbiota and autism spectrum disorders. *Front. Cell Neurosci.* **2017**, *11*, 120. [CrossRef] [PubMed]

222. Ashwood, P.; Krakowiak, P.; Hertz-Picciotto, I.; Hansen, R.; Pessah, I.; Van de Water, J. Elevated plasma cytokines in autism spectrum disorders provide evidence of immune dysfunction and are associated with impaired behavioral outcome. *Brain Behav. Immun.* **2011**, *25*, 40–45. [CrossRef]

223. Siniscalco, D.; Schultz, S.; Brigida, A.L.; Antonucci, N. Inflammation and neuro-immune dysregulations in autism spectrum disorders. *Pharmaceuticals* **2018**, *11*, 56. [CrossRef] [PubMed]

224. Quan, L.J.; Yi, J.P.; Zhao, Y.; Zhang, F.; Shi, X.T.; Feng, Z.; Miller, H.L. Plasma trimethylamine N-oxide, a gut microbe generated phosphatidylcholine metabolite, is associated with autism spectrum disorders. *Neurotoxicology* **2020**, *76*, 93–98. [CrossRef] [PubMed]

225. Liu, A.; Lv, H.; Wang, H.; Yang, H.; Li, Y.; Qian, J. Aging increases the severity of colitis and the related changes to the gut barrier and gut microbiota in humans and mice. *J. Gerontol. A Biol. Sci. Med. Sci.* **2020**, *75*, 1284–1292. [CrossRef]

226. Qi, Z.; Qi, S.; Gui, L.; Shen, L.; Feng, Z. Daphnetin protects oxidative stress-induced neuronal apoptosis via regulation of MAPK signaling and HSP70 expression. *Oncol. Lett.* **2016**, *12*, 1959–1964. [CrossRef]

227. Reeg, S.; Jung, T.; Castro, J.P.; Davies, K.J.A.; Henze, A.; Grune, T. The molecular chaperone Hsp70 promotes the proteolytic removal of oxidatively damaged proteins by the proteasome. *Free Radic. Biol. Med.* **2016**, *99*, 153–166. [CrossRef]

228. Uversky, V.N.; Li, J.; Fink, A.L. Trimethylamine-N-oxide-induced folding of alpha-synuclein. *FEBS Lett.* **2001**, *509*, 31–35. [CrossRef]

229. Yoshida, H.; Yoshizawa, T.; Shibasaki, F.; Shoji, S.; Kanazawa, I. Chemical chaperones reduce aggregate formation and cell death caused by the truncated Machado-Joseph disease gene product with an expanded polyglutamine stretch. *Neurobiol. Dis.* **2002**, *10*, 88–99. [CrossRef]

230. Paul, S. Polyglutamine-mediated neurodegeneration: Use of chaperones as prevention strategy. *Biochemistry* **2007**, *72*, 359–366. [CrossRef]

231. Bartolini, M.; Andrisano, V. Strategies for the inhibition of protein aggregation in human diseases. *Chembiochem* **2010**, *11*, 1018–1035. [CrossRef] [PubMed]

232. Lisabeth, E.M.; Falivelli, G.; Pasquale, E.B. Eph receptor signaling and ephrins. *Cold Spring Harb. Perspect. Biol.* **2013**, *5*, a009159. [CrossRef]

233. Chang, Q.; Jorgensen, C.; Pawson, T.; Hedley, D.W. Effects of dasatinib on EphA2 receptor tyrosine kinase activity and downstream signalling in pancreatic cancer. *Br. J. Cancer* **2008**, *99*, 1074–1082. [CrossRef] [PubMed]

234. Konstantinova, I.; Nikolova, G.; Ohara-Imaizumi, M.; Meda, P.; Kucera, T.; Zarbalis, K.; Wurst, W.; Nagamatsu, S.; Lammert, E. EphA-Ephrin-A-mediated beta cell communication regulates insulin secretion from pancreatic islets. *Cell* **2007**, *129*, 359–370. [CrossRef] [PubMed]

235. Li, D.; Ke, Y.; Zhan, R.; Liu, C.; Zhao, M.; Zeng, A.; Shi, X.; Ji, L.; Cheng, S.; Pan, B.; et al. Trimethylamine-N-oxide promotes brain aging and cognitive impairment in mice. *Aging Cell* **2018**, *17*, e12768. [CrossRef] [PubMed]

236. D'Orio, B.; Fracassi, A.; Ceru, M.P.; Moreno, S. Targeting PPARalpha in alzheimer's disease. *Curr. Alzheimer Res.* **2018**, *15*, 345–354. [CrossRef] [PubMed]

237. Barrea, L.; Annunziata, G.; Muscogiuri, G.; Di Somma, C.; Laudisio, D.; Maisto, M.; de Alteriis, G.; Tenore, G.C.; Colao, A.; Savastano, S. Trimethylamine-N-oxide (TMAO) as novel potential biomarker of early predictors of metabolic syndrome. *Nutrients* **2018**, *10*, 1917. [CrossRef]

238. Erickson, M.L.; Malin, S.K.; Wang, Z.; Brown, J.M.; Hazen, S.L.; Kirwan, J.P. Effects of lifestyle intervention on plasma Trimethylamine N-Oxide in obese adults. *Nutrients* **2019**, *11*, 179. [CrossRef] [PubMed]

239. Dumas, M.E.; Rothwell, A.R.; Hoyles, L.; Aranias, T.; Chilloux, J.; Calderari, S.; Noll, E.M.; Pean, N.; Boulange, C.L.; Blancher, C.; et al. Microbial-Host Co-metabolites are prodromal markers predicting phenotypic heterogeneity in behavior, obesity, and impaired glucose tolerance. *Cell Rep.* **2017**, *20*, 136–148. [CrossRef]

240. Parks, B.W.; Nam, E.; Org, E.; Kostem, E.; Norheim, F.; Hui, S.T.; Pan, C.; Civelek, M.; Rau, C.D.; Bennett, B.J.; et al. Genetic control of obesity and gut microbiota composition in response to high-fat, high-sucrose diet in mice. *Cell Metab.* **2013**, *17*, 141–152. [CrossRef]

241. Veeravalli, S.; Omar, B.A.; Houseman, L.; Hancock, M.; Gonzalez Malagon, S.G.; Scott, F.; Janmohamed, A.; Phillips, I.R.; Shephard, E.A. The phenotype of a flavin-containing monooyxgenase knockout mouse implicates the drug-metabolizing enzyme FMO1 as a novel regulator of energy balance. *Biochem. Pharm.* **2014**, *90*, 88–95. [CrossRef] [PubMed]

242. Gao, X.; Liu, X.; Xu, J.; Xue, C.; Xue, Y.; Wang, Y. Dietary trimethylamine N-oxide exacerbates impaired glucose tolerance in mice fed a high fat diet. *J. Biosci. Bioeng.* **2014**, *118*, 476–481. [CrossRef] [PubMed]

243. Ahima, R.S.; Flier, J.S. Adipose tissue as an endocrine organ. *Trends Endocrinol. Met.* **2000**, *11*, 327–332. [CrossRef]

244. Kanda, H.; Tateya, S.; Tamori, Y.; Kotani, K.; Hiasa, K.-i.; Kitazawa, R.; Kitazawa, S.; Miyachi, H.; Maeda, S.; Egashira, K.; et al. MCP-1 contributes to macrophage infiltration into adipose tissue, insulin resistance, and hepatic steatosis in obesity. *J. Clin. Investig.* **2006**, *116*, 1494–1505. [CrossRef]

245. Hotamisligil, G.S.; Shargill, N.S.; Spiegelman, B.M. Adipose expression of tumor necrosis factor-alpha: Direct role in obesity-linked insulin resistance. *Science* **1993**, *259*, 87–91. [CrossRef]

246. Cani, P.D.; Amar, J.; Iglesias, M.A.; Knauf, C.; Neynhck, A.; Alessi, M.C.; Burcelin, R.; Delzenne, N. Metabolic endotoxemia initiates obesity and insulin resistance. *Ann. Nutr. Metab.* **2007**, *51*, 79. [CrossRef]

247. Lagathu, C.; Bastard, J.P.; Auclair, M.; Maachi, M.; Capeau, J.; Caron, M. Chronic interleukin-6 (IL-6) treatment increased IL-6 secretion and induced insulin resistance in adipocyte: Prevention by rosiglitazone. *Biochem. Biophys. Res. Commun.* **2003**, *311*, 372–379. [CrossRef]

248. Yancey, P.H.; Gerringer, M.E.; Drazen, J.C.; Rowden, A.A.; Jamieson, A. Marine fish may be biochemically constrained from inhabiting the deepest ocean depths. *Proc. Natl. Acad. Sci. USA* **2014**, *111*, 4461–4465. [CrossRef]

249. Yancey, P.H. Organic osmolytes as compatible, metabolic and counteracting cytoprotectants in high osmolarity and other stresses. *J. Exp. Biol.* **2005**, *208*, 2819–2830. [CrossRef]

250. Baskakov, I.; Wang, A.; Bolen, D.W. Trimethylamine-N-oxide counteracts urea effects on rabbit muscle lactate dehydrogenase function: A test of the counteraction hypothesis. *Biophys. J.* **1998**, *74*, 2666–2673. [CrossRef]

251. Baskakov, I.; Bolen, D.W. Time-dependent effects of trimethylamine-N-oxide/urea on lactate dehydrogenase activity: An unexplored dimension of the adaptation paradigm. *Biophys. J.* **1998**, *74*, 2658–2665. [CrossRef]

252. Ortiz-Costa, S.; Sorenson, M.M.; Sola-Penna, M. Counteracting effects of urea and methylamines in function and structure of skeletal muscle myosin. *Arch. Biochem. Biophys.* **2002**, *408*, 272–278. [CrossRef]

253. Li, Z.; Wu, Z.; Yan, J.; Liu, H.; Liu, Q.; Deng, Y.; Ou, C.; Chen, M. Gut microbe-derived metabolite trimethylamine N-oxide induces cardiac hypertrophy and fibrosis. *Lab. Investig.* **2019**, *99*, 346–357. [CrossRef]

254. Li, X.; Geng, J.; Zhao, J.; Ni, Q.; Zhao, C.; Zheng, Y.; Chen, X.; Wang, L. Trimethylamine N-Oxide exacerbates cardiac fibrosis via activating the NLRP3 inflammasome. *Front. Physiol.* **2019**, *10*, 866. [CrossRef]

255. Chen, K.; Zheng, X.; Feng, M.; Li, D.; Zhang, H. Gut microbiota-dependent metabolite Trimethylamine N-Oxide contributes to cardiac dysfunction in western diet-induced obese mice. *Front. Physiol.* **2017**, *8*, 139. [CrossRef]

256. Savi, M.; Bocchi, L.; Bresciani, L.; Falco, A.; Quaini, F.; Mena, P.; Brighenti, F.; Crozier, A.; Stilli, D.; Del Rio, D. Trimethylamine-N-Oxide (TMAO)-induced impairment of cardiomyocyte function and the protective role of urolithin B-Glucuronide. *Molecules* **2018**, *23*, 549. [CrossRef]

257. Piacentino, V., 3rd; Weber, C.R.; Chen, X.; Weisser-Thomas, J.; Margulies, K.B.; Bers, D.M.; Houser, S.R. Cellular basis of abnormal calcium transients of failing human ventricular myocytes. *Circ. Res.* **2003**, *92*, 651–658. [CrossRef]

258. Oakley, C.I.; Vallejo, J.A.; Wang, D.; Gray, M.A.; Tiede-Lewis, L.M.; Shawgo, T.; Daon, E.; Zorn, G., 3rd; Stubbs, J.R.; Wacker, M.J. Trimethylamine-N-oxide acutely increases cardiac muscle contractility. *Am. J. Physiol. Heart Circ. Physiol.* **2020**, *318*, H1272–H1282. [CrossRef]

259. Videja, M.; Vilskersts, R.; Korzh, S.; Cirule, H.; Sevostjanovs, E.; Dambrova, M.; Makrecka-Kuka, M. Microbiota-derived metabolite Trimethylamine N-Oxide protects mitochondrial energy metabolism and cardiac functionality in a rat model of right ventricle heart failure. *Front. Cell Dev. Biol.* **2020**, *8*, 622741. [CrossRef]

260. Jaworska, K.; Bielinska, K.; Gawrys-Kopczynska, M.; Ufnal, M. TMA (trimethylamine), but not its oxide TMAO (trimethylamine-oxide), exerts haemodynamic effects: Implications for interpretation of cardiovascular actions of gut microbiome. *Cardiovasc. Res.* **2019**, *115*, 1948–1949. [CrossRef]

261. Frey, N.; Olson, E.N. Cardiac hypertrophy: The good, the bad and the ugly. *Annu. Rev. Physiol.* **2003**, *65*, 45–79. [CrossRef]

262. Richter, K.; Kietzmann, T. Reactive oxygen species and fibrosis: Further evidence of a significant liaison. *Cell Tissue Res.* **2016**, *365*, 591–605. [CrossRef] [PubMed]

263. Ryu, Y.; Jin, L.; Kee, H.J.; Piao, Z.H.; Cho, J.Y.; Kim, G.R.; Choi, S.Y.; Lin, M.Q.; Jeong, M.H. Gallic acid prevents isoproterenol-induced cardiac hypertrophy and fibrosis through regulation of JNK2 signaling and Smad3 binding activity. *Sci. Rep.* **2016**, *6*, 34790. [CrossRef]

264. Makrecka-Kuka, M.; Volska, K.; Antone, U.; Vilskersts, R.; Grinberga, S.; Bandere, D.; Liepinsh, E.; Dambrova, M. Trimethylamine N-oxide impairs pyruvate and fatty acid oxidation in cardiac mitochondria. *Toxicol. Lett.* **2017**, *267*, 32–38. [CrossRef] [PubMed]

265. Hiraoka, M. A novel action of insulin on cardiac membrane. *Circ. Res.* **2003**, *92*, 707–709. [CrossRef] [PubMed]

266. Doenst, T.; Pytel, G.; Schrepper, A.; Amorim, P.; Farber, G.; Shingu, Y.; Mohr, F.W.; Schwarzer, M. Decreased rates of substrate oxidation ex vivo predict the onset of heart failure and contractile dysfunction in rats with pressure overload. *Cardiovasc. Res.* **2010**, *86*, 461–470. [CrossRef]

267. Kato, T.; Niizuma, S.; Inuzuka, Y.; Kawashima, T.; Okuda, J.; Tamaki, Y.; Iwanaga, Y.; Narazaki, M.; Matsuda, T.; Soga, T.; et al. Analysis of metabolic remodeling in compensated left ventricular hypertrophy and heart failure. *Circ. Heart Fail.* **2010**, *3*, 420–430. [CrossRef]

268. Mori, J.; Basu, R.; McLean, B.A.; Das, S.K.; Zhang, L.; Patel, V.B.; Wagg, C.S.; Kassiri, Z.; Lopaschuk, G.D.; Oudit, G.Y. Agonist-induced hypertrophy and diastolic dysfunction are associated with selective reduction in glucose oxidation: A metabolic contribution to heart failure with normal ejection fraction. *Circ. Heart Fail.* **2012**, *5*, 493–503. [CrossRef]

269. Osorio, J.C.; Stanley, W.C.; Linke, A.; Castellari, M.; Diep, Q.N.; Panchal, A.R.; Hintze, T.H.; Lopaschuk, G.D.; Recchia, F.A. Impaired myocardial fatty acid oxidation and reduced protein expression of retinoid X receptor-alpha in pacing-induced heart failure. *Circulation* **2002**, *106*, 606–612. [CrossRef]

270. Seldin, M.M.; Meng, Y.; Qi, H.; Zhu, W.; Wang, Z.; Hazen, S.L.; Lusis, A.J.; Shih, D.M. Trimethylamine N-Oxide promotes vascular inflammation through signaling of mitogen-activated protein kinase and nuclear factor-kappaB. *J. Am. Heart Assoc.* **2016**, *5*, e002767. [CrossRef]

271. Querio, G.; Antoniotti, S.; Levi, R.; Gallo, M.P. Trimethylamine N-Oxide does not impact viability, ROS production, and mitochondrial membrane potential of adult rat cardiomyocytes. *Int. J. Mol. Sci.* **2019**, *20*, 3045. [CrossRef]

272. Huc, T.; Drapala, A.; Gawrys, M.; Konop, M.; Bielinska, K.; Zaorska, E.; Samborowska, E.; Wyczalkowska-Tomasik, A.; Paczek, L.; Dadlez, M.; et al. Chronic, low-dose TMAO treatment reduces diastolic dysfunction and heart fibrosis in hypertensive rats. *Am. J. Physiol. Heart Circ. Physiol.* **2018**, *315*, H1805–H1820. [CrossRef]

273. Steiner, D.J.; Kim, A.; Miller, K.; Hara, M. Pancreatic islet plasticity: Interspecies comparison of islet architecture and composition. *Islets* **2010**, *2*, 135–145. [CrossRef] [PubMed]

274. Ionescu-Tirgoviste, C.; Gagniuc, P.A.; Gubceac, E.; Mardare, L.; Popescu, I.; Dima, S.; Militaru, M. A 3D map of the islet routes throughout the healthy human pancreas. *Sci. Rep.* **2015**, *5*, 14634. [CrossRef] [PubMed]

275. Wang, X.; Misawa, R.; Zielinski, M.C.; Cowen, P.; Jo, J.; Periwal, V.; Ricordi, C.; Khan, A.; Szust, J.; Shen, J.; et al. Regional differences in islet distribution in the human pancreas–preferential beta-cell loss in the head region in patients with type 2 diabetes. *PLoS ONE* **2013**, *8*, e67454. [CrossRef] [PubMed]

 *nutrients*

Article

Obesogenic and Ketogenic Diets Distinctly Regulate the SARS-CoV-2 Entry Proteins ACE2 and TMPRSS2 and the Renin-Angiotensin System in Rat Lung and Heart Tissues

Daniel Da Eira, Shailee Jani and Rolando B. Ceddia *

Muscle Health Research Center, School of Kinesiology and Health Science, York University, Toronto, ON M3J1P3, Canada; DanielDaEira@hotmail.com (D.D.E.); janishailee@gmail.com (S.J.)
* Correspondence: roceddia@yorku.ca; Tel.: +1-(416)-736-2100 (ext. 77204); Fax: +1-(416)-736-5774

Abstract: Background: Obesity increases the severity of SARS-CoV-2 outcomes. Thus, this study tested whether obesogenic and ketogenic diets distinctly affect SARS-CoV-2 entry proteins and the renin-angiotensin system (RAS) in rat pulmonary and cardiac tissues. Methods: Male Sprague-Dawley rats were fed either standard chow (SC), a high-fat sucrose-enriched diet (HFS), or a ketogenic diet (KD) for 16 weeks. Afterwards, levels of angiotensin converting enzyme 2 (ACE2), transmembrane protease serine 2 (TMPRSS2), RAS components, and inflammatory genes were measured in the lungs and hearts of these animals. Results: In the lungs, HFS elevated ACE2 and TMPRSS2 levels relative to SC diet, whereas the KD lowered the levels of these proteins and the gene expressions of toll-like receptor 4 and interleukin-6 receptor relative to HFS. The diets did not alter ACE2 and TMPRSS2 in the heart, although ACE2 was more abundant in heart than lung tissues. Conclusion: Diet-induced obesity increased the levels of viral entry proteins in the lungs, providing a mechanism whereby SARS-CoV-2 infectivity can be enhanced in obese individuals. Conversely, by maintaining low levels of ACE2 and TMPRSS2 and by exerting an anti-inflammatory effect, the KD can potentially attenuate the severity of infection and migration of SARS-CoV-2 to other ACE2-expressing tissues.

Keywords: obesity; ketones; inflammation; TLR4; IL6; TNF-alpha; MasR

Citation: Da Eira, D.; Jani, S.; Ceddia, R.B. Obesogenic and Ketogenic Diets Distinctly Regulate the SARS-CoV-2 Entry Proteins ACE2 and TMPRSS2 and the Renin-Angiotensin System in Rat Lung and Heart Tissues. *Nutrients* **2021**, *13*, 3357. https://doi.org/10.3390/nu13103357

Academic Editors: Anna M. Giudetti and Anna Tagliabue

Received: 7 July 2021
Accepted: 23 September 2021
Published: 25 September 2021

Publisher's Note: MDPI stays neutral with regard to jurisdictional claims in published maps and institutional affiliations.

1. Introduction

Obesity is characterized by an expansion in white adipose tissue (WAT) mass and is linked to a number of comorbidities, including hypertension and diabetes [1]. More recently, obesity has emerged as a major risk factor for severe illness related to COVID-19 infection [2,3]. Infection with the severe acute respiratory syndrome-coronavirus-2 (SARS-CoV-2) occurs mainly through the airway [4]. At the cellular level, the virus binds to its receptor, ACE2 [4,5]. The interaction between ACE2 and SARS-CoV-2 occurs via the virus's spike (S) protein; however, an additional crucial step, catalysed by transmembrane protease serine 2 (TMPRSS2), leads to cleavage of the S protein [4,5]. This latter step allows for the fusion of the virus and host cell membranes, so endocytosis of the virus can occur. Once inside the cell, the virus can utilize the cell's ribosomes to transcribe its own RNA and undergo replication [6]. The enhanced susceptibility of obese individuals to severe COVID-19 outcomes has been linked to elevations in ACE2 and TMPRSS2 in lung tissues [4,7]. In fact, Sarver and Wong showed that ACE2 expression was elevated in the lungs and tracheas of male, obese mice [4]. Similarly, Batchu et al. reported elevations in ACE2 and TMPRSS2 protein contents in the lungs of HF-fed, diabetic mice [7]. In this context, the diet-induced elevations in SARS-CoV-2 entry proteins provide a plausible mechanism that explains the increased susceptibility of obese individuals to severe outcomes. It is important to note that ACE2 is also a component of the renin-angiotensin aldosterone system (RAS), which has emerged as another possible mechanism mediating COVID-19-related illness [8].

The RAS is classically responsible for the regulation of fluid balance and blood pressure [8]. In short, the kidney releases renin, which converts liver-derived angiotensinogen into angiotensin I (Ang-I) [9]. Ang-I is then processed into angiotensin-II (Ang-II), mainly in the pulmonary circulation [10], by the angiotensin converting enzyme (ACE1) [9]. Ang-II can then bind to the G-protein coupled receptor angiotensin-II, type 1 receptor (AT$_1$R) [9] where it induces vasoconstriction, water retention and sympathetic nervous system activation [9]. Under conditions of hypertension and obesity, hyperactivation of the RAS leads to inflammation, fibrosis and oxidative damage [8]. In this context, therapies aimed at increasing flux through the counterregulatory arm of the RAS have gained attention as a potential treatment for hypertension and obesity [2,8]. These effects are mainly mediated by ACE2, which converts Ang-II into Ang(1,7) [8]. Ang(1,7) binds to the Mas receptor (MasR), which exerts protective, anti-inflammatory effects [9]. Moreover, Ang-II can also bind to angiotensin II, type 2 receptor (AT$_2$R) [2]. AT$_2$R exerts similar effects to MasR, however it is mainly expressed under conditions of AT$_1$R hyperactivation and injury to suppress inflammation and oxidative stress [2,11]. Following COVID infection, ACE2 becomes downregulated [2]. Consequently, the obesity-induced elevation in Ang-II increases signalling through AT$_1$R, which promotes inflammation and tissue damage [2]. Thus, the role of ACE2 in COVID-19-related illness extends beyond serving as a viral entry protein. Rather, it is also implicated in the inflammatory response that ensues, which carries a profound impact for tissue damage and organ failure [2]. In this context, shifting obese individuals from the ACE1/Ang-II/AT$_1$R arm to the ACE2/Ang(1,7)/MasR and AT$_2$R arm is of great importance to reduce the susceptibility of obese individuals to severe COVID-19 outcomes.

Recently, the ketogenic diet (KD) has sparked interest as a potential therapeutic tool for the management of obesity. In fact, it has been shown that the KD improved blood pressure, glycemia and body weight in Type 2 diabetic patients [12] and carries potential as an anti-inflammatory diet [13]. However, to our knowledge, little is known regarding the effect of the KD on the RAS components in lung and heart tissues. Moreover, despite there being evidence that obesity enhances the expression of ACE2 and TMPRSS2 in lung tissue, to the best of our knowledge, no studies have evaluated whether a KD can affect molecular steps involved in SARS-CoV-2 infection. In this context, our study aims to compare the effects of different dietary interventions on the contents of viral entry proteins and RAS components in lung and heart tissues. Here, we provide a detailed analysis of how an obesogenic diet regulates ACE2 and TMPRSS2 contents, as well as AT$_1$R and AT$_2$R levels in lung and cardiac tissues. Additionally, we test whether a KD can alter the levels of these proteins and reduce the expression of inflammatory genes in the lungs.

2. Materials and Methods

2.1. Reagents

Protease (cOmplete Ultra Tablets) and phosphatase (PhosSTOP) inhibitors—were obtained from Roche Diagnostics GmbH (Mannheim, Germany). The Insulin ELISA kit was purchased from Millipore-Sigma (Burlington, MA, USA). The ACE1 (ab254222), ACE2 (ab239924), AT$_1$R (ab124734), AT$_2$R (ab92445) and TMPRSS2 (ab92323) antibodies were purchased from Abcam (Toronto, ON, Canada) and the β-actin (cat. no. 4967) antibody was purchased from Cell Signaling (Danvers, MA, USA).

2.2. Animals

Male albino rats from the Sprague-Dawley strain (Envigo, Indianapolis, IN, USA) weighing 200–250 g (initial weight)—were maintained in a constant-temperature (23 °C), with a fixed 12-h light/12-h dark cycle and fed for 16 weeks ad libitum either a standard chow diet (SC, 27.0%, 13.0%, and 60.0% of calories provided by protein, fat, and carbohydrates, respectively) a HF, sucrose-enriched (HFS) diet (20.0%, 60.0%, and 20.0% of calories provided by protein, fat, and carbohydrates [sucrose], respectively) or a KD (20%, 80% and 0% of calories provided by protein, fat and carbohydrates, respectively). The energy densities for the SC, HFS and ketogenic diets were 3.43, 5.24 and 6.14 kcal/g, respectively. The SC diet (standard rat chow, catalog # 5012) was purchased from TestDiet (Richmond,

IN, USA). The HFS and ketogenic diets (catalog # D12492 and D03022101, respectively) were purchased from Research Diets Inc. (New Brunswick, NJ, USA). Energy intake was not significantly different between the three groups at weeks 0, 4 and 8 (Table 1). At week 16, the KD-fed animals had 17.9% and 18.1% lower energy intake than the SC and HFS-fed animals, respectively (Table 1).

Table 1. Energy intake at weeks 0, 4, 8, and 16 in the SC-, HFS-, and KD-fed animals.

	Week	SC	HFS	KD
Energy intake (kcal/rat/day)	0	69.23 ± 3.8	66.49 ± 2.3	66.49 ± 2.9
	4	72.60 ± 4.2	68.37 ± 4.0	63.83 ± 0.76
	8	65.83 ± 2.9	66.94 ± 3.2	59.82 ± 1.1
	16	69.65 ± 2.1	69.75 ± 6.0	57.15 ± 5.0 *

* $p < 0.05$ vs. SC and HFS within the respective week. Data are expressed as mean ± SEM. Two-way ANOVA. $n = 3$–7. SC, standard chow; HFS, high-fat sucrose-enriched diet; KD, ketogenic diet.

2.3. Ethics Approval

The protocol containing all animal procedures described in this study was specifically approved by the Committee on the Ethics of Animal Experiments of York University (York University Animal Care Committee, YUACC, permit number: 2021-03) and performed strictly in accordance with the YUACC guidelines. All tissue extraction procedures were performed under ketamine/xylazine anaesthesia, and all efforts were made to minimize suffering [14]. All experiments in this study were carried out in compliance with the ARRIVE guidelines [15].

2.4. Glucose Monitoring and Determination of Plasma Insulin Concentrations

Blood from all animals in a fed state was collected between 15:00 and 16:00 h by saphenous vein bleeding and used to determine plasma glucose, using the OneTouch UltraMini blood glucose monitoring system from LifeScan Canada Ltd. (Vancouver, BC, Canada). Insulin was measured using a commercially available kit listed in the reagents section. All procedures were performed according to instructions provided by the manufacturer of the kit.

2.5. Western Blotting Analysis of ACE1, ACE2, AT_1R, AT_2R, and TMPRSS2 Protein Levels in Lung and Cardiac Tissues

Lung and heart tissues were homogenized in a buffer containing 25 mM Tris-HCl, 25 mM NaCl (pH 7.4), 1 mM $MgCl_2$, 2.7 mM KCl, 1% Triton X-100 and protease and phosphatase inhibitors (Roche Diagnostics GmbH, Mannheim, Germany). Homogenates were centrifuged, the supernatant was collected, and an aliquot was used to measure protein by the Bradford method. Samples were diluted 1:1 (vol:vol) with 2× Laemmli sample buffer and heated to 95 °C for 5 min. Then, 25 µg of protein were loaded in each well. Samples were then subjected to SDS-PAGE, transferred to PVDF membrane, and probed for the proteins of interest. All primary antibodies were used at a dilution of 1:1000. All densitometry analyses were performed using the ImageJ program.

2.6. RNA Isolation and Quantitative PCR

Primers were designed using the software PrimerQuest (IDT) based on probe sequences available at the Affymetrix database (NetAffx™ Analysis Centre, http://www.affymetrix.com/analysis, accessed on 3 May 2020) for each given gene. RNA was isolated from lung and heart tissues using Trizol™ (ThermoFisher Scientific, Waltham, MA, USA). Complimentary DNA (cDNA) was made from 2 µg of extracted RNA using the ABM OneScript cDNA Synthesis kit (Diamed, Mississauga, ON, Canada), according to the manufacturer's instructions. Samples were run in duplicates on 96-well plates, and each 20 µL reaction contained 4 µL of cDNA, 0.4 µL of primer, 10 µL of Brightgreen 2× qPCR Mastermix (Diamed, Mississauga, ON, Canada) and 5.6 µL of RNase-free water. Real-time

PCR analysis was performed using a Bio-Rad CFX96 Real Time PCR Detection System (Bio-Rad, Mississauga, ON, Canada) using the following amplification conditions: 95 °C (10 min); 40 cycles of 95 °C (15 s), 60 °C (60 s). All genes were normalized to the control gene β-actin, and values are expressed as fold increases relative to control [16]. Primer sequences utilized are shown in Table 2.

Table 2. Primer sequences for qPCR.

	Forward	Reverse
mas1	5′-CGCCAACCCTTTCATCTACT-3′	5′-CCTAGGTTGCATCTCGTCTTT-3′
tlr4	5′-ACCTAAGGAGAGGAGGCTAAG-3′	5′-GGTAACTGCAGCACACTACA-3′
il6r	5′-TGGAGCAGACAGAGAGACTT-3′	5′-AGCTTACAGGTAACAGAGCATAAA-3′
tnfr1	5′-CCCTGTGAACCTCCTCTTTG-3′	5′-CTATGTACACCAAGTCGGTAGC-3′
β-actin	5′-GTGAAAAGATGACCCAGATC-3′	5′-CACCGCCTGGATGGCTACGT-3′

2.7. Statistical Analyses

Data were expressed as Mean ± SE. Statistical analyses were performed by using mixed-model analyses, and one-way ANOVAs and two way ANOVAs with Bonferroni multiple comparison post-hoc tests, as indicated in the figure legends. The GraphPad Prism software version 9.1.12 was used for all statistical analyses and for the preparation of all graphs. The level of significance was set to $p < 0.05$.

3. Results

Body weight (BW), glycemia, and insulinemia—HFS-fed animals gained significantly more weight than the SC and KD-fed animals (Table 3). This difference in BW was statistically significant at weeks 8 (409.77 ± 14.6 g in the HFS group vs. 370.21 ± 14.3 and 374.24 ± 6.6 g in the SC and KD groups, respectively), 12 (452.3 ± 15.7 g in the HFS group vs. 400.47 ± 15 and 404.69 ± 6.5 g in the SC and KD groups, respectively), and 16 (470.81 ± 17.1 g in the HFS group vs. 421.29 ± 15.7 and 425.54 ± 4.7 g in the SC and KD groups, respectively). With respect to blood glucose levels, there was no significant difference among the three groups at weeks 0 and 16 (Table 4). Insulinemia increased in both the HFS (1.90 ± 0.33 to 7.68 ± 1.54 ng/mL) and KD groups (2.18 ± 0.34 to 5.85 ± 0.66 ng/mL) from week 0 to week 16 (Table 4). However, at week 16, only the HFS had significantly higher blood insulin levels than the SC group (Table 4). In fact, the HFS diet-induced elevation in blood insulin concentration was 79%, relative to the SC animals (Table 4), which is an indication of insulin-resistance in these animals. In contrast, the SC and KD groups did not have statistically significantly different blood insulin levels at week 16 (Table 4).

Table 3. Body weight of SC-, HFS-, and KD-fed rats at the beginning of the study (week 0) and after 4, 8, 12, and 16 weeks of dietary intervention.

		Duration of Study (Weeks)				
	Groups	**0**	**4**	**8**	**12**	**16**
Body mass (g)	SC	231.97 ± 7.1	317.21 ± 10.4	370.21 ± 14.3	400.47 ± 15.0	421.29 ± 15.7
	HFS	236.18 ± 5.4	345.97 ± 10.7	409.77 ± 14.6 *	452.3 ± 15.7 *	470.81 ± 17.1 *
	KD	230.12 ± 3.6	321.21 ± 4.5	374.24 ± 6.6	404.69 ± 6.5	425.54 ± 4.7

* $p < 0.05$ vs. SC and KD within the respective weeks. Data are expressed as mean ± SEM. Mixed-effects model analysis. $n = 7$–9.

Table 4. Glycemia and insulinemia levels at the beginning (week 0) and at the end of the dietary intervention (week 16) in SC-, HFS-, and KD-fed rats.

Blood Parameters	Groups	Duration of Study (Weeks)	
		0	16
Glucose (mmol/L)	SC	6.85 ± 0.12	5.72 ± 0.18 [#]
	HFS	6.77 ± 0.21	5.91 ± 0.10 [#]
	KD	6.59 ± 0.12	5.71 ± 0.07 [#]
Insulin (ng/mL)	SC	2.16 ± 0.50	4.30 ± 0.43
	HFS	1.90 ± 0.33	7.68 ± 1.54 [*,#]
	KD	2.18 ± 0.34	5.85 ± 0.66 [#]

[*] $p < 0.05$ vs. SC within the respective week; [#] $p < 0.05$ vs. week 0 measurement. Data are expressed as mean $\pm$ SEM. Mixed-effects model analysis, $n = 6$–7.

3.1. ACE2 and TMPRSS2 Levels in Lung and Heart Tissues

We extracted lung and heart tissues from the animals for measurement of SARS-CoV-2 cellular entry proteins ACE2 and TMPRSS2. The HFS diet increased lung ACE2 content 3.8- and 6-fold, when compared to the SC and KD groups, respectively (Figure 1A). A similar pattern was observed for TMPRSS2 where, for the HFS diet, the levels of this protein increased by 5.1- and 3.4-fold, relative to the SC and KD groups, respectively (Figure 1B). However, in cardiac tissue, no significant differences were detected for ACE2 or TMPRSS2 levels when comparing the different dietary interventions (Figure 1C,D, respectively). These data indicate that the HFS diet increased SARS-CoV-2 entry proteins in lung tissue, but not in the heart.

3.2. ACE1, AT₁R, and AT₂R Levels in Lung and Heart Tissues

In pulmonary tissue, the KD attenuated ACE1 protein content by 56%, relative to the SC group (Figure 2A). Furthermore, KD also reduced AT_1R protein content in this tissue; however, it was not statistically significant (Figure 2C). Pulmonary AT_2R levels were not altered by diet either (Figure 2E). In cardiac tissue, ACE1 was not significantly different among the groups (Figure 2B). However, the HFS diet induced 2.6 and 4.9-fold increases in AT_1R (Figure 2D) and AT_2R (Figure 2F) levels, respectively, when compared to the SC group. The contents of these proteins were not significantly different between the SC- and KD-fed animals.

3.3. Comparison of ACE1, ACE2, AT₁R, and AT₂R in lung and Heart Tissues

Upon observing the diet-induced changes in RAS proteins and the tissue-specific nature of these changes, we then decided to compare the contents of ACE1, ACE2, AT_1R, and AT_2R in the lungs and heart. Pulmonary tissue displayed higher levels of ACE1 than cardiac tissue (Figure 3), whereas more ACE2, AT_1R, and AT_2R was found in the heart than in the lungs. (Figure 3).

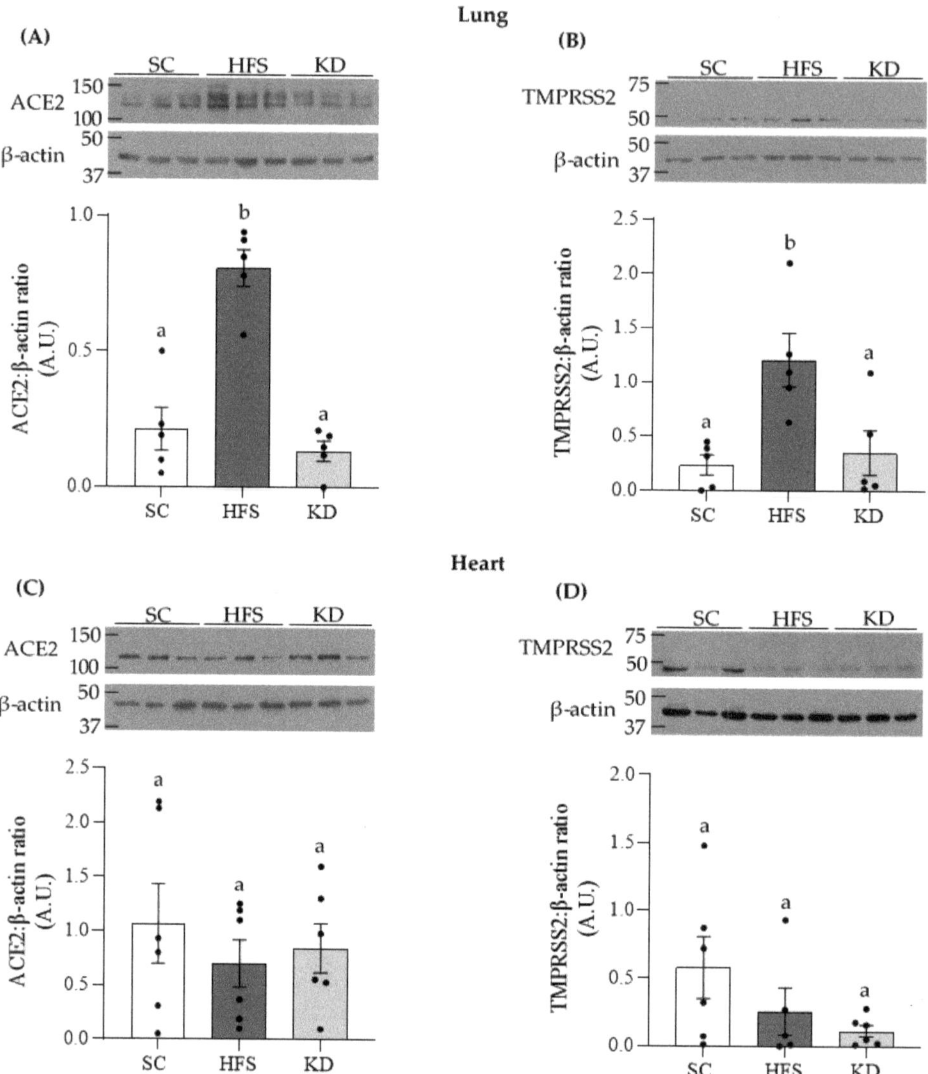

Figure 1. Lung contents of angiotensin converting enzyme 2 (ACE2) (**A**) and transmembrane protease serine 2 (TMPRSS2) (**B**) were significantly elevated by the HFS, but not by the KD. In contrast, diet did not alter the contents of these proteins in heart tissue (**C,D**). Statistical significance is denoted by different letters $p < 0.05$. Bars represent mean ± SEM. One-way ANOVA, $n = 5$–6. SC, standard chow; HFS, high-fat sucrose-enriched diet; KD, ketogenic diet.

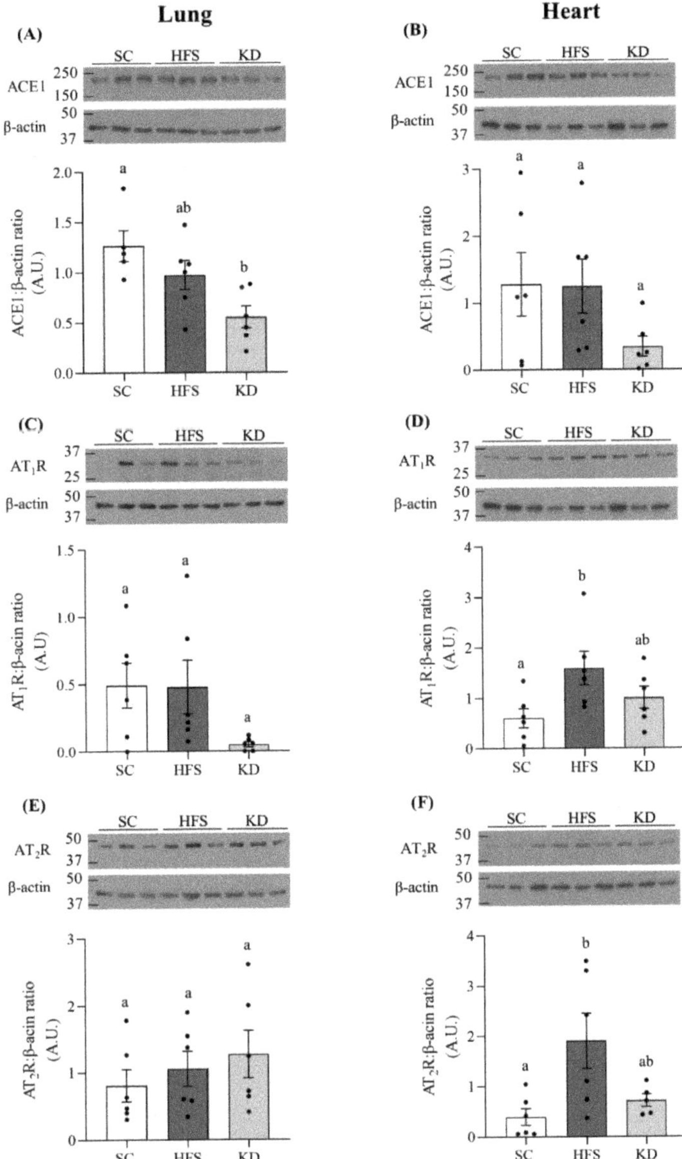

Figure 2. The KD attenuated angiotensin converting enzyme 1 (ACE1) and angiotensin II, type I receptor (AT$_1$R) contents in pulmonary tissue (**A,C**, respectively), however only the former was found to be statistically significant. Angiotensin II, type 2 receptor (AT$_2$R) was unaltered in the lungs, regardless of diet (**E**). On the other hand, dietary intervention did not influence heart ACE1 content (**B**), however AT$_1$R and AT$_2$R were significantly elevated in the cardiac tissues of HFS-fed rats (**D,F**, respectively). Significant differences are denoted by different letters $p < 0.05$. Bars represent mean $\pm$ SEM. One-way ANOVA, $n = 5$–6.

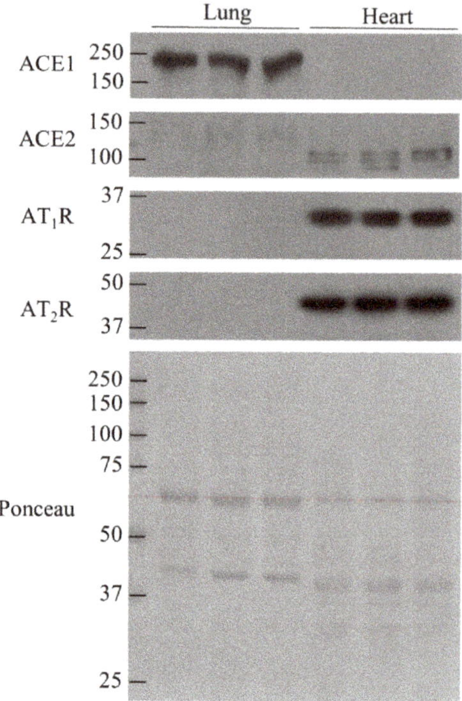

Figure 3. AT_1R, AT_2R and ACE2 are more abundant in heart tissue than in lung tissue. In contrast, lung tissue contains more ACE1. $n = 3$.

3.4. Gene Expressions of Mas1, tlr4, il6r, and tnfr1 in Lung and Heart Tissues

The gene expression of *mas1* was not significantly different among any of the three dietary interventions in either lung (Figure 4A) or heart tissue (Figure 4E). However, the gene expressions of *tlr4* and *il6r* were significantly reduced by 74% and 79%, respectively, in the lungs of KD-fed animals, when compared to the HFS animals (Figure 4B,C, respectively). KD also suppressed *tnfr1* gene expression in this tissue by ~73% (Figure 4D), although this was not statistically significant ($p = 0.059$). In the heart, no significant differences were found for *tlr4*, *il6r*, and *tnfr1* gene expressions among the three diet groups (Figure 4F–H, respectively). Altogether, these results indicate that the KD exerted an anti-inflammatory effect in the lungs, but not in the heart.

Lung

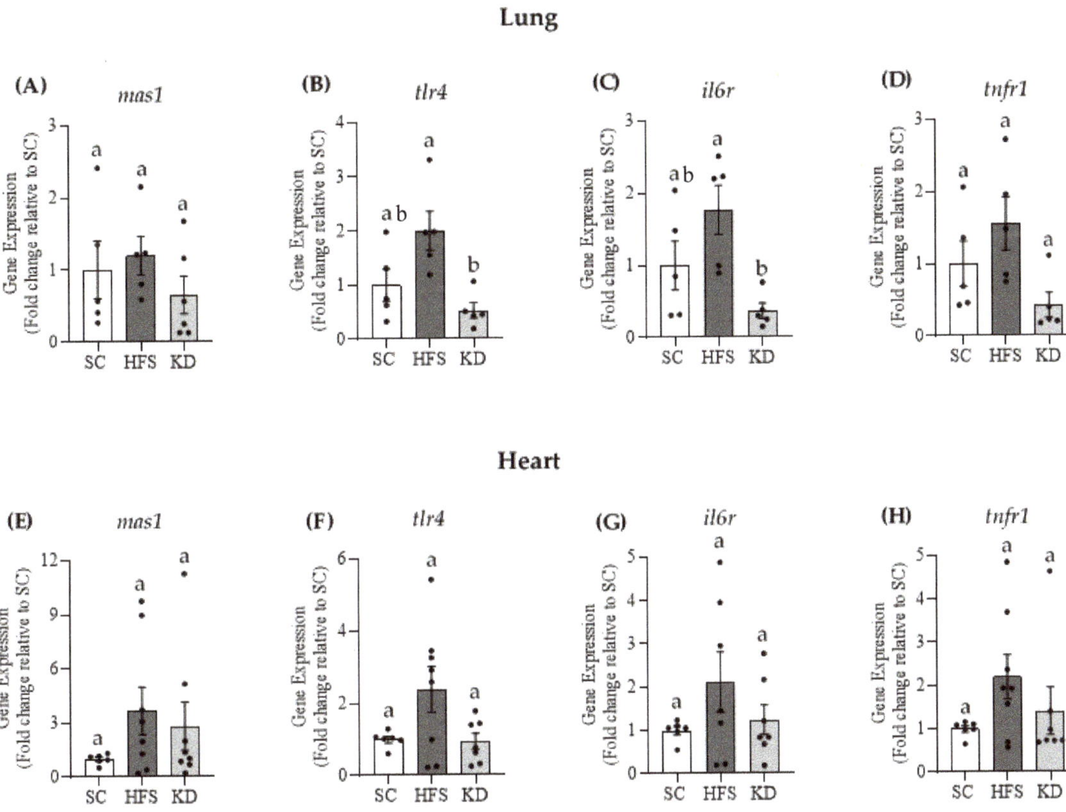

Figure 4. The KD does not affect *mas1* gene expression in lung (**A**) or heart (**E**) tissues. Interestingly, KD-feeding significantly reduces toll-like receptor 4 (*tlr4*) and interleukin-6 receptor (*il6r*) gene expression in pulmonary tissues (**B,C**, respectively), when compared to the HFS diet. The KD also reduced tumour necrosis factor receptor 1 (*tnfr1*) gene expression in lungs (**D**), though this was not statistically significant. *Tlr4, il6r and tnfr1* were not significantly altered by diet in cardiac tissues (**F–H**, respectively). Statistical significance is denoted by different letters $p < 0.05$. Bars represent mean $\pm$ SEM. One-way ANOVA, $n = 5–8$.

4. Discussion

Here, we show that obesogenic and ketogenic diets regulate in a distinct and tissue-specific manner proteins of the RAS, as well as major components of the molecular machinery involved in cellular infection by SARS-CoV-2. In fact, rats fed the HFS obesogenic diet displayed elevated insulinemia and much higher levels of ACE2 and TMPRSS2 in the lungs than animals fed either SC or a KD. This is consistent with previous findings of increased lung levels of ACE2 and TMPRSS2 in diabetic mice [7], and with reports linking insulin resistance and Type 2 diabetes to severe COVID outcomes [2,7]. However, to our knowledge, this is the first study to show that besides attenuating weight gain and insuline-mia, a KD maintained lower ACE2 and TMPRSS2 protein levels in the lungs in comparison to a HFS diet. This is relevant for the prevention of severe COVID-19-related illness, given that ACE2 and TMPRSS2 have been identified as crucial components of the molecular machinery used by SARS-CoV-2 to infects cells [5]. In fact, use of the specific TMPRSS2 inhibitor, camostat mesylate, has been shown to inhibit SARS-CoV-2 entry in Caco-2 cells, and anti-ACE2 antibody incubation with Vero cells blocked SARS-CoV-1 and SARS-CoV-2 entry [5]. Thus, the observed elevation in these two proteins in the lungs of HFS-fed rats provides a plausible mechanism whereby obese individuals are more susceptible to severe infection with SARS-CoV-2. The underlying mechanisms that explain this diet-induced elevation in ACE2 have not been fully elucidated [17]. However, under conditions of obesity, adipose tissue secretes angiotensinogen, which can eventually be converted to Ang-II, leading to hyperactivation of the ACE1/Ang-II/AT$_1$R arm of the RAS and subsequent tissue damage [2,8]. Hyperinsulinemia may also cause RAS dysregulation [8]. In this context, the observed elevation in ACE2 is likely a counterregulatory mechanism to attenuate the detrimental effects induced by obesity-mediated RAS hyperactivation [17]. Furthermore, non-alcoholic fatty liver disease (NAFLD) is a major comorbidity of obesity, and it has been linked to increased hepatic expression of ACE2 and TMPRSS2 [18], which could contribute to the aggravate morbidity in SARS-CoV-2 patients. Importantly, ketogenic diets have been shown to be very effective in combating NAFLD [19–21] and attenuate the potential deleterious effects of SARS-CoV-2 in the liver. However, we did not assess any variable associated with NAFLD in our studies; therefore, we could not make any inferences about hepatic expression of SARS-CoV-2 critical entry points.

Although there were no significant alterations in ACE2 and TMPRSS2 in the heart with any of the dietary interventions, ACE2 was found to be more abundant in cardiac than in lung tissue. Thus, because the obesogenic diet significantly increased ACE2 in the lungs, COVID-19 infection still poses a significant risk for cardiac health in obese individuals, since it is one of the primary points of infection. This could potentially enhance the severity of the infection in pulmonary tissues, leading to viral spill over, whereby the virus can access other tissues, including the heart, which is already abundant in ACE2 and, as a result, predisposed to infection and damage [2,6]. The roles of ACE2 in SARS-CoV-2 infection, as well as the RAS, are not independent. Thus, the other RAS components are also emerging as therapeutic targets to treat illness related to COVID-19. In fact, the use of angiotensin receptor blockers (ARB) or ACE inhibitors have been proposed as treatments for SARS-CoV-2 patients [2]. This serves to increase flux through the ACE2/Ang(1,7)/MasR and AT$_2$R arm of the RAS and to promote anti-inflammatory mechanisms that counteract the COVID-induced cytokine storm [2,6,8]. In the present study, we observed no alteration in AT$_1$R or AT$_2$R in lung tissues with any of the three dietary interventions. However, the KD did significantly reduce ACE1 levels. This indicates that the RAS system was shifted away from the ACE1/Ang-II/AT$_1$R arm and towards the counterregulatory, anti-inflammatory RAS arm, given that we also observed reductions in *tlr4*, and *il6r* gene expressions, as well as a trend to decrease *tnfr1* gene expression. The observed reduction in the expression of these inflammatory genes is relevant, as IL-6 and TNFα are two major cytokines that facilitate the coronavirus-induced cytokine storm [2]. Furthermore, Zhao et al. demonstrated that TLR4 inhibition suppressed interleukin-1β release in cells treated with the SARS-CoV-2 S protein [22].

Our findings are in line with other studies that report the anti-inflammatory nature of the KD [13,23], and suggest that this diet could reduce the risk of severe infection and inflammatory response related to COVID-19. We have not investigated the mechanism by which the KD led to lower levels of expression of inflammatory genes in comparison to the obesogenic diet. However, the KD is characterized by increased liver production of ketones [24]. Thus, the anti-inflammatory effect of the KD could be at least partially attributed to increased levels of circulating ketones, particularly β-hydroxybutyrate (β-HB), upon KD feeding [25]. In fact, β-HB has been demonstrated to block the nucleotide-binding domain leucine-rich-containing family pyrin domain-containing-3 (NLRP3) inflammasome and NLRP3 inflammasome-mediated interleukin (IL)-1β and IL-18 production by monocytes [26]. Additionally, the KD has been shown to exert its anti-inflammatory effects by inhibiting nuclear factor kappa-light-chain-enhancer of activated B cells (NF-kB) activation, suppressing the activity of histone deacetylases (HDACs), reducing the production of reactive oxygen species (ROS), and by improving mitochondrial respiration [24]. Thus, all these KD effects can ameliorate chronic low-grade inflammation and its associated diseases. It is important to mention that the KD used in this study was completely devoid of carbohydrates, which differs from the commonly used KD that recommends low carbohydrate content (less than 50 g/day) and not the complete elimination of this macronutrient from the diet [27]. Thus, caution should be taken when extrapolating our findings to human subjects. However, the absence of carbohydrates in the diet allowed us to identify that it is the combination of HF with sucrose, as opposed to HF alone, that promotes the deleterious metabolic effects associated with increased levels of SARS-CoV-2 entry proteins and inflammation in lung and heart tissues in obesity.

We did not find any significant difference in the cardiac gene expressions of Mas1 or the aforementioned inflammatory genes, although a trend to increase the latter in the HFS-fed rats was detected. However, the HFS diet did increase AT_1R content, which was also met by a similar increase in AT_2R. This likely served to offset the proinflammatory response mediated by AT_1R. Indeed, it has been reported that AT_2R increases in response to AT_1R hyperactivation and injury [2,11]. Interestingly, AT_1R and AT_2R are more abundant in the heart than in the lungs. In this context, it is possible that the elevated expression of these receptors allows for an enhanced flexibility of this tissue to regulate its inflammatory status. It is important to note that AT_1R and AT_2R levels did not change in the hearts of KD-fed animals, relative to the SC animals, suggesting that this diet did not provoke RAS activation. Unlike the lungs, heart ACE1 content was not altered by diet. This could be attributed to the lower ACE1 content in the heart, compared to the lungs, suggesting that the reliance of the former on ACE1 for RAS regulation is lower than the latter, where there was a diet-induced alteration in the content of this protein. This is consistent with the notion that the ACE1-dependent conversion of Ang-I to Ang-II takes place predominantly in pulmonary circulation [10].

In conclusion, these data indicate that the HFS diet increases body weight, insulinemia RAS hyperactivation and lung ACE2 and TMPRSS2, which could explain the predisposition of obese individuals to severe COVID-19 outcomes. In contrast, the KD did not elicit the same elevations in weight gain and insulinemia as the HFS diet did. Our data also suggests that there is not a hyperactivation of the RAS in KD-fed animals, as indicated by the reductions in pulmonary ACE1 and the expressions of inflammatory genes, as well as unaltered cardiac AT_1R and AT_2R contents. Thus, a counterregulatory elevation in lung ACE2 was not observed in KD animals. Unlike the lungs, the heart does not experience a diet-induced alteration in ACE2, however the abundance of this protein, relative to the lungs, makes it a susceptible organ to SARS-CoV-2 damage, following viral spill over from pulmonary tissues. In this context, despite altering RAS proteins in a tissue-specific manner, it is clear that the KD offers a promising potential therapeutic tool for its reduction in SARS-CoV-2 entry proteins and its induction of an anti-inflammatory profile.

Author Contributions: D.D.E. conducted experiments, analyzed the results, and prepared figures, and wrote the manuscript. S.J. conducted experiments and revised the manuscript. R.B.C. designed the experimental approach, conducted experiments, and revised the manuscript. All authors have read and agreed to the published version of the manuscript.

Funding: This research was funded by a Natural Science and Engineering Research Council of Canada Discovery Grant (RBC, #311818-2011) and infrastructure grants from the Canada Foundation for Innovation and the Ontario Research Fund awarded to RBC. DE was supported by the Natural Sciences and Engineering Research Council of Canada Alexander Graham Bell Canada Graduate Scholarship-Doctoral.

Institutional Review Board Statement: The study was conducted according to the guidelines of the Declaration of Helsinki. The protocol containing all animal procedures described in this study was specifically approved by the Committee on the Ethics of Animal Experiments of York University (York University Animal Care Committee, YUACC, permit number: 2021-03, approved 16 March 2021) and performed strictly in accordance with the YUACC guidelines. All tissue extraction procedures were performed under ketamine/xylazine anesthesia, and all efforts were made to minimize suffering [14]. All experiments in this study were carried out in compliance with the ARRIVE guidelines [15].

Informed Consent Statement: Not applicable.

Data Availability Statement: All data are reported in the manuscript.

Conflicts of Interest: No conflict of interest are declared by the authors.

References

1. Longo, M.; Zatterale, F.; Naderi, J.; Parrillo, L.; Formisano, P.; Raciti, G.A.; Beguinot, F.; Miele, C. Adipose tissue dysfunction as determinant of obesity-associated metabolic complications. *Int. J. Mol. Sci.* **2019**, *20*, 2358. [CrossRef]
2. Luzi, L.; Bucciarelli, L.; Ferrulli, A.; Terruzzi, I.; Massarini, S. Obesity and COVID-19: The Ominous duet affecting the Renin-Angiotensin System Livio. *Minerva Endocrinol.* **2021**, *46*, 193–201. [CrossRef]
3. Palaiodimos, L.; Kokkinidis, D.G.; Li, W.; Karamanis, D.; Ognibene, J.; Arora, S.; Southern, W.N.; Mantzoros, C.S. Severe obesity is associated with higher in-hospital mortality in a cohort of patients with COVID-19 in the Bronx, New York. *Metabolism* **2020**, *108*, 154262. [CrossRef] [PubMed]
4. Sarver, D.C.; Wong, G.W. Obesity alters Ace2 and Tmprss2 expression in lung, trachea, and esophagus in a sex-dependent manner: Implications for COVID-19. *Biochem. Biophys. Res. Commun.* **2021**, *538*, 92–96. [CrossRef] [PubMed]
5. Hoffmann, M.; Kleine-Weber, H.; Schroeder, S.; Krüger, N.; Herrler, T.; Erichsen, S.; Schiergens, T.S.; Herrler, G.; Wu, N.H.; Nitsche, A.; et al. SARS-CoV-2 Cell Entry Depends on ACE2 and TMPRSS2 and Is Blocked by a Clinically Proven Protease Inhibitor. *Cell* **2020**, *181*, 271–280.e8. [CrossRef] [PubMed]
6. Machhi, J.; Herskovitz, J.; Senan, A.M.; Dutta, D.; Nath, B.; Oleynikov, M.D.; Blomberg, W.R.; Meigs, D.D.; Hasan, M.; Patel, M.; et al. The Natural History, Pathobiology, and Clinical Manifestations of SARS-CoV-2 Infections. *J. Neuroimmune Pharmacol.* **2020**, *15*, 359–386. [CrossRef] [PubMed]
7. Batchu, S.N.; Kaur, H.; Yerra, V.G.; Advani, S.L.; Golam Kabir, M.; Liu, Y.; Klein, T.; Advani, A. Lung and Kidney ACE2 and TMPRSS2 in Renin-Angiotensin System Blocker–Treated Comorbid Diabetic Mice Mimicking Host Factors That Have Been Linked to Severe COVID-19. *Diabetes* **2021**, *70*, 759–771. [CrossRef]
8. Akoumianakis, I.; Filippatos, T. The renin-angiotensin-aldosterone system as a link between obesity and coronavirus disease 2019 severity. *Obes. Rev.* **2020**, *21*, 1–8. [CrossRef]
9. Ames, M.K.; Atkins, C.E.; Pitt, B. The renin-angiotensin-aldosterone system and its suppression. *J. Vet. Intern. Med.* **2019**, *33*, 363–382. [CrossRef]
10. Marcus, Y.; Shefer, G.; Stern, N. Adipose tissue renin-angiotensin-aldosterone system (RAAS) and progression of insulin resistance. *Mol. Cell. Endocrinol.* **2013**, *378*, 1–14. [CrossRef]
11. Schulman, I.H.; Raij, L. The angiotensin II type 2 receptor: What is its clinical significance? *Curr. Hypertens. Rep.* **2008**, *10*, 188–193. [CrossRef]
12. Walton, C.M.; Perry, K.; Hart, R.H.; Berry, S.L.; Bikman, B.T. Improvement in Glycemic and Lipid Profiles in Type 2 Diabetics with a 90-Day Ketogenic Diet. *J. Diabetes Res.* **2019**. [CrossRef]
13. Asrih, M.; Altirriba, J.; Rohner-Jeanrenaud, F.; Jornayvaz, F.R. Ketogenic diet impairs FGF21 signaling and promotes differential inflammatory responses in the liver and white adipose tissue. *PLoS ONE* **2015**, *10*, e0126364. [CrossRef]
14. Pinho, R.A.; Sepa-Kishi, D.M.; Bikopoulos, G.; Wu, M.V.; Uthayakumar, A.; Mohasses, A.; Hughes, M.C.; Perry, C.G.R.; Ceddia, R.B. High-fat diet induces skeletal muscle oxidative stress in a fiber type-dependent manner in rats. *Free Radic. Biol. Med.* **2017**, *110*. [CrossRef]
15. du Sert, N.P.; Ahluwalia, A.; Alam, S.; Avey, M.T.; Baker, M.; Browne, W.J.; Clark, A.; Cuthill, I.C.; Dirnagl, U.; Emerson, M.; et al. Reporting animal research: Explanation and elaboration for the arrive guidelines 2.0. *PLoS Biol.* **2020**, *18*, e3000411.
16. Livak, K.J.; Schmittgen, T.D. Analysis of Relative Gene Expression Data Using Real-Time Quantitative PCR and the $2^{-\Delta\Delta CT}$ Method. *Methods* **2001**, *25*, 402–408. [CrossRef] [PubMed]
17. Wijnant, S.R.A.; Jacobs, M.; Van Eeckhoutte, H.; Lapauw, B.; Joos, G.F.; Bracke, K.R.; Brusselle, G.G. Expression of ACE2, the SARS-CoV-2 Receptor, in Lung Tissue of Patients with Type 2 Diabetes. *Diabetes* **2020**, *69*, 2691–2699. [CrossRef]
18. Fondevila, M.F.; Mercado-Gómez, M.; Rodríguez, A.; Gonzalez-Rellan, M.J.; Iruzubieta, P.; Valentí, V.; Escalada, J.; Schwaninger, M.; Prevot, V.; Dieguez, C.; et al. Obese patients with NASH have increased hepatic expression of SARS-CoV-2 critical entry points. *J. Hepatol.* **2021**, *74*, 469–471. [CrossRef]
19. D'Abbondanza, M.; Ministrini, S.; Pucci, G.; Nulli Migliola, E.; Martorelli, E.E.; Gandolfo, V.; Siepi, D.; Lupattelli, G.; Vaudo, G. Very Low-Carbohydrate Ketogenic Diet for the Treatment of Severe Obesity and Associated Non-Alcoholic Fatty Liver Disease: The Role of Sex Differences. *Nutrients* **2020**, *12*, 2748. [CrossRef]
20. Watanabe, M.; Tozzi, R.; Risi, R.; Tuccinardi, D.; Mariani, S.; Basciani, S.; Spera, G.; Lubrano, C.; Gnessi, L. Beneficial effects of the ketogenic diet on nonalcoholic fatty liver disease: A comprehensive review of the literature. *Obes. Rev.* **2020**, *21*. [CrossRef] [PubMed]
21. Luukkonen, P.K.; Dufour, S.; Lyu, K.; Zhang, X.M.; Hakkarainen, A.; Lehtimäki, T.E.; Cline, G.W.; Petersen, K.F.; Shulman, G.I.; Yki-Järvinen, H. Effect of a ketogenic diet on hepatic steatosis and hepatic mitochondrial metabolism in nonalcoholic fatty liver disease. *Proc. Natl. Acad. Sci. USA* **2020**, *117*, 7347–7354. [CrossRef] [PubMed]
22. Zhao, Y.; Kuang, M.; Li, J.; Zhu, L.; Jia, Z.; Guo, X.; Hu, Y.; Kong, J.; Yin, H.; Wang, X.; et al. SARS-CoV-2 spike protein interacts with and activates TLR41. *Cell Res.* **2021**, 1–3. [CrossRef]
23. Forsythe, C.E.; Phinney, S.D.; Fernandez, M.L.; Quann, E.E.; Wood, R.J.; Bibus, D.M.; Kraemer, W.J.; Feinman, R.D.; Volek, J.S. Comparison of low fat and low carbohydrate diets on circulating fatty acid composition and markers of inflammation. *Lipids* **2008**, *43*, 65–77. [CrossRef]

24. Pinto, A.; Bonucci, A.; Maggi, E.; Corsi, M.; Businaro, R. Anti-oxidant and anti-inflammatory activity of ketogenic diet: New perspectives for neuroprotection in alzheimer's disease. *Antioxidants* **2018**, *7*, 63. [CrossRef] [PubMed]
25. Streijger, F.; Plunet, W.T.; Lee, J.H.T.; Liu, J.; Lam, C.K.; Park, S.; Hilton, B.J.; Fransen, B.L.; Matheson, K.A.J.; Assinck, P.; et al. Ketogenic diet improves forelimb motor function after spinal cord injury in rodents. *PLoS ONE* **2013**, *8*, e78765. [CrossRef]
26. Youm, Y.-H.; Nguyen, K.Y.; Grant, R.W.; Goldberg, E.L.; Bodogai, M.; Kim, D.; D'Agostino, D.; Planavsky, N.; Lupfer, C.; Kanneganti, T.D.; et al. The ketone metabolite β-hydroxybutyrate blocks NLRP3 inflammasome–mediated inflammatory disease. *Nat. Med.* **2015**, *21*, 263–269. [CrossRef]
27. Batch, J.T.; Lamsal, S.P.; Adkins, M.; Sultan, S.; Ramirez, M.N. Advantages and Disadvantages of the Ketogenic Diet: A Review Article. *Cureus* **2020**, *12*. [CrossRef]

MDPI
St. Alban-Anlage 66
4052 Basel
Switzerland
Tel. +41 61 683 77 34
Fax +41 61 302 89 18
www.mdpi.com

Nutrients Editorial Office
E-mail: nutrients@mdpi.com
www.mdpi.com/journal/nutrients